Contact in Structural Mechanics

To my parents
To Nicole and Younnik
To Mai

Series Editor
Noël Challamel

Contact in Structural Mechanics

A Weighted Residual Approach

Anh Le van

iSTE

WILEY

First published 2024 in Great Britain and the United States by ISTE Ltd and John Wiley & Sons, Inc.

ISTE Ltd
27-37 St George's Road
London SW19 4EU
UK

www.iste.co.uk

John Wiley & Sons, Inc.
111 River Street
Hoboken, NJ 07030
USA

www.wiley.com

Library of Congress Control Number: 2023949724

British Library Cataloguing-in-Publication Data
A CIP record for this book is available from the British Library
ISBN 978-1-78630-906-8

Contents

Preface **ix**

1 Introduction to Contact Problems in Structural Mechanics **1**
- 1.1 Solving a contact problem numerically via the penalty method . . . 3
- 1.2 Numerical solution of a contact problem using the multiplier method . . . 8
  - 1.2.1 Preliminaries: problems with equality constraints . . . 8
  - 1.2.2 Problems with inequality constraints . . . 10
- 1.3 Numerical solution of a contact problem by the augmented Lagrangian method . 15
- 1.4 Book synopsis . . . 21

2 Contact Kinematics **23**
- 2.1 Motions and strains . . . 23
- 2.2 Potential contact surfaces . . . 25
- 2.3 Normal contact kinematics . . . 26
- 2.4 Variation of kinematic quantities with respect to time . . . 29
- 2.5 Tangential contact kinematics – Relative velocity . . . 34

3 Sthenics of Contact **37**
- 3.1 Stresses in bodies . . . 37
- 3.2 Contact stress vector . . . 38

4 The Constitutive Law **39**
- 4.1 Hyperelastic materials . . . 39
- 4.2 Elastoplastic materials with isotropic hardening . . . 41

5 Contact Laws **45**
- 5.1 Normal contact law . . . 45
- 5.2 Tangential contact law . . . 47

6 Strong Formulation of the Contact Problem **51**
- 6.1 Field equations . . . 51
- 6.2 Boundary conditions . . . 52
- 6.3 Initial conditions . . . 53
- 6.4 Remarks . . . 53

7 Weak Formulation of the Contact Problem **55**
7.1 Transforming the contact laws into equalities . . . 55
7.2 Preliminary ideas for the weak form . . . 59
7.3 Weak form of the contact problem . . . 60
7.4 Equivalence between the strong and the weak forms . . . 62
7.5 Final remarks . . . 66

8 Matrix Equations of the Contact Problem **69**
8.1 Introduction . . . 69
8.2 Meshes . . . 70
8.3 Matrix notation in finite elements . . . 72
8.4 The element nodal vectors . . . 73
8.5 Interpolation of positions, displacements and virtual velocities . . . 75
8.5.1 Interpolation on the contactor surface . . . 75
8.5.2 Interpolation on the target surface . . . 75
8.6 Interpolation of multipliers . . . 76
8.6.1 Definition of the vector λ . . . 76
8.6.2 Interpolation of λ . . . 78
8.6.3 Interpolation of λ^* . . . 78
8.7 Discretization of the element virtual contact power $(\mathcal{P}^*_{\text{contact}})^{e^{(1)}}$ . . . 78
8.7.1 Explicit expressions for $\{\mathbf{\Phi}^{e^{(1)}}_{\text{contact}}\}$, $\{\mathbf{\Phi}^{e^{(2)}}_{\text{contact}}\}$ and $\{\mathbf{R}^{e^{(1)}}_{\Lambda}\}$ in the three cases: algorithmic *gap*, algorithmic *slip* and algorithmic *stick* . . . 85
8.8 System of matrix equations for the contact problem . . . 88
8.8.1 Global nodal vectors . . . 88
8.8.2 Discretization of the classical terms . . . 90
8.8.3 Assembly of element virtual contact powers . . . 91
8.8.4 System of matrix equations . . . 94
8.9 Abnormal contact stresses . . . 96
8.9.1 First cause of abnormal contact stresses . . . 96
8.9.2 Second cause of abnormal contact stresses . . . 98
8.9.3 Third cause of abnormal contact stresses . . . 98
8.10 Projection calculation: contact detection . . . 99
8.11 Discrete expression of the slip $\mathbf{V}_T\Delta t$ . . . 101
8.12 Physical units . . . 106
8.13 Chapter summary . . . 107

9 Solution of the Quasi-static Contact Problem **109**
9.1 System of equations for the static contact problem . . . 109
9.2 Incremental loop initialization: the vectors $\mathbf{U}_0, \mathbf{\Lambda}_0$ . . . 111
9.3 Calculation of step $n \geq 1$: calculating $\mathbf{U}_n, \mathbf{\Lambda}_n$ . . . 111
9.3.1 Principle of the iterative Newton–Raphson scheme . . . 111
9.3.2 Tangent matrix . . . 113
9.3.3 Block matrix inversion . . . 114
9.3.4 Iterative loop initialization: the vectors $\mathbf{U}^0_n$ and $\mathbf{\Lambda}^0_n$ . . . 115
9.4 Solution algorithm . . . 115
9.5 Calculation method for the tangent matrix . . . 117
9.5.1 Direct method . . . 117
9.5.2 Indirect method . . . 118
9.5.3 Restriction to the contact tangent matrix . . . 121

9.6 Calculation of the contact tangent matrix . . . 123
9.6.1 Variations of the arguments of the functional $\mathcal{P}^*_{\text{contact}}$ . . . 123
9.6.2 Calculation of the variation $\delta\mathcal{P}^*_{\text{contact}}$ . . . 126
9.6.3 Calculation of the variation $(\delta\mathcal{P}^*_{\text{contact}})^{e^{(1)}}$ . . . 127
9.6.4 Discretization of the variation $(\delta\mathcal{P}^*_{\text{contact}})^{e^{(1)}}$ – Element contact tangent matrix $[\mathbf{K}^e_{\text{contact}}]$ . . . 133
9.6.5 Discretization of the variation $\delta\mathcal{P}^*_{\text{contact}}$ – Contact tangent matrix $[\mathbf{K}_{\text{contact}}]$ 135
9.6.6 Explicit expression for the element contact tangent matrix $[\mathbf{K}^e_{\text{contact}}]$ . . . 138
9.6.7 $[\mathbf{K}^e_{\text{contact}}]$ in the case of the algorithmic *gap* at the considered integration point . . . 144
9.6.8 $[\mathbf{K}^e_{\text{contact}}]$ in the case of algorithmic contact with *slip* at the considered integration point . . . 144
9.6.9 $[\mathbf{K}^e_{\text{contact}}]$ in the case of algorithmic contact with *stick* at the considered integration point . . . 146
9.6.10 Symmetry of the contact tangent matrix $[\mathbf{K}_{\text{contact}}]$ . . . 147
9.7 Particular case of two non-contacting bodies . . . 148
9.8 Particular case of the frictionless problem . . . 149
9.8.1 Algorithmic *gap* case at the considered integration point . . . 150
9.8.2 Algorithmic contact with *slip* case at the considered integration point . . . 152
9.9 Solution via the arc-length method . . . 152
9.10 Physical units . . . 154
9.11 Summary of the chapter . . . 155

10 Numerical Examples of Quasi-static Contact **157**
10.1 Contact patch test . . . 157
10.2 Hertzian contact problem . . . 159
10.2.1 Frictionless contact case . . . 160
10.2.2 Case of frictional contact with $\mu = 0.3$ . . . 163
10.3 Rolling disk . . . 167
10.4 Contact between two beams . . . 171
10.4.1 Dead load . . . 171
10.4.2 Follower load . . . 176
10.5 Contact of two pressurized membranes . . . 176
10.5.1 Centered membranes . . . 180
10.5.2 Membranes staggered along x . . . 182
10.6 Extrusion of an elastoplastic cylinder . . . 184
10.7 Interference fit problem . . . 189
10.7.1 Abnormal contact stresses . . . 192
10.7.2 Influence of the mesh . . . 194
10.8 Conclusion . . . 194

11 Solution of the Dynamic Contact Problem **197**
11.1 A brief review of the computational methods in dynamic contact . . . 197
11.2 Solution of the dynamic contact problem via Newmark's algorithm . . . 200
11.2.1 Initializing the incremental loop: the vectors $\mathbf{U}_0, \mathbf{V}_0, \mathbf{A}_0$ and $\mathbf{\Lambda}_0$ . . . 202
11.2.2 Calculation for a step $n \geq 1$: calculating $\mathbf{U}_n, \mathbf{V}_n, \mathbf{A}_n, \mathbf{\Lambda}_n$ . . . 202
11.2.3 Initializing the iterative loop: the vectors $\mathbf{U}^0_n, \mathbf{V}^0_n, \mathbf{A}^0_n, \mathbf{\Lambda}^0_n$ . . . 207
11.3 Solution algorithm . . . 208
11.4 Summary . . . 210

12 Numerical Examples of Dynamic Contact **213**
12.1 Impact of two elastic rods . . . 213
12.1.1 Analytical solution . . . 214
12.1.2 Numerical applications . . . 217
12.1.3 Numerical solution . . . 218
12.2 Disk impacting a rigid plane . . . 220
12.2.1 Frictionless case . . . 222
12.2.2 Case with friction $\mu = 0.3$ . . . 224
12.3 Disk falling into a funnel . . . 228
12.3.1 Frictionless case . . . 231
12.3.2 Case with friction $\mu = 0.4$ . . . 234
12.4 Final remarks . . . 236

Appendix A: Variations of Kinematic Quantities **239**

References **247**

Index **257**

Preface

Contact is a phenomenon that must be taken into account in structure mechanics when dealing with several bodies that are connected, or that are likely to come into contact with one another. When compared to contactless mechanical problems, the novelty concerns the appearance of nonclassical boundary conditions, which are not expressed simply in terms of prescribed displacements or stresses. These contact boundary conditions, which include both equations and also inequalities, impose the non-interpenetration of the bodies and other more or less complex relations between the relative slipping and stresses at the boundary. Contact thus brings two difficulties:

(i) The first comes from the fact that the contact laws are written along portions of the boundary of each body. It is therefore necessary to master how to calculate the geometry of these surfaces (in the 3D case). Moreover, in general, contact surfaces evolve throughout the deformations of the bodies and are the unknowns of the problem themselves.

(ii) The second difficulty is due to the inequalities contained within the contact laws. The mechanical problem is therefore necessarily nonlinear, even when assuming only small perturbations. From a numerical point of view, inequalities are known to be more difficult to deal with than equalities.

There are many works in the literature which deal with both theoretical and numerical questions of the contact problem and which are carried out in different contexts, small or large deformations, static or dynamic, with or without friction, with Coulomb friction or with other more complex interface laws.

Theoretical studies on the existence and uniqueness of the solution or analytical solutions are difficult, even in the absence of friction and with only small deformations. A numerical solution is also difficult to obtain when friction is taken into account, especially when using a more sophisticated friction law than Coulomb's friction law, or when working within the framework of large deformations with large relative slips. For several decades, many numerical formulations, along with many variations, have been proposed to solve the various complicated contact problems in structural mechanics, and in efficient and robust ways.

This work presents a method to solve the contact problem numerically, with large deformations and with Coulomb friction. The proposed formulation differs from the others, insofar as it is based upon a weak form generalizing the principle of virtual power, which is well known in noncontact problems. The weak form is a relation of weighted residuals, just like the principle of virtual power. Consequently, it can be discretized using the finite element method by following the same typical approach as in a contactless problem. The difference with a contactless problem is that here we obtain a system of coupled equations with unknown displacements and unknown contact stresses.

Apart from the weak form of the weighted residuals, which is specific to this work, the other parts are essentially shared with the other contact formulations in the literature. Thus, for example, contact kinematics has the same formulation here as is in the other ones. Regardless of the approach chosen, the numerical solution of the nonlinear matrix system of equations calls upon an iterative Newton-type scheme, and the contact tangent matrix is obtained using the linearization or the variation techniques, which will be described in this work.

The author hopes that this book can serve as a preliminary reading that allows the reader to acquire the necessary basic notions, before delving into the beautiful and immense existing literature on contact problems.

December 2023

1

Introduction to Contact Problems in Structural Mechanics

Often, a body is not isolated in space. Whether motionless or moving, it is generally brought into contact with other bodies, over an essentially long period of time. In the case involving contact between several solid bodies (fluids are excluded here), physical observation reveals that the bodies cannot interpenetrate, and that their interaction takes place along their external boundaries. New mechanical connections are then formed; stresses develop on the contact surface and modify their shapes more or less strongly, as well as the trajectory of each body. Therefore, mechanics naturally include the study of contact with the aim of modeling this type of interaction between the bodies and calculating their resulting movement.

The first mathematical relation to express consists of interpreting the non-interpenetrability condition that can be observed, which appears as an inequality to be satisfied. Other relations are then to be added, the number and complexity of which vary according to the physical nature of the contact (with or without friction, with or without stick, static or dynamic, etc.). In general, a contact problem is always difficult to solve, insofar as it involves not only equations but also inequations.

In 1881, the frictionless contact problem between two spheres was solved analytically by Hertz, thus giving rise to the contact mechanics of structures. Hertz assumed that the contact portion is elliptical in shape, and that the bodies in contact behave in a linearly elastic way under small perturbations. Since then, other work has been carried out, allowing Hertz's results to be extended to more complex configurations such as the solutions of Mindlin [94] and Spence [127, 128]. The reader can consult the works of Johnson [67] and Goldsmith [45], which provide a collection of analytical solutions of contact problems with or without friction, in statics or dynamics, and for elastic or elastoplastic bodies. More recently, Storakers [130] succeeded in solving Hertz's problem with friction for various contact profiles by combining both analytical and numerical methods. In general, theoretical solutions of contact problems are difficult to construct; they are rare, and have often been obtained with simplifying assumptions, in particular that of small perturbations.

The development of computers since the 1970s has enabled researchers to solve more complicated and realistic contact problems numerically. Besides the typical nonlinearities – whether it be material or geometric – a numerical contact method must be able to deal with the non-linearity of the boundary interactions between two or more bodies. Difficulties stem from the fact that the contact laws reflecting these interactions are often irregular and multi-valued.

In the following, we present a brief review of the key numerical formulations that are used to solve contact problems. Some underlying mathematical results will be recalled succinctly and without proof. In order to understand how the numerical solving methods operate, we will rely on the following simple example of a mass-spring system:

EXAMPLE.– Consider a particle of mass m connected to a spring of stiffness k. In the reference state, there is no gravity, the spring is at rest, and the particle is at a distance h from a fixed plane (Figure 1.1(a)). Under the effect of gravity g, the particle moves by an amount u measured in the direction of gravity and calculated relative to the reference position of the particle (Figure 1.1(b)). The proximity between the particle and the rigid plane, denoted $g(u)$ (no possible confusion with the acceleration of gravity g), is defined by $g(u) \equiv u - h$. This quantity must always be either negative or zero.

The problem therefore consists of determining the equilibrium position of the particle, that is to say the displacement u, under the force of gravity in the presence of an obstacle, which here is the rigid plane, and thus imposes the inequality constraint $g(u) = u - h \leq 0$.

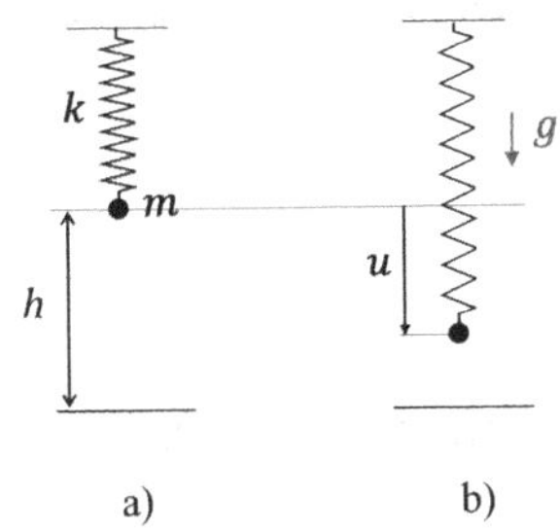

Figure 1.1. *Mass-spring system. (a) Reference position (no gravity); (b) equilibrium position under gravity. For a color version of this figure, see www.iste.co.uk/levan/contact.zip*

We easily obtain the exact solution of this problem:

1. If $mg < kh$, then the displacement is $u = \dfrac{mg}{k} < h$: the particle is displaced without touching the rigid plane. The reaction from the rigid plane on the particle, denoted λ, is zero. The condition $mg < kh$ signifies that the particle weight must be low enough so that no contact occurs.

2. If $mg \geq kh$, then $u = h$: the particle enters into contact with the rigid plane. The contact reaction equals $\lambda = mg - kh = mg - ku \geq 0$.

The total potential energy of the system is

$$\Pi(u) = \frac{1}{2}ku^2 - mgu \qquad [1.1]$$

In the $h > mg/k$ case (Figure 1.2(a)), the solution u corresponds to the absolute minimum of Π. In the other case (Figure 1.2(b)), the solution $u = h$ does not correspond to the absolute minimum of Π, but instead to the minimum of Π on the space of admissible solutions defined by $g(u) = u - h \leq 0$.

The example above shows that the proximity $g(u)$ and the contact reaction λ must satisfy the inequalities $g \leq 0$ and $\lambda \geq 0$, such that:

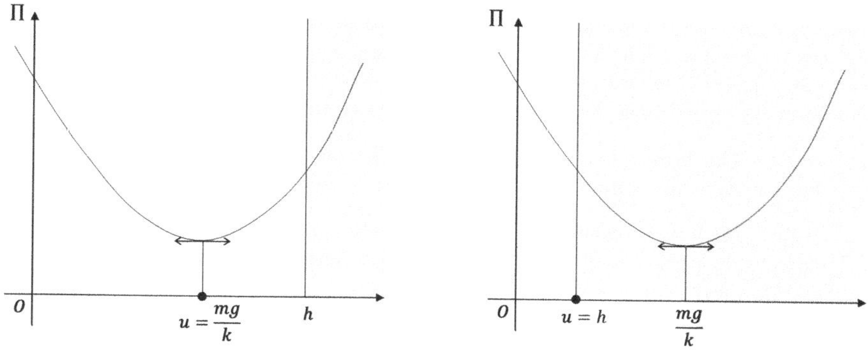

(a) Case $h > mg/k$ (b) Case $h \leq mg/k$

Figure 1.2. *The displacement u which solves the mass-spring problem, represented by the thick black point ●*

- if contact does not take place, that is to say if $g(u) < 0$, then $\lambda = 0$;
- if contact does take place, that is to say if $g(u) = 0$, then $\lambda \geq 0$.

We can group these two cases together using the following mathematical relations, called the Hertz–Signorini–Moreau conditions:

$$g(u) \leq 0 \qquad \lambda \geq 0 \quad \text{and} \quad \lambda g(u) = 0 \qquad [1.2]$$

Figure 1.3 shows the special relationship between g and λ which is imposed by the Hertz–Signorini–Moreau conditions.

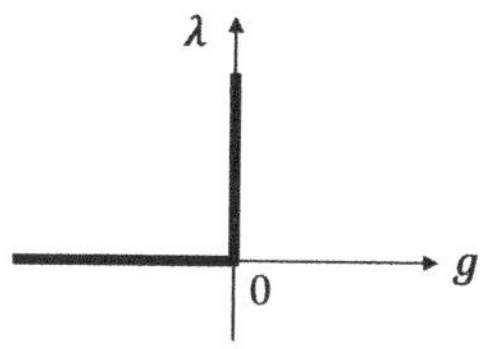

Figure 1.3. *Hertz–Signorini–Moreau conditions*

In the general case of the contact problem between two 3D solids, it becomes difficult to solve it, due to the fact that the portions of the surfaces, which are likely to come into contact, are not a priori known, and can have complicated shapes, especially when the bodies undergo large deformations. The relations [1.2] remain valid, however, depending on a suitably defined proximity g and the normal contact reaction that we denote by λ, and must be written for any point of the surface at which contact may occur.

1.1. Solving a contact problem numerically via the penalty method

A classical method to solve the contact problem, and which is easy to implement, is the penalty method. This method can be presented within the framework of optimization theory, but it is easier to explain in terms of mechanics.

EXAMPLE.– Let us go back to the mass-spring system considered above, where we had to calculate the displacement u of the particle while respecting the inequality constraint $g(u) = u - h \leq 0$. The idea of the penalty method is to abandon the constraint $g \leq 0$ and, at the level of the rigid plane, to add an artificial spring of stiffness ϵ, in such a way that:

- if the particle does not make contact with the rigid plane, then this spring does not come into play (Figure 1.4(a));
- if the particle exceeds the rigid plane, then the artificial spring is compressed and opposes the particle with a resistive force proportional to the penetration $g(u) = u - h$ (Figure 1.4(b)).

Thus, the artificial spring introduced makes it possible to oppose the penetration, and the larger its stiffness ϵ, the stronger the artificial spring. The condition $g \leq 0$ is not completely satisfied, although intuitively, it should be expected that it should become more so as ϵ tends to infinity.

By employing the penalty method, we have to solve a classical problem found in structural mechanics, instead of having to deal with an inequality constraint.

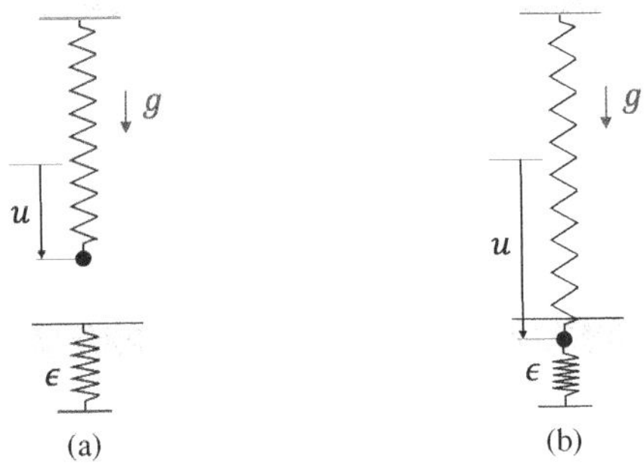

Figure 1.4. *Mass-spring system solution via the penalty method. For a color version of this figure, see www.iste.co.uk/levan/contact.zip*

To solve the mass-spring problem, we therefore seek a u which minimizes the following function, and which is equal to the sum of the total potential energy [1.1] and the strain energy of the artificial spring:

$$\Pi^{pen}(u) = \frac{1}{2}ku^2 - mgu + \frac{1}{2}\epsilon\,\langle g(u)\rangle^2$$

where $\langle g\rangle$ stands for the positive part of g: $\langle g\rangle = g$ if $g \geq 0$, $= 0$ otherwise. In optimization theory, the additional $\frac{1}{2}\epsilon\,\langle g(u)\rangle^2$ term is called the penalty term, and expresses the fact that the artificial spring will only work when compressed. The ϵ term itself is called the penalty parameter.

The extremum of Π^{pen} is obtained by setting the derivative with respect to u of this function equal to zero. Using $\frac{\partial}{\partial g}\langle g\rangle^2 = 2\langle g\rangle H(g) = 2\langle g\rangle$ (where H is the Heaviside function), we have

$$0 = \frac{d\Pi^{pen}}{du} = ku - mg + \epsilon\langle g\rangle\frac{dg}{du} \quad \text{with} \quad \frac{dg}{du} = 1 \qquad \Rightarrow \qquad ku - mg + \epsilon\langle u - h\rangle = 0 \qquad [1.3]$$

We distinguish between two exclusive cases:

1. If $u < h$, then equation [1.3] gives $u = \frac{mg}{k}$. The condition $u < h$ to be satisfied can then be written as $\frac{mg}{k} < h$, namely $mg < kh$.

2. If $u \geq h$, then equation [1.3] gives $u = \frac{mg + \epsilon h}{k + \epsilon}$. The condition $u \geq h$ can then be written as $\frac{mg + \epsilon h}{k + \epsilon} \geq h$, namely $mg \geq kh$, which is exactly the condition complementary to that of the first case.

We observe that

– if $mg < kh$, then the obtained solution u is equal to the exact solution;

– if $mg \geq kh$, then the obtained solution u is not exact. The proximity equals $g(u) = u - h = \frac{mg - kh}{k + \epsilon} \geq 0$: there is a certain penetration of the particle into the rigid obstacle, which corresponds to the spring of stiffness ϵ being compressed (Figure 1.4(b)). It is only by letting $\epsilon \to \infty$ that we obtain the exact solution $u = h$ and $g(u) = 0$.

Figure 1.5 represents the function Π^{pen} compared to the total potential energy Π of the contactless problem, given by [1.1]. The two functions are identical on the interval $u \leq h$ and differ when $u > h$. Depending on whether $kh > mg$ or not, the solution u that minimizes Π^{pen} may coincide with the solution that minimizes Π and, consequently, is either exact or not.

• We will generalize the idea presented above to three-dimensional problems by considering a system of two solids, denoted 1 and 2. The unknown field u is not a scalar one as in the previous example, but rather a vector field that represents the displacement field of each body in the mechanical system. At a given moment, the two bodies are likely to come into contact along two surfaces, denoted $S_c^{(1)}$ and $S_c^{(2)}$, belonging to the boundaries of bodies 1 and 2, respectively (see Figure 1.6). Let us choose one of the two bodies arbitrarily, say body number 1, and now let us try to describe its motion relative to body number 2. Body 2 is called the target, body 1 is called the contactor, the potential contact surface $S_c^{(2)}$ is called the target surface and potential contact surface $S_c^{(1)}$ is called the contactor surface.

Consider a particle located on the contact surface; its current position is the point $\mathbf{x}$ and its orthogonal projection onto the target surface is $\mathbf{y}$. The antecedents of $\mathbf{x}$ and $\mathbf{y}$ are two points $\mathbf{X}$ and $\mathbf{Y}$, respectively, belonging to the reference positions on the contact surfaces, denoted as $S_{0c}^{(1)}$ and $S_{0c}^{(2)}$. We define the proximity g at the point $\mathbf{x}$ by:

$$g = -\boldsymbol{\nu}.(\mathbf{x} - \mathbf{y})$$

where $\boldsymbol{\nu}$ is the normal at $\mathbf{y}$ to the surface $S_c^{(2)}$. According to its definition, the proximity g is negative if there is a gap between the two bodies, and positive if there is interpenetration.

As we will work in the total Lagrangian formulation, we choose to consider any contact quantity that varies in space as a function of the current point $\mathbf{X} \in S_{0c}^{(1)}$. So, for example, the proximity g is the real function $S_{0c}^{(1)} \ni \mathbf{X} \mapsto g(\mathbf{X}) \in \mathbb{R}$. That said, as part of the optimization, any dependence on $\mathbf{X}$ will be implied so as not to make the notation too cumbersome; therefore, we will write $g(u)$ instead, emphasizing the dependence of g on the displacement fields u of the bodies.

In the absence of friction, the penalty method consists of covering the target surface with a distribution of artificial springs, each with a sufficiently high stiffness ϵ, similar to what we

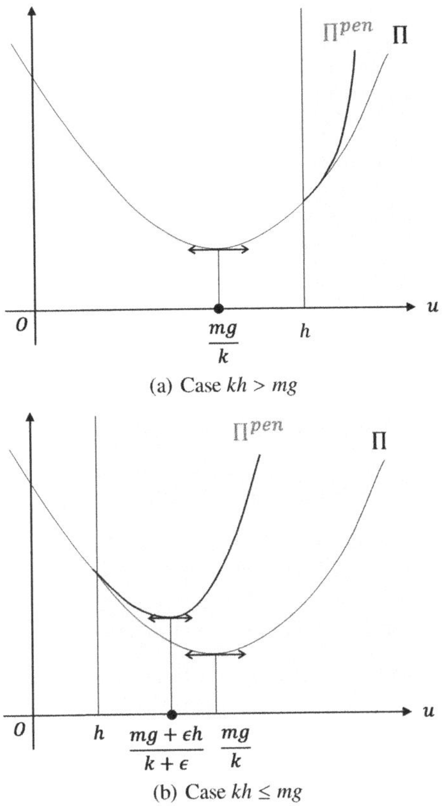

Figure 1.5. *The function Π^{pen} used in the penalty method, by comparing it to the total potential energy Π. The displacement solution u of the problem is represented by the thick black point ●. For a color version of this figure, see www.iste.co.uk/levan/contact.zip*

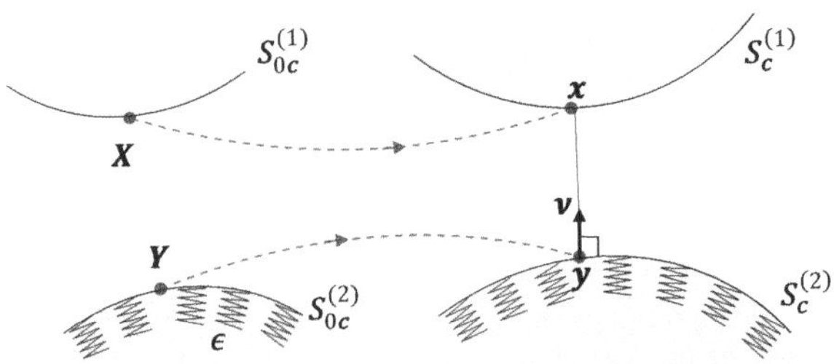

Figure 1.6. *3D problem – artificial springs used in the penalty method. For a color version of this figure, see www.iste.co.uk/levan/contact.zip*

have just seen in the example of the mass-spring system (Figure 1.6). We are then reduced to considering a mechanical system, supplemented by a network of artificial springs, whose solution is that of a classical problem without any special treatment regarding the contact.

To solve the contact problem, we seek the field u which minimizes the functional given by the sum of the total potential energy Π of the system and the strain energy of the artificial springs. In other words, we are looking for the solution field u of:

$$\min_{v} \left[\Pi(v) + \frac{1}{2} \int_{S_{0c}^{(1)}} \epsilon \, \langle g(v) \rangle^2 dS_0 \right] \qquad [1.4]$$

where, we recall, $\epsilon > 0$ is a given fixed number. Thus, we have replaced the non-regular contact law [1.2] with a single-valued and differentiable law (the response of springs of stiffness ϵ), such that when the stiffness ϵ tends toward infinity, the regularized law tends toward the original non-regular law.

Let u be the displacement field of the two bodies and the solution of the problem, and let δu be an arbitrary variation of u. The corresponding variation of the functional in [1.4] is zero and, by using $\delta \langle g \rangle^2 = 2\langle g \rangle H(g)\, \delta g = 2\langle g \rangle\, \delta g$, we have to solve the following variational equation:

$$\forall \delta u, \qquad \delta \Pi(u) + \int_{S_{0c}^{(1)}} \epsilon \, \langle g(u) \rangle \, \delta g \, dS_0 = 0 \qquad [1.5]$$

where the variation δg is expressed as a function of the displacement variations δu on the contact surfaces (see equation [A.7] in the Appendix).

This approach can also be extended to contact problems with friction by considering artificial springs that are arranged tangentially to the target surface. However, tangential contact laws are more difficult to translate than the normal contact law.

The penalty method has the following advantages:

1. It is straightforward to formulate and program. This leads to solving equation [1.5] without any inequality constraints. Numerical solutions can be obtained by discretizing this equation by means of the finite element method in a classical way.

2. The penalty method was presented in statics and leads to an energy minimization problem, but in fact, nothing prevents it from being applied to a dynamic contact problem for which no functional exists to be minimized. To do this, artificial springs are added to the target surface as we did in the static problem; however, this time we consider the principle of virtual power that includes the virtual power from the added springs and that from the inertial terms. We then discretize the principle of virtual power in the classical way – as in a non-contact problem – using the finite element method, before solving the resultant equation as usual.

However, the penalty method has a big drawback: it does not allow for the contact inequalities to be satisfied exactly. In order to approach the exact solution, one must allocate very large values to the penalty parameters, with the risk that the tangent matrix may become ill-conditioned and the calculation can become unstable and divergent.

The first work that applied the penalty method to contact problems considered small deformations. In Kikuchi and Oden [71], theoretical results can be found on this method in both the friction and frictionless cases, whereas in Alart and Curnier [1], examples of contact problems with friction and small perturbations can be found.

Other work with large deformations was then carried out by Hallquist et al. [50], Parisch [108], Papadopoulos and Taylor [106], Puso and Laursen [114] (by combining the penalty method with the mortar method), Konyukhov and Schweizerhof [74] and Harnau et al. [52] in the frictionless case. Contact problems with friction have been treated by Oden and Pires [102], Wriggers et al. [142], Perić and Owen [110], Parisch and Lubbing [109], Bittencourt and Creus [18] and later by Fischer and Wriggers [39] in combination with the mortar method. In order to ensure that the numerical calculation scheme converges when the penalty parameter is large, several stabilization techniques have been developed in Alart and Curnier [1] and Papadopoulus and Taylor [106].

1.2. Numerical solution of a contact problem using the multiplier method

Contrary to the penalty method, the multiplier method allows us to satisfy the contact conditions exactly.

1.2.1. *Preliminaries: problems with equality constraints*

Although a contact problem always involves inequality conditions, let us begin by describing the multiplier method in the case where there are only equality conditions. This simple case is a good introduction to the multiplier method in the presence of inequality constraints. On the other hand, the results of this simple case will be used to construct the augmented Lagrangian, which we will see later.

We are interested in the following problem:

$$\text{minimize } J(v) \text{ subject the constraint } h(v) = 0 \qquad [1.6]$$

where $J : U \rightarrow \mathbb{R}$ and $h : U \rightarrow \mathbb{R}$ are real-valued functions defined on an open subset U of $\mathbb{R}$.

To simplify the presentation, we have considered functions of a single real variable $v \in \mathbb{R}$. The problem can easily be generalized to functions of several variables $v \in \mathbb{R}^n$, in which case $\frac{\partial J}{\partial v}$ below should be thought of as the gradient vector $\mathbf{grad}J$, and the multiplier μ below is a vector in $\mathbb{R}^n$.

Furthermore, we will not recall every mathematical hypothesis required to obtain the results. For example, we will not always therefore specify the properties of the subset U (convexity, compactness, etc.) or the regularity conditions on the various functions that we encounter.

Lagrange multiplier theorem.– Assume that the functions J and h are continuously differentiable. If u is a local extremum of $J(v)$, then there exists a number η, called the Lagrange multiplier, such that

$$\frac{\partial J}{\partial v}(u) + \eta \frac{\partial h}{\partial v}(u) = 0 \qquad [1.7]$$

According to [1.7], to find a solution to the problem [1.6], we have to solve the following two (generally nonlinear) equations for the two unknowns, u and η:

$$\frac{\partial J}{\partial v}(u) + \eta \frac{\partial h}{\partial v}(u) = 0 \qquad \text{and} \qquad h(u) = 0 \qquad [1.8]$$

As [1.7] is only a necessary condition, it must be checked that the solution obtained from [1.8] is indeed the desired minimum of problem [1.6].

We can solve the nonlinear system [1.8] numerically by using the iterative Newton–Raphson method.

The problem can also be reformulated into another equivalent way by using the Lagrangian, which is defined as follows:

DEFINITION.– The Lagrangian associated with the problem [1.6], denoted L, is defined by

$$L(v,\mu) \equiv J(v) + \mu h(v) \tag{1.9}$$

The equations [1.8] can then be written in terms of the Lagrangian L:

$$\frac{\partial L}{\partial v}(u,\eta) = 0 \qquad \frac{\partial L}{\partial \mu}(u,\eta) = 0$$

The following theorem shows that we can reduce the minimization problem [1.6] to the problem of finding a saddle point of the Lagrangian $L(v,\mu)$:

THEOREM.– If u is a solution of the problem $\min_v J(v)$ with $h(v) = 0$, then there exists a multiplier η such that the pair (u,η) is a saddle point of the Lagrangian $L(v,\mu)$:

$$\inf_{v\in U, h(v)=0} J(v) = \inf_{v\in U} \sup_{\mu\in\mathbb{R}} L(v,\mu) = \sup_{\mu\in\mathbb{R}} \inf_{v\in U} L(v,\mu)$$

Recall that the saddle point (u,η) in question is, by definition, the point that satisfies the double inequality:

$$\forall v, \forall \mu, \qquad L(u,\mu) \le L(u,\eta) \le L(v,\eta)$$

Hence, we can replace the constrained minimization problem [1.6] with the unknown u and with the inequality constraint $h(u) = 0$, with the problem of finding the saddle point of the Lagrangian $L(v,\mu)$ (assuming that there exists a saddle point):

$$\text{find the solution } (u,\eta) \text{ of } \max_{\mu} \min_{v} L(v,\mu) \tag{1.10}$$

This new problem has two unknowns, u and η. On the other hand, it no longer includes an equality constraint $h(u) = 0$. The solution of the problem [1.10] is obtained by following these steps:

Search algorithm for the Lagrangian L saddle point. [1.11]

1. Determine the dual function $H(\mu) \equiv \min_v L(v,\mu)$.

   Minimization is performed by considering μ as a fixed parameter. We know that H is concave.

2. Determine the solution η of $\max_{\mu} H(\mu)$.

3. Once η is known, look for the solution u of $\min_v L(v,\eta) = \min_v \left[J(v) + \eta h(v)\right]$.

   The search is facilitated by the fact that we know the optimal value $L(u,\eta)$, which is equal to $H(\eta)$.

The saddle point problem can be solved numerically by Uzawa's algorithm (1958) (as in the case of the inequality constraint below).

1.2.2. *Problems with inequality constraints*

Now let us move on to problems with inequality conditions. We are interested in the following problem:

$$\text{minimize } J(v) \text{ subject to the constraint } g(v) \leq 0 \qquad [1.12]$$

where $J : U \to \mathbb{R}$ and $g : U \to \mathbb{R}$ are real-valued functions defined on an open subset U of $\mathbb{R}$.

Karush's (1939), and Kuhn and Tucker's (1951) theorem.– Assume that the functions J and g are continuously differentiable. If u is a local minimum of $J(v)$, then there exists a number $\lambda \geq 0$, called the Lagrange multiplier (or the Karush–Kuhn–Tucker multiplier), such that

$$\frac{\partial J}{\partial v}(u) + \lambda \frac{\partial g}{\partial v}(u) = 0 \quad \text{and } \lambda g(u) = 0 \qquad [1.13]$$

In [1.7], the sign of the multiplier η is not constrained, whereas, in [1.13], the multiplier λ should be ≥ 0 and, in addition, there is the complementarity condition $[1.13]_2$, which does not exist for equality constraints. The complementarity condition means that:

- if the constraint is not active for the solution u, that is, $g(u) < 0$, then necessarily $\lambda = 0$;
- if $\lambda > 0$, then the constraint is necessarily active for u: $g(u) = 0$.

According to [1.13], to find a solution to the problem [1.12], we have to solve the following system of equations for the two unknowns, u and λ:

$$\frac{\partial J}{\partial v}(u) + \lambda \frac{\partial g}{\partial v}(u) = 0 \qquad \text{and} \qquad \lambda g(u) = 0 \quad \text{subject to the inequalities } g(u) \leq 0 \text{ and } \lambda \geq 0 \qquad [1.14]$$

The three relations $\lambda g(u) = 0$, $g(u) \leq 0$ and $\lambda \geq 0$, are called *the Karush–Kuhn–Tucker conditions* in optimization theory, and they coincide with the Hertz–Signorini–Moreau conditions [1.2] in contact theory.

Since [1.13] is only a necessary condition, it must be checked that the solution obtained from [1.14] is indeed the desired minimum for problem [1.12].

EXAMPLE.– Let us take the mass-spring system considered above, and solve it via the multiplier method. We must minimize the total potential energy [1.1] subject to the constraint $g(u) = u - h \leq 0$. The system [1.14] to solve is written as

$$ku - mg + \lambda = 0 \qquad \lambda(u - h) = 0 \qquad \text{with} \qquad \lambda \geq 0 \qquad \text{and} \qquad u - h \leq 0$$

There are two exclusive cases:

1. If $u - h < 0$, then the equation $\lambda(u - h) = 0$ implies $\lambda = 0$, and then the equation $ku - mg + \lambda = 0$ gives $u = \dfrac{mg}{k}$.

   The condition $u - h < 0$ to satisfy then becomes $\dfrac{mg}{k} < h$, namely, $mg < kh$.

2. If $u - h = 0$, that is to say $u = h$, then the equation $\lambda(u - h) = 0$ implies that λ can be non-zero, with $\lambda \geq 0$ here. The equation $ku - mg + \lambda = 0$ then gives $\lambda = mg - ku = mg - kh$.

   The condition $\lambda \geq 0$ is then written as $mg - kh \geq 0$, namely, $mg \geq kh$, and is the complementary condition to that of the first case.

The displacement u obtained is exact. On the other hand, the obtained multiplier λ is equal to the exact value of the contact force applied on the particle by the rigid plane.

In the general case, we can solve the system [1.14] numerically by Wilson's method (1963), which is an extension of Newton's method and which can be used in cases that involve inequality constraints. However, its numerical implementation is rather difficult.

- The solution of the system [1.14] requires taking into account two inequality constraints, $g(u) \leq 0$ and $\lambda \geq 0$. We prefer another approach that allows us to reduce the problem [1.12] into an optimization one without the constraint $g(u) \leq 0$. To do this, define the Lagrangian L analogously to [1.9]:

DEFINITION.– *The Lagrangian L* associated to the problem [1.12] is defined by

$$L(v, \mu) \equiv J(v) + \mu g(v) \quad [1.15]$$

The system [1.14] can be written in terms of the Lagrangian L as

$$\frac{\partial L}{\partial v}(u, \lambda) = 0 \qquad \lambda \frac{\partial L}{\partial \mu}(u, \lambda) = 0 \qquad \frac{\partial L}{\partial \mu}(u, \lambda) \leq 0 \qquad \lambda \geq 0$$

We can transform the minimization problem [1.12] into the problem of finding the saddle point of the Lagrangian $L(v, \mu)$, thanks to the following theorem.

THEOREM.– If (u, λ) is a saddle point of the Lagrangian $L(v, \mu)$ on the set of the admissible v and on the set where $\mu \geq 0$, then u is a solution of $\min\limits_{v} J(v)$ subject to $g(v) \leq 0$.

The saddle point (u, λ) in question is, by definition, the point which satisfies the double inequality:

$$\forall v, \forall \mu \geq 0, \qquad L(u, \mu) \leq L(u, \lambda) \leq L(v, \lambda)$$

In the case of convex functions, we have the following result, analogous to the Karush–Kuhn–Tucker theorem:

THEOREM.– [1.16]

Assume that J and g are convex and that the regularity conditions hold.

If u is a solution of the problem $\min\limits_{v} J(v)$ subject to $g(v) \leq 0$, then there exists a multiplier $\lambda \geq 0$ such that the pair (u, λ) is a saddle point of the Lagrangian $L(v, \mu)$:

$$\inf_{v \in U, g(v) \leq 0} J(v) = \inf_{v \in U} \sup_{\mu \geq 0} L(v, \mu) = \sup_{\mu \geq 0} \inf_{v \in U} L(v, \mu)$$

Hence, we can replace the constrained minimization problem [1.12] with the unknown u and with the inequality constraint $g(u) \leq 0$, with the problem of finding a saddle point for the Lagrangian $L(v, \mu)$ (assuming there is a saddle point):

$$\text{look for a solution } (u, \lambda \geq 0) \text{ of } \max_{\mu \geq 0} \min_{v} L(v, \mu) \quad [1.17]$$

This new problem has two unknowns, u and λ. On the other hand, it no longer includes the inequality constraint $g(u) \leq 0$. The constraint $\mu \geq 0$ is easy to deal with. The solution of the problem [1.17] is found by the following steps:

Algorithm for finding a saddle point of the Lagrangian L (inequality constraint case). [1.18]

1. On the set where $\mu \geq 0$, determine the dual function $H(\mu) \equiv \min_v L(v,\mu)$.

   The minimization is performed by considering $\mu \geq 0$ as a fixed parameter. We know that H is concave.

2. Determine a solution λ of $\max_{\mu \geq 0} H(\mu)$.

   It is necessary to maximize $H(\mu)$, while remaining in the set where $\mu \geq 0$.

3. Once λ is known, we look for the solution u of $\min_v L(v,\lambda) = \min_v \left[J(v) + \lambda g(v)\right]$.

   This search is facilitated by the fact that we know the optimal value $L(u,\lambda)$, which is equal to $H(\lambda)$.

EXAMPLE.– Let us solve the problem with the mass-spring system above, by following the algorithm [1.18]. The Lagrangian [1.15] for the considered problem is

$$L(v,\mu) = \frac{1}{2}kv^2 - mgv + \mu(v-h)$$

1. Let us construct the dual function $H(\mu) \equiv \min_v L(v,\mu)$. We have:

$$H(\mu) = \min_v \left[\frac{1}{2}kv^2 - mgv + \mu(v-h)\right] = \min_v \left[\frac{1}{2}kv^2 + (\mu - mg)v - \mu h\right]$$

   Since the derivative with respect to v of the expression in the square brackets is $kv+\mu-mg$, the minimum is reached at $v = (mg-\mu)/k$. Hence, for the H function:

$$H(\mu) = \frac{1}{2k}\left[-\mu^2 + 2\mu(mg-kh) - m^2g^2\right] \qquad (H \text{ is concave})$$

2. Let us determine the solution λ of $\max_{\mu \geq 0} H(\mu)$. The derivative of H, equal to $\frac{1}{k}[-\mu + mg - kh]$, vanishes at $\mu = mg - kh$. From it, we obtain the maximum of H on the set where $\mu \geq 0$, which depends on the sign of $mg - kh$ (see Figure 1.7):

   (a) if $mg - kh \geq 0$, then the maximum of H is attained at the value of μ which we have just found: $\lambda = mg - kh$;

   (b) if $mg - kh < 0$, then the maximum of H is attained at $\lambda = 0$.

3. Search for a solution u of $\min_v L(v,\lambda)$:

   (a) if $mg - kh \geq 0$, then $\lambda = mg - kh$ and so $L(v,\lambda) = \frac{1}{2}kv^2 - khv - mgh + kh^2$. The minimum of $L(v,\lambda)$ is attained at $u = h$;

(b) if $mg - kh < 0$, then $\lambda = 0$ and so $L(v, \lambda) = \frac{1}{2}kv^2 - mgv$. The minimum of $L(v, \lambda)$ is attained at $u = \frac{mg}{k}$.

We obtain the exact solution.

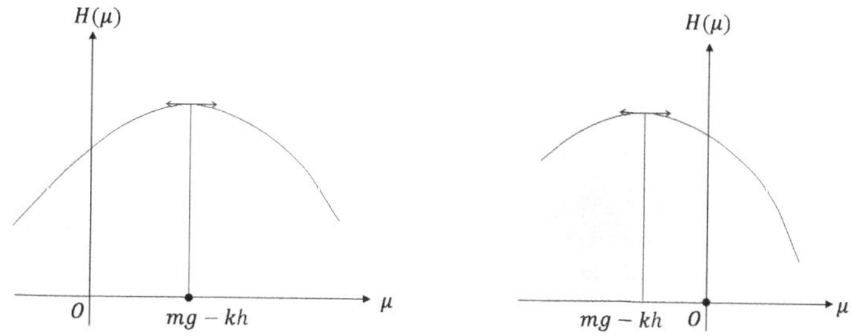

(a) Case $mg - kh \geq 0$ (b) Case $mg - kh < 0$

Figure 1.7. *Solution maximizing the dual function $H(\mu)$ on the set of $\mu \geq 0$. The solution λ is represented by the thick black point* ●

In the previous analytical solution, tasks 1 and 2 of algorithm [1.18] are carried out successively. However, for a numerical solution, task 1 is not straightforward: the function $H(\mu)$ cannot be constructed numerically. For this reason, in programming, we combine the two tasks, tasks 1 and 2, into a single one called *the dual problem*:

$$\text{look for a solution } \lambda \geq 0 \text{ for } \max_{\mu \geq 0} H(\mu) = \max_{\mu \geq 0} \min_{v} L(v, \mu) \qquad [1.19]$$

In contrast, the original optimization problem [1.12] is then called the primal problem.

We may solve the dual problem [1.19] numerically by an iterative method, where we construct a sequence of values μ^k that converges to the desired solution λ. Maximizing H is easy if the derivative of H can be computed (or a sub-gradient in the non-differentiable case). The positivity constraint $\mu \geq 0$ is easy to take into account using a projection. A commonly used algorithm for finding a saddle point is Uzawa's algorithm (1958), which is a gradient (or sub-gradient) algorithm.

• The results obtained from the multiplier method recalled above are valid for an objective function $J(v)$ and a constraint function $g(v)$, which are real-valued functions whose variable v is a scalar. These results can be generalized to the framework of three-dimensional mechanical systems, where v is a kinematically admissible vector field and the objective function $J(v)$ is the potential energy functional $\Pi(v)$. Consider the frictionless contact problem of two solids, denoted as 1 and 2. With the notation introduced in the penalty method above and summarized in Figure 1.6, we seek to find the displacement field solution u of the following problem:

$$\text{minimize } \Pi(v) \text{ subject to the constraint } g(v) \leq 0 \text{ at every point of } S_{0c}^{(1)} \qquad [1.20]$$

We then define the Lagrangian associated with this problem by

$$L(v, \mu) \equiv \Pi(v) + \int_{S_{0c}^{(1)}} \mu g(v)\, dS_0$$

where μ is the multiplier function defined on $S^{(1)}_{0c}$: $S^{(1)}_{0c} \ni \mathbf{X} \to \mu(\mathbf{X}) \in \mathbb{R}^+$ (which is abbreviated as $\mu \geq 0$).

THEOREM.– If (u, λ) is a saddle point of the Lagrangian $L(v, \mu)$ on the set of admissible fields v and the set of fields $\mu \geq 0$, then u is a solution of the problem [1.20].

On the set where $\mu \geq 0$, we define the dual functional $H(\mu) \equiv \min_v L(v, \mu)$. If a saddle point exists, then this can be determined by solving the dual problem (where there is no $g \leq 0$ constraint), as in [1.19]:

$$\text{find a solution } \lambda \geq 0 \text{ of } \max_{\mu \geq 0} H(\mu)$$

We can formulate a contact problem with friction as a constrained minimization problem. However, the inequality constraints of the tangential contact laws are difficult to translate, and the functional to be minimized contains non-differentiable potentials, one of which is non-classical, since it is parameterized by the field of normal contact constraints.

The multiplier method allows us to obtain a solution that satisfies the problem equations exactly, as well as the inequalities. However, it has some drawbacks:

1. The min-max approach seen above works well when the Lagrangian $L(v, \mu)$ admits a saddle point, since then the Lagrangian is stable in v and the dual function $H(\mu)$ is differentiable. However, there are simple examples in mathematical programming where $L(v, \mu)$ is concave in v (for any fixed μ) and there is no saddle point. On the other hand, in a large number of optimization problems (especially in linear problems), the function $J(v)$ is not strictly convex, and it may happen that the stability and differentiability conditions are not satisfied.

   When the dual function $H(\mu)$ is not differentiable, use sub-gradient algorithms in order to maximize H.

2. In mechanical terms, the displacement field solution is sought by solving an optimization problem, namely that of finding a saddle point. This is not the usual approach in structural mechanics where the principle of virtual power is discretized using the finite element method, and where the resulting matrix system is solved. It is not straightforward to incorporate an optimization program into a classical finite element code in mechanics.

3. The multiplier method, as presented previously, cannot be generalized to include dynamic contact, where the functional to be minimized does not exist.

In mechanics, the proximity g is in general a complex function of the displacement field v, and it is difficult to solve the dual problem [1.19] and hence determine the contact stress field λ. In particular, we do not know at which points the contact is active or not.

Chan and Tuba [21], and Hughes et al. [64] used the multiplier method to solve the contact problem numerically in the context of small perturbations and in the absence of friction. In order to exploit the method, the authors had to carry out a strategy in which the contact condition is activated, where one distinguishes between two states: in contact and out of contact. If contact at a point is detected, then the normal contact law is expressed as an equality constraint, which forces the proximity to zero ($g = 0$) at this point, and the equality constraint is ensured by a Lagrange multiplier λ, denoted λ_N, equal to the contact pressure at this point. Belgacem et al. [10] combined the mortar method with that of the multipliers. Crisfield [28] proposed a patch test to evaluate the quality of the linear and quadratic contact elements. Focusing solely on small deformations, Jones and Papadopoulos [68] proposed a patch test in the case of a contact problem with friction.

Extensions of this method to the framework of large deformations have been carried out by Papadopoulos and Solberg [107], and Hüeber and Wohlmuth [61], with the latter work being carried out in combination with the mortar method. That said, the Lagrange multiplier method can only be easily used in contact problems with small perturbations, where the contact zone is known a priori. It is poorly suited to the framework involving large deformations, where the contact zones are a priori unknown. Using this method in the presence of friction is difficult.

1.3. Numerical solution of a contact problem by the augmented Lagrangian method

The augmented Lagrangian method was first introduced in mathematical programming with the aim of combining the advantages of both the penalty and the multiplier methods, while avoiding their respective drawbacks.

We are going to explain how to solve a contact problem using the augmented Lagragian method by putting ourselves in the frictionless framework of a contact problem in order to simplify the exposition. With the notations introduced in the penalty method above, and summarized in Figure 1.6, we seek the displacement field solution u of the following problem:

$$\min_{v} \left[\Pi(v) + \frac{1}{2} \int_{S_{0c}^{(1)}} \epsilon \, \langle g(v) \rangle^2 dS_0 \right] \text{ subject to the condition } g(v) \leq 0 \text{ at every point on } S_{0c}^{(1)} \quad [1.21]$$

where $\epsilon > 0$ is a fixed chosen number and v denotes a kinematically admissible field. This expression combines the two penalty and multiplier methods:

(i) The $\frac{1}{2}\epsilon \, \langle g(v) \rangle^2$ term in the objective function arises from the artificial springs added onto the target surface, as we did in the penalty method.

(ii) The $g(v) \leq 0$ condition means that we want to respect the non-penetration condition exactly, as we did in the multiplier method, instead of only roughly checking it as had been done in the penalty method.

If the solution is exact, then we must find a possible solution u that corresponds to the case involving effective contact, that is, a solution that satisfies $g = 0$. However, the function $g \mapsto \langle g \rangle^2$ is precisely not differentiable at $g = 0$. To avoid this differentiability problem, the idea is to modify the objective function in [1.21] using the proposition stated below.

In what follows, s denotes a real-valued function defined on $S_{0c}^{(1)}$ of the same nature as g: $S_{0c}^{(1)} \ni \mathbf{X} \to s(\mathbf{X}) \in \mathbb{R}$.

PROPOSITION.– With v fixed, we have:

$$\min_{s} \int_{S_{0c}^{(1)}} \left[g(v) + s^2 \right]^2 dS_0 = \int_{S_{0c}^{(1)}} \langle g(v) \rangle^2 \, dS_0 \quad [1.22]$$

PROOF. Let r be the argument function on the left-hand side of [1.22], and δr an arbitrary variation of r. The corresponding variation of the functional on the left-hand side of [1.22] is zero:

$$0 = \delta \int_{S_{0c}^{(1)}} \left[g(v) + s^2 \right]^2 dS_0 = 4 \int_{S_{0c}^{(1)}} \left[g(v) + r^2 \right] r \, \delta r \, dS_0$$

As the variation δr is arbitrary, the previous relation implies that

$$\left[g(v) + r^2\right] r = 0 \text{ at every point } \mathbf{X} \in S_{0c}^{(1)} \qquad [1.23]$$

(a) At every point $\mathbf{X} \in S_{0c}^{(1)}$ where $g(v) > 0$, we have $g(v) + r^2 > 0$, and the relation [1.23] implies therefore that $r = 0$.

(b) At every point $\mathbf{X} \in S_{0c}^{(1)}$ where $g(v) \leq 0$, the relation [1.23] implies that $r^2 = -g(v)$ or $r = 0$. We can easily check that the second solution $r = 0$ should be discarded, because it provides a value of $\int_{S_{0c}^{(1)}} \left[g(v) + r^2\right]^2 dS_0$ greater than the value corresponding to the first solution, $r^2 = -g(v)$.

To combine the two cases, at any point of $S_{0c}^{(1)}$ we write that $g(v) + r^2 = \langle g(v) \rangle$. ■

Using [1.22], we can recast the minimum in [1.21] as follows:

$$\begin{aligned}
\min_{v} \left[\Pi(v) + \frac{1}{2} \int_{S_{0c}^{(1)}} \epsilon \langle g(v) \rangle^2 dS_0 \right] &= \min_{v} \left\{\Pi(v) + \min_{s} \frac{1}{2} \int_{S_{0c}^{(1)}} \epsilon \left[g(v) + s^2\right]^2 dS_0 \right\} \\
&= \min_{v} \min_{s} \left\{\Pi(v) + \frac{1}{2} \int_{S_{0c}^{(1)}} \epsilon \left[g(v) + s^2\right]^2 dS_0 \right\} \\
&= \min_{v,s} \left\{\Pi(v) + \frac{1}{2} \int_{S_{0c}^{(1)}} \epsilon \left[g(v) + s^2\right]^2 dS_0 \right\}
\end{aligned}$$

The second idea, due to Rockafellar (1973), consists of rewriting the inequality constraint $g(v) \leq 0$ in [1.21] into the form of an equality constraint $h(v, s) \equiv g(v) + s^2 = 0$.

By combining the two previous ideas, the problem [1.21] is transformed into the problem of finding a pair of functions (u, r) that solves

$$\min_{v,s} \left\{\Pi(v) + \frac{1}{2} \int_{S_{0c}^{(1)}} \epsilon \left[g(v) + s^2\right]^2 dS_0 \right\} \text{ subject to } h(v, s) \equiv g(v) + s^2 = 0 \qquad [1.24]$$

The Lagrangian associated with the problem [1.24] is given by [1.9]:

$$L(v, s, \mu) \equiv \Pi(v) + \int_{S_{0c}^{(1)}} \left\{\frac{1}{2} \epsilon \left[g(v) + s^2\right]^2 + \mu \left[g(v) + s^2\right]\right\} dS_0 \qquad [1.25]$$

where μ is the multiplier function, defined on $S_{0c}^{(1)} : S_{0c}^{(1)} \ni \mathbf{X} \to \mu(\mathbf{X}) \in \mathbb{R}$.

REMARK. In fact, we should have written $h(v, s') \equiv g(v) + s'^2 = 0$, where s' is a function independent of s, and thus the two functions s and s' should actually appear in the Lagrangian [1.25]. However, in the construction of the augmented Lagrangian that we have presented here, we chose to set $s' = s$. ■

From [1.10], we can replace the minimization problem [1.24] with the problem of finding a saddle point of this Lagrangian L:

$$\text{find a solution } (u, r, \eta) \text{ of } \max_{\mu} \min_{v,s} L(v, s, \mu) \qquad [1.26]$$

It is straightforward to determine $\min_s L(v, s, \mu)$ with v, μ fixed. Indeed, we have:

$$\min_s L(v, s, \mu) = \Pi(v) + \min_s \int_{S_{0c}^{(1)}} \left\{ \frac{1}{2}\epsilon \left[g(v) + s^2\right]^2 + \mu \left[g(v) + s^2\right] \right\} dS_0 \qquad [1.27]$$

Let r be the function that minimizes the integral and δr an arbitrary variation of r. The corresponding variation of the functional $\int_{S_{0c}^{(1)}}$ is zero:

$$0 = \delta \int_{S_{0c}^{(1)}} \{\cdots\} dS_0 = 2 \int_{S_{0c}^{(1)}} \left[\epsilon r^2 + \mu + \epsilon g(v)\right] r \, \delta r \, dS_0$$

We continue to reason analogously here as we did in the proof of [1.22]. As the variation δr is arbitrary, the previous relation implies that:

$$\left[\epsilon r^2 + \mu + \epsilon g(v)\right] r = 0 \text{ at every point } \mathbf{X} \in S_{0c}^{(1)} \qquad [1.28]$$

(a) At every point $\mathbf{X} \in S_{0c}^{(1)}$ where $\mu + \epsilon g(v) > 0$, we have $\epsilon r^2 + \left[\mu + \epsilon g(v)\right] > 0$, and the relation [1.28] implies therefore that $r = 0$. Consequently, the integrand of [1.27] can be written at these points as

$$\frac{1}{2}\epsilon \left[g(v) + r^2\right]^2 + \mu \left[g(v) + r^2\right] = \frac{1}{2}\epsilon g^2(v) + \mu g(v)$$

(b) At every point $\mathbf{X} \in S_{0c}^{(1)}$ where $\mu + \epsilon g(v) \leq 0$, the relation [1.28] implies that $r^2 = -\frac{1}{\epsilon}\left[\mu + \epsilon g(v)\right]$ (the solution $r = 0$ is eliminated). The integrand of [1.27] can be written at these points as

$$\frac{1}{2}\epsilon \left[g(v) + r^2\right]^2 + \mu \left[g(v) + r^2\right] = -\frac{\mu^2}{2\epsilon}$$

The two previous cases can be grouped together according to the following expression, valid at any point of $S_{0c}^{(1)}$:

$$\frac{1}{2}\epsilon \left[g(v) + r^2\right]^2 + \mu \left[g(v) + r^2\right] = \frac{1}{2\epsilon} \left[\langle \mu + \epsilon g(v) \rangle^2 - \mu^2\right]$$

Thus, we have arrived at a new Lagrangian, called the augmented Lagrangian, and which replaces the Lagrangian in [1.25]:

Definition.– The augmented Lagrangian associated with the problem [1.21], denoted L^{aug}, is defined by:

$$L^{aug}(v, \mu) \equiv \Pi(v) + \int_{S_{0c}^{(1)}} \frac{1}{2\epsilon} \left[\langle \mu + \epsilon g(v) \rangle^2 - \mu^2\right] dS_0 \qquad [1.29]$$

This allows us to reformulate the problem [1.26] into the terms of a saddle point problem for the augmented Lagrangian:

$$\text{look for a solution } (u, \eta) \text{ of } \max_\mu \min_v L^{aug}(v, \mu) \qquad [1.30]$$

As the saddle point (u,η) is a stationary point of L^{aug}, we have to solve the following variational equations:

$$\forall \delta u, \forall \delta \eta, \qquad \begin{cases} \delta_v L^{aug}(u,\eta) = 0 \\ \delta_\mu L^{aug}(u,\eta) = 0 \end{cases} \Leftrightarrow \begin{cases} \delta \Pi(u) + \displaystyle\int_{S_{0c}^{(1)}} \langle \eta + \epsilon g(u) \rangle \, \delta g(u) \, dS_0 = 0 \\ \displaystyle\int_{S_{0c}^{(1)}} \left[\langle \eta + \epsilon g(u) \rangle - \eta \right] \delta \eta \, dS_0 = 0 \end{cases} \qquad [1.31]$$

Equation $[1.31]_2$ implies that $\langle \eta + \epsilon g(u) \rangle = \eta$ at every point of $S_{0c}^{(1)}$, and we can check that this relation is equivalent to the Karush–Kuhn–Tucker conditions:

$$\langle \eta + \epsilon g(u) \rangle = \eta \qquad \Leftrightarrow \qquad \eta g(u) = 0, g(u) \leq 0, \eta \geq 0$$

On the other hand, by setting $\langle \eta + \epsilon g(u) \rangle = \eta$ in equation $[1.31]_1$, we arrive at

$$\delta_v \Pi(u) + \int_{S_{0c}^{(1)}} \eta \, \delta_v g(u) \, dS_0 = 0$$

We see that the system [1.31] is the same as the system [1.14], which is to be solved in the multiplier method and which provides the exact solution. Moreover, here we obtain the exact solution regardless of the chosen value for the penalty parameter ϵ.

EXAMPLE.– Let us return to the example of the mass-spring system considered above, and solve it using the augmented Lagrangian method. Here, the augmented Lagrangian [1.29] is written as

$$\Pi^{aug}(v,\mu) = \frac{1}{2}kv^2 - mgv + \frac{1}{2\epsilon}\left[\langle \mu + \epsilon(v-h) \rangle^2 - \mu^2 \right] \qquad (\epsilon > 0)$$

The system [1.31] is written as

$$\begin{aligned} 0 &= \frac{\partial \Pi^{aug}}{\partial v}(u,\eta) = ku - mg + \langle \eta + \epsilon(u-h) \rangle \\ 0 &= \frac{\partial \Pi^{aug}}{\partial v}(u,\eta) = \frac{1}{\epsilon}\left[\langle \eta + \epsilon(u-h) \rangle - \eta \right] \end{aligned} \qquad [1.32]$$

We can distinguish between two exclusive cases:

1. Suppose $\eta + \epsilon(u-h) < 0$, then the system [1.32] gives:

$$\begin{cases} ku - mg = 0 \\ -\eta = 0 \end{cases} \qquad \Rightarrow \qquad \begin{cases} u = mg/k \\ \eta = 0 \end{cases}$$

The condition $\eta + \epsilon(u-h) < 0$ that we have to satisfy is then written as $u - h < 0$, namely $mg < kh$;

2. Suppose $\eta + \epsilon(u-h) \geq 0$, then the system [1.32] gives:

$$\begin{cases} ku - mg + \eta + \epsilon(u-h) = 0 \\ \eta + \epsilon(u-h) - \eta = 0 \end{cases} \qquad \Rightarrow \qquad \begin{cases} u = h \\ \eta = mg - kh \end{cases}$$

The condition $\eta + \epsilon(u-h) \geq 0$ is then written as $\eta \geq 0$, namely $mg \geq kh$, which is the complementary condition to that of the first case.

We get the exact solution.

As we had hoped for, the augmented Lagrangian method allows us to combine the advantages of both the penalty and multiplier methods, while simultaneously avoiding their disadvantages:

1. We have seen that the multiplier method does not work in certain optimization problems, namely, when the Lagrangian $L(v, \mu)$ defined by [1.15] does not possess a saddle point, or when the Lagrangian is unstable with respect to v, or when the dual function $H(\mu)$ is not differentiable.

   The augmented Lagrangian method avoids these pitfalls. It regularizes the contact laws as in the penalty method, and it turns out that the obtained augmented Lagrangian indeed admits a saddle point, is generally stable, and gives rise to a differentiable dual functional.

2. With the multiplier method, we have to either solve the system of equations [1.14] which contains the inequality constraint $g(u) \leq 0$, or the saddle point problem [1.17] which contains the $\mu \geq 0$ condition.

   With the augmented Lagrangian method, the saddle point problem [1.30], as well as the system of equations [1.31], does not contain any inequality constraints. This is an advantage in mechanics, since, while inequality constraints are common in mathematical programming, they are not easy to deal with in finite element code for structural mechanics.

3. Like the multiplier method, the augmented Lagrangian method allows us to obtain the exact solution, in particular, it makes it possible to satisfy the inequality contact laws exactly. Moreover, we obtain the exact solution regardless of the chosen value for the penalty parameter ϵ.

   In this regard, the augmented Lagrangian method is an improvement over the penalty method, insofar as we are not obliged to use penalty parameters that are too large, and also that we consequently avoid having an ill-conditioned tangent matrix.

   In practice, the independence of the result with respect to the choice of the parameter ϵ is not entirely verified. We note that, in complicated problems, it is necessary to remain within a specific range of values for ϵ, otherwise the numerical computation converges very slowly or may not even converge at all.

4. Notice, as in the penalty method, that even if the augmented Lagrangian method was described in a static problem with the aim of minimizing a functional, it also applies to contact problems in dynamics.

   Indeed, one can transform the system of equations [1.31] into a weak form which generalizes the principle of virtual power in statics, by formally making any perturbations into a virtual velocity. The weak form obtained is mixed, in the sense that it involves both the displacement fields $\mathbf{U}$ of the two bodies and the multiplier field $\mathbf{\Lambda}$ defined on the contactor surface. The form weak in dynamics is deduced from this by simply adding the virtual power from the accelerations. It is advisable to check that the weak form hence obtained is indeed equivalent to the local movement equation of each body and also the contact laws. Then, by discretizing this weak form using the finite element method, we arrive at a coupled matrix system with the unknowns $\mathbf{U}$ and $\mathbf{\Lambda}$. Solving this system of equations produces the solution of the dynamical contact problem.

   If we stay in statics, we can numerically solve the saddle point problem [1.30] using Uzawa's algorithm. That being said, in mechanics, we prefer to work with a weak form that resembles the principle of virtual power as we have just explained above, rather than work with mathematical programming tools.

In view of its many advantages, the augmented Lagrangian method is recognized as one of the most effective ways to solve the contact problem in mechanics where we must deal with highly nonlinear contact conditions. However, this is not as easy to implement as in the frictionless case seen above.

Indeed, in optimization theory, when faced with a given minimization problem such as [1.12] – where the functional to be minimized and the constraint are real-valued functions, depending on a real variable – there may be different ways of creating the augmented Lagrangian; consequently, there can be several possible expressions for the augmented Lagrangian.

The situation is of course more difficult in a contact problem with friction, where we work with vector fields, and where the inequalities to be respected are more complicated. There are various different ways to construct an augmented Lagrangian, and implementing them numerically is much more complicated.

Several authors have sought to apply the augmented Lagrangian method to mechanics. In the absence of contact, the problem of solids and incompressible fluids was solved by Glowinski and Le Tallec [44] using this method.

For the contact problem with small deformations, the theoretical question regarding the convergence of various formulations involving the augmented Lagrangian was studied by Stadler [129], in the context of linear elasticity. Numerical solutions were handled by Alart and Curnier [2] and Zavarise et al. [145].

Glowinski and Le Tallec [44] were among the first to solve the frictionless contact problem with large deformations and with a rigid obstacle plane using the same approach, before being followed by Heegard and Curnier [54], who considered the contact problem between two elastic bodies, by Zavarise and Wriggers [144] and later by Harnau et al. [52]. The frictional contact problem with large deformations between deformable bodies has been successfully solved by Simo and Laursen [121, 83], including the algorithmic symmetrization of the contact tangent matrix by Laursen and Oancea [81], Wriggers [137], and Pietrzak and Curnier [111, 112], by Feng et al. [37], for the hyperelastic bodies of Blatz-Ko, Mijar and Arora [93], Puso and Laursen [115] (who combined the augmented Lagrangian method with the mortar method) and, later, by Konyukhov and Schweizerhof [75], who generalized Uzawa's symmetrization technique used in [121] to the anisotropic friction-adhesion model.

In most works, the tangential contact stresses are obtained by integrating the contact laws locally, following the radial-return algorithm in elastoplasticity (see Giannakopoulos [43]), then the contact tangent matrix is deduced in a consistent way from the contact stresses. Curnier and his colleagues [2, 54, 112] chose another approach, by directly incorporating the contact laws into their augmented Lagrangian formulations, allowing them to obtain the tangential contact stresses correctly, without having to resort to the local integration.

To globally solve the discrete system of equations in the $(\mathbf{U}, \mathbf{\Lambda})$, Uzawa's method has often been used, owing to its simplicity: the $\mathbf{\Lambda}$ multipliers are updated in an outer augmentation loop, while the displacements are calculated in a standard way in an inner loop, where the multipliers are fixed. However, the major drawback of Uzawa's method is that its global convergence speed is not quadratic due to the augmentation loop. To avoid this problem, it suffices to apply the Newton–Raphson method to the entire coupled system $(\mathbf{U}, \mathbf{\Lambda})$, even if it means more complex programming (see Zavarise and Wriggers [144], for example). Even better, other authors, such as Alart and Curnier [2] and Pietrzak and Curnier [112] used a modified Newton scheme to take into account the non-differentiability of the contact equations and obtained the results with an even better rate of convergence.

In this chapter, we voluntarily limited ourselves to describe the penalty method, the Lagrange multiplier method and the augmented Lagrangian method, which are the three basic methods used to solve contact problems. There are, however, other methods that solve the contact problem numerically, for example, the perturbed Lagrangian method, which resembles that of the augmented Lagrangian, since it includes both penalty terms and multipliers (see Simo et al. [125]). We can also exploit Nitsche's method when dealing with frictional contact conditions, as seen in Wriggers and Zavarise [140] and Oliver et al. [103] and Hartmann et al. [53].

1.4. Book synopsis

The aim of this textbook is to propose a new formulation for either the static contact problem or the dynamic one with Coulomb friction, and which provides the same advantages as those with augmented Lagrangian-type formulations mentioned above. The content of this book is organized as follows.

Chapter 2 is devoted to the kinematics of bodies, that is to say their movement and their strains, and in particular the kinematics of contact between two bodies. Within the framework of finite deformations, we will present the first essential contact notions such as the parameterization of the target surface and the contactor surface, the projection point on the target surface and the proximity. We will study the derivatives of the kinematic quantities with respect to time, in particular the slip vector, which is an important quantity when we study contact with friction.

In Chapter 3, we will be interested in forces and stresses: the external forces being exerted on each body within the system under consideration, the internal forces in each body, called stresses, and in particular the interaction forces between the bodies that appear in a contact problem. The essential notions of this chapter are the contact stress vector, contact pressure and frictional stresses.

Chapter 4 is devoted to describing the constitutive laws of the bodies. As these are present in any study of mechanics, this chapter does not contain anything specific to contact problems.

Chapter 5 deals with the contact laws. These are the special boundary conditions that are included with the classical ones (prescribed displacements and/or stresses), and which are essential in a contact problem.

In this work, we will only consider normal contact laws without adhesion, and tangential contact laws with Coulomb's law of dry friction. The contact laws are presented as relations between the kinematic contact quantities (studied in Chapter 2) and the contact stresses (studied in Chapter 3) similar to the material constitutive laws of the materials, which relate strains to stresses in the bodies.

Chapter 6 describes the strong form of the contact problem. It collects the various relations seen in the preceding chapters together, which are needed for the solution of the contact problem with finite perturbations. This collection forms a system of coupled partial derivatives equations in time and space, accompanied by the inequalities that are specific to the contact problem.

The strong problem is important for the mechanical problem to be well-posed but it is already difficult enough to solve as it is. Thus, in Chapter 7, we will introduce the weak formulation with the aim of solving the problem numerically via the finite element method. Analogously to what is done for the principle of virtual power, we will proceed in two steps:

1. First, we will first state the weak form as a postulate.
2. Then, we will prove that this weak form is equivalent to the strong form seen in Chapter 6. In particular, the weak form restores the contact laws on the contact surfaces.

The proposed weak form has certain analogies with that of Pietrzak and Curnier [112], but it is not obtained using the augmented Lagrangian method. It generalizes the principle of virtual power to contact problems and is therefore a weighted residuals relation. However, it is mixed in the sense that it involves both the displacement fields of the bodies and the multiplier fields on the contactor surface, with the field of virtual velocities and the field of virtual multipliers as virtual fields. As the weak form is written in the form of a weighted residuals relation, it lends itself well to the spatial discretization used in the finite element method.

In Chapter 8, we will discretize the weak form of the previous chapter by means of the finite element method, from which we will obtain a system of coupled nonlinear differential equations in time, whose unknowns are both the nodal displacements and the nodal multipliers.

In Chapter 9, we will restrict ourselves to the static (or quasi-static) framework, in order to reduce the equations coming from the previous chapter to a system of nonlinear algebraic equations. We will explain how this system can be solved using the iterative Newton–Raphson scheme, and we will detail the expression for the contact tangent matrix, which is the essential ingredient for each iteration of the computation.

Chapter 10 contains the static numerical examples obtained by applying the method described in the previous chapter, and allowing for the formulation performance to be tested.

In Chapter 11, we will solve the contact problem in the context of dynamics. The discretization of the system of differential equations that govern the problem with respect to time will be carried out using the standard Newmark scheme. We will thus be led to deal with a new nonlinear algebraic system, which in turn can be solved by an iterative Newton–Raphson type scheme.

The Newmark scheme is chosen because of how simple it is to implement, but there are other more sophisticated and efficient schemes. The search for a good temporal integration scheme is important when it comes to processing the dynamic contact problem, but doing so would exceed the scope of this work.

Examples of dynamic contact problems will be presented in Chapter 12.

The Appendix at the end of this book deals with calculating the variation in the different kinematic quantities that are defined on the contact surfaces. The results obtained are used to establish the expression of the contact tangent matrix in Chapter 9.

2

Contact Kinematics

Kinematics involves the study of the movement and strains of bodies, independently of what causes the motion. For a problem where certain bodies are likely to enter into contact with each other, it is particularly interesting to study their positions and motions relative to one another.

This chapter is devoted to kinematics and, in particular, to the kinematics of contact. We will present a mechanical problem in the three-dimensional and dynamic framework, given that either the two-dimensional or static problems can be easily obtained from these with a few obvious adjustments.

In general, a contact problem may involve several bodies. However, since contact between several bodies is equivalent to multiple interactions between several pairs of bodies, we can restrict our focus to the study of contact between just two bodies.

The main concepts introduced in this chapter are the following:

- the target surface, the contactor surface and their parameterizations;
- the orthogonal projection of a point from the contactor surface onto the target surface. The proximity g, which is an essential scalar quantity, letting us know whether two bodies are in contact at any given point or not;
- the time derivatives of the kinematic quantities. The results obtained from this will inform us of how the kinematic quantities evolve with respect to time when the contact surfaces evolve, during the motions of the bodies;
- certain time derivatives that have been studied previously allow us to introduce the relative velocity, which is an important quantity when studying contact problems with friction.

2.1. Motions and strains

We work in a Galilean reference frame R_0. At a given time, the position of a particle in physical space is represented by a point in the three-dimensional affine space $\mathcal{E}$. If a point in $\mathcal{E}$ has been chosen to serve as the origin, then we can identify the affine space $\mathcal{E}$ with its corresponding vector space, should this be necessary.

We will describe the mechanical problem according to the Lagrangian point of view. Consider two moving bodies in R_0, which we shall call body no. 1 and body no. 2 as in Figure 2.1. Their reference positions are $\Omega_0^{(1)}$ and $\Omega_0^{(2)} \subset \mathcal{E}$, and their current positions at the instant t, where t lies within the time interval $[0, t_{max}]$, are $\Omega^{(1)}$ and $\Omega^{(2)} \subset \mathcal{E}$. The reference position of the current particle of each body $i \in \{1,2\}$ is denoted by $\mathbf{X}^{(i)} \in \Omega_0^{(i)}$, and its current position at time t is denoted $\mathbf{x}^{(i)} \in \Omega^{(i)}$. The deformation of each body $i \in \{1,2\}$ at time t is defined by

$$\begin{aligned} \boldsymbol{\phi}^{(i)}(.,t) : \ \Omega_0^{(i)} &\rightarrow \mathcal{E} \\ \mathbf{X}^{(i)} &\mapsto \mathbf{x}^{(i)} = \boldsymbol{\phi}^{(i)}(\mathbf{X}^{(i)},t) \end{aligned} \qquad [2.1]$$

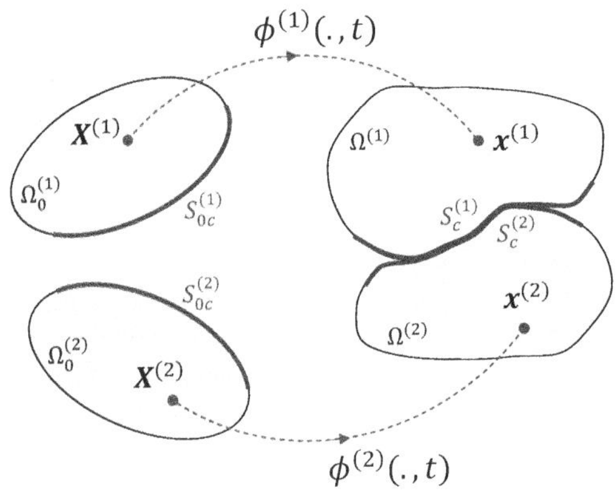

Figure 2.1. *Definition of the unilateral contact problem. For a color version of this figure, see www.iste.co.uk/levan/contact.zip*

The displacement of a particle of each body $i \in \{1,2\}$ at time t is defined by

$$\mathbf{U}^{(i)}(\mathbf{X}^{(i)},t) = \mathbf{x}^{(i)} - \mathbf{X}^{(i)} \qquad [2.2]$$

The velocity and the acceleration of the same particle at time t are defined by

$$\mathbf{V}^{(i)}(\mathbf{X}^{(i)},t) \equiv \frac{d\mathbf{U}^{(i)}}{dt}(\mathbf{X}^{(i)},t) \qquad \boldsymbol{\Gamma}(\mathbf{X}^{(i)},t) \equiv \frac{d\mathbf{V}^{(i)}}{dt}(\mathbf{X}^{(i)},t) = \frac{d^2\mathbf{U}^{(i)}}{dt^2}(\mathbf{X}^{(i)},t) \qquad [2.3]$$

The deformation gradient tensor $\mathbf{F}^{(i)}$ of body i is defined by:

$$\mathbf{F}^{(i)}(\mathbf{X}^{(i)},t) = \frac{\partial \boldsymbol{\phi}^{(i)}(\mathbf{X}^{(i)},t)}{\partial \mathbf{X}^{(i)}} \qquad [2.4]$$

We will also use the right Cauchy stretch tensor $\mathbf{C}^{(i)}$ and the Green strain tensor $\mathbf{E}^{(i)}$ of body i, respectively:

$$\mathbf{C}^{(i)} = \mathbf{F}^{(i)T}.\mathbf{F}^{(i)} \qquad [2.5]$$

$$\mathbf{E}^{(i)} = \frac{1}{2}(\mathbf{C}^{(i)} - \mathbf{I}) \qquad [2.6]$$

where $\mathbf{I}$ is the unit tensor.

2.2. Potential contact surfaces

Throughout their motions, the bodies may end up coming into contact with one another for any period of time lying within the interval $[0, t_{max}]$. For body number i, its reference boundary $S_0^{(i)}$ is partitioned into three disjoint sets, denoted by $S_{0U}^{(i)}$, $S_{0T}^{(i)}$, and $S_{0c}^{(i)}$, where

- $S_{0U}^{(i)}$ and $S_{0T}^{(i)}$ are the boundary portions where the displacement and the stress vectors are prescribed, respectively.

- $S_{0c}^{(i)}$ is the boundary portion where contact is most likely to take place.

The effective contact regions between the bodies evolve over time and are not known beforehand; therefore, their determination is part of the solution of the contact problem. The reference potential contact surfaces for $S_{0c}^{(1)}$ and $S_{0c}^{(2)}$ must also be chosen and be large enough to include any particle likely to come into contact during our analysis.

The current positions of $S_0^{(i)}, S_{0U}^{(i)}, S_{0T}^{(i)}$ and $S_{0c}^{(i)}$ are denoted by $S^{(i)}, S_U^{(i)}, S_T^{(i)}$ and $S_c^{(i)}$, respectively, meaning that $S_c^{(i)} = \boldsymbol{\phi}^{(i)}(S_{0c}^{(i)}, t)$, for example. Along the sections of $S_c^{(i)}$ where contact takes place, the stresses are not known and generally neither are the displacements. We say "generally" since, in some exceptional cases, $S_c^{(i)}$ may contain points with prescribed displacements.

To describe the relative motion of the two bodies, it is convenient to pick one of the two bodies, say body number 2, to act as the reference and to describe the proximity of the other body, in this case body number 1, relative to it. Body number 2 will be called *the target* (or *the master*), and body number 1 will be called *the contactor* (or *the slave*). Similarly, the potential contact surface $S_c^{(2)}$ will be called *the target surface*, and that of $S_c^{(1)}$ *the contactor surface*.

We assume that the reference contact surface $S_{0c}^{(1)}$ is parameterized by a mapping that we denote by $\boldsymbol{\theta}^{(1)}$, defined on some known domain of $\mathbb{R}^2$ as in Figure 2.2. Hence at time t, the current contact surface $S_c^{(1)}$ is parameterized by the mapping $\boldsymbol{\psi}^{(1)}(.,t) = \boldsymbol{\phi}^{(1)}(.,t) \circ \boldsymbol{\theta}^{(1)}$:

$$\begin{array}{lll} \text{a domain of } \mathbb{R}^2 & \rightarrow S_{0c}^{(1)} & \rightarrow S_c^{(1)} \\ \boldsymbol{\xi}^{(1)} = (\xi^{(1)1}, \xi^{(1)2}) & \mapsto \mathbf{X}^{(1)} = \boldsymbol{\theta}^{(1)}(\boldsymbol{\xi}^{(1)}) & \mapsto \mathbf{x}^{(1)} = \boldsymbol{\phi}^{(1)}(\mathbf{X}^{(1)}, t) = \boldsymbol{\psi}^{(1)}(\boldsymbol{\xi}^{(1)}, t) \end{array} \quad [2.7]$$

Similarly, we assume that the reference target surface $S_{0c}^{(2)}$ is parameterized by a mapping that we denote by $\boldsymbol{\theta}^{(2)}$. Thus at time t, the current target surface $S_c^{(2)}$ is parameterized by the mapping $\boldsymbol{\psi}^{(2)}(.,t) = \boldsymbol{\phi}^{(2)}(.,t) \circ \boldsymbol{\theta}^{(2)}$:

$$\begin{array}{lll} \text{a domain of } \mathbb{R}^2 & \rightarrow S_{0c}^{(2)} & \rightarrow S_c^{(2)} \\ \boldsymbol{\xi}^{(2)} = (\xi^{(2)1}, \xi^{(2)2}) & \mapsto \mathbf{X}^{(2)} = \boldsymbol{\theta}^{(2)}(\boldsymbol{\xi}^{(2)}) & \mapsto \mathbf{x}^{(2)} = \boldsymbol{\phi}^{(2)}(\mathbf{X}^{(2)}, t) = \boldsymbol{\psi}^{(2)}(\boldsymbol{\xi}^{(2)}, t) \end{array} \quad [2.8]$$

The coordinates $(\xi^{(1)1}, \xi^{(1)2})$ and $(\xi^{(2)1}, \xi^{(2)2})$ are *the curvilinear coordinates* defined on the contactor and target surfaces, respectively.

The map $\boldsymbol{\psi}^{(2)}$ is assumed to be sufficiently regular, so that a natural basis $(\mathbf{a}_1, \mathbf{a}_2)$ at any point $\mathbf{x}^{(2)}$ of the current target surface $S_c^{(2)}$ can be defined as follows:

$$\mathbf{a}_1 = \frac{\partial \boldsymbol{\psi}^{(2)}}{\partial \xi^{(2)1}}(\boldsymbol{\xi}^{(2)}, t) \qquad \mathbf{a}_2 = \frac{\partial \boldsymbol{\psi}^{(2)}}{\partial \xi^{(2)2}}(\boldsymbol{\xi}^{(2)}, t) \quad [2.9]$$

The unit normal at $\mathbf{x}^{(2)}$ to $S_c^{(2)}$ is denoted by $\mathbf{n}$:

$$\mathbf{n} = \frac{\mathbf{a}_1 \times \mathbf{a}_2}{\|\mathbf{a}_1 \times \mathbf{a}_2\|} \quad [2.10]$$

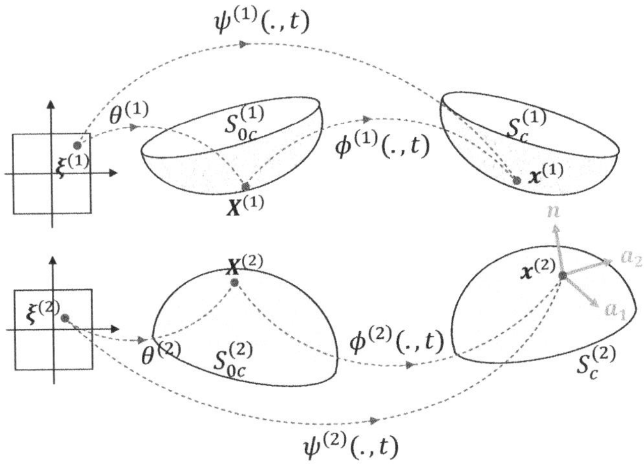

Figure 2.2. *Parameterization of the contact surfaces. For a color version of this figure, see www.iste.co.uk/levan/contact.zip*

If the target is a solid body, then $S_c^{(2)}$ will be parameterized in such a way that its normal $\mathbf{n}$ is directed outwards from the body. On the other hand, if the target is a membrane, then it will be parameterized in such a way that $\mathbf{n}$ is directed toward the contactor surface $S_c^{(1)}$.

Let us define a 2 × 2 matrix, denoted $[a_{..}]$, by $a_{\alpha\beta} = \mathbf{a}_\alpha.\mathbf{a}_\beta$, where the Greek indices equal either 1 or 2. It is therefore the matrix representative of the metric tensor at $\mathbf{x}^{(2)}$. Let $[a^{..}]$ be the inverse matrix of $[a_{..}]$. Then we can define the dual basis $(\mathbf{a}^1, \mathbf{a}^2)$ in the tangent plane to the target surface by

$$\forall \alpha \in \{1,2\}, \mathbf{a}^\alpha = a^{\alpha\beta}\mathbf{a}_\beta \tag{2.11}$$

where a summation is implied over any repeated index.

2.3. Normal contact kinematics

DEFINITION.– Given a point $\mathbf{x} \in S_c^{(1)}$, we define its projection onto $S_c^{(2)}$ to be the point $\mathbf{y} \in S_c^{(2)}$ closest to $\mathbf{x}$ in the Euclidean sense (Figure 2.3):

$$\boxed{\mathbf{y} = \arg \min_{\mathbf{x}^{(2)} \in S_c^{(2)}} \|\mathbf{x} - \mathbf{x}^{(2)}\|} \tag{2.12}$$

REMARK.– Existence and uniqueness of the projection point. The projection point $\mathbf{y}$, defined in [2.12], exists, provided that the point $\mathbf{x}$ is close enough to the target surface $S_c^{(2)}$. Uniqueness of the projection point depends on the convexity of the target surface $S_c^{(2)}$, and on which side of $S_c^{(2)}$ that $\mathbf{x}$ belongs to. Figure 2.4 shows two cases for which we do not have a unique projection.

In theory, non-uniqueness of the projection point is not such a big disadvantage, insofar as it vanishes once there is effective contact (since a particle of $S_c^{(1)}$ can only touch one single particle of $S_c^{(2)}$). The interested reader can consult [77] page 115, [138] page 48 and [30] for more details regarding this question. On the other hand, non-uniqueness may involve difficulties when computing a solution numerically, as we will see in Chapter 8 when considering the discretized contact problem using the finite element method. ■

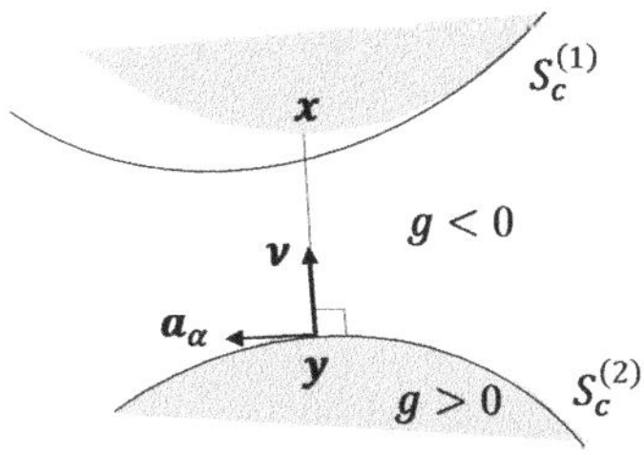

Figure 2.3. *Projection of* $\boldsymbol{x} \in S_c^{(1)}$ *onto the target surface*

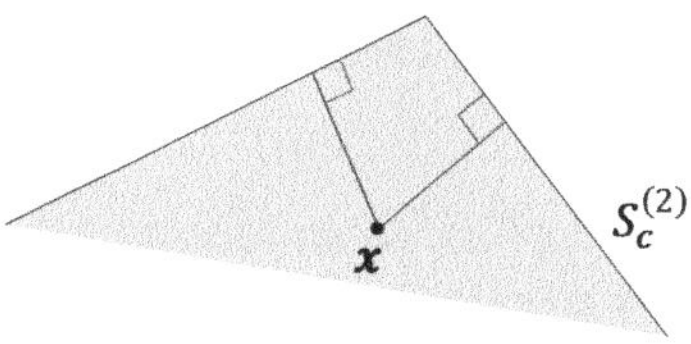

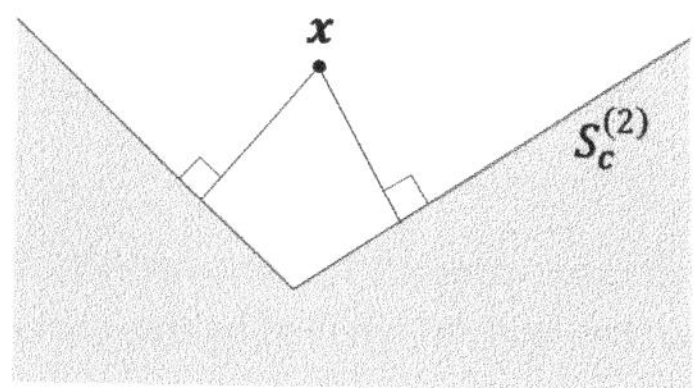

Figure 2.4. *Non-uniqueness of the projection*

DEFINITION.– If the point $\mathbf{x}$ has (at least) one projection point $\mathbf{y} \in S_c^{(2)}$ associated with it, then we define its *proximity* g to the point $\mathbf{x}$ by

$$\boxed{g = -\boldsymbol{\nu}.(\mathbf{x} - \mathbf{y})} \quad \Leftrightarrow \quad \boxed{\mathbf{x} - \mathbf{y} = -g\boldsymbol{\nu}} \qquad [2.13]$$

where $\boldsymbol{\nu}$ is the unit normal at $\mathbf{y}$ to the surface $S_c^{(2)}$.

The negative sign chosen in [2.13] means that the proximity g is negative if there is a gap between the two bodies, and that it is positive should interpenetration occur. Therefore, the proximity g is the opposite of the usual quantity called *the gap*.

We can check that g is an *objective* quantity.

As we adopt the Lagrangian description, we will transform the Eulerian quantities introduced previously into quantities that are defined relative to the reference configuration, which is achieved using the deformations $\boldsymbol{\phi}^{(1)}$ and $\boldsymbol{\phi}^{(2)}$ of [2.1]. Let $\mathbf{X} \in S_{0c}^{(1)}$ be the reference position of the point $\mathbf{x} \in S_c^{(1)}$, and $\mathbf{Y} \in S_{0c}^{(2)}$ the reference position of the projection point $\mathbf{y} \in S_c^{(2)}$ of $\mathbf{x}$ as in Figure 2.5. Then we can recast [2.12] into an equation involving $\mathbf{X}$ and $\mathbf{Y}$:

$$\boxed{\mathbf{Y} = \arg \min_{X^{(2)} \in S_{0c}^{(2)}} \|\boldsymbol{\phi}^{(1)}(\mathbf{X}, t) - \boldsymbol{\phi}^{(2)}(\mathbf{X}^{(2)}, t)\|} \qquad [2.14]$$

Thus, to the point $\mathbf{X}$ we have associated (at least) one point $\mathbf{Y}$, and we can rewrite definition [2.13] of the proximity g in terms of the Lagrangian variables:

$$g = -\boldsymbol{\nu}.(\boldsymbol{\phi}^{(1)}(\mathbf{X}, t) - \boldsymbol{\phi}^{(2)}(\mathbf{Y}, t)) \qquad [2.15]$$

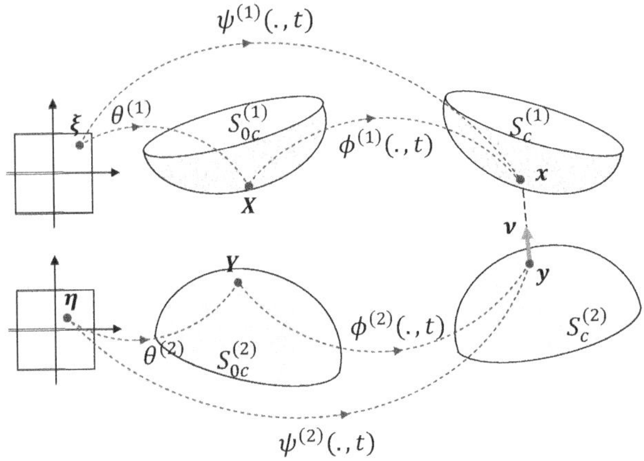

Figure 2.5. *The kinematic variables associated with the point* **x** *and its projection point* **y***. For a color version of this figure, see www.iste.co.uk/levan/contact.zip*

Finally, by using [2.12] $\mathbf{x} = \boldsymbol{\phi}^{(1)}(\mathbf{X}, t)$ and $\mathbf{x}^{(2)} = \boldsymbol{\psi}^{(2)}(\boldsymbol{\xi}^{(2)}, t)$, we arrive at the following equation that must be satisfied by the inverse image $\boldsymbol{\eta}$ of $\mathbf{Y}$ under the parameterization $\boldsymbol{\theta}^{(2)}$:

$$\boxed{\boldsymbol{\eta} = \arg\min_{\boldsymbol{\xi}^{(2)}} \|\boldsymbol{\phi}^{(1)}(\mathbf{X}, t) - \boldsymbol{\psi}^{(2)}(\boldsymbol{\xi}^{(2)}, t)\|} \qquad [2.16]$$

To shorten the writing, we will henceforth write the preceding kinematic quantities as functions of $(\mathbf{X}, t)$ only, omitting the dependencies on the parameterizations $\boldsymbol{\phi}^{(1)}$, $\boldsymbol{\phi}^{(2)}$ and $\boldsymbol{\psi}^{(2)}$:

$$\mathbf{Y} = \mathbf{Y}(\mathbf{X}, t) \qquad \boldsymbol{\eta} = \boldsymbol{\eta}(\mathbf{X}, t) \qquad g = g(\mathbf{X}, t)$$

Before continuing, it is worth summarizing the notation introduced so far:

– For $i \in \{1, 2\}$, the symbol $\mathbf{x}^{(i)}$ denotes a *current* point of $S_c^{(i)}$. The inverse images of $\mathbf{x}^{(i)}$ are $\mathbf{X}^{(i)}$ and $\boldsymbol{\xi}^{(i)}$, as shown in the following sequence:

$$\begin{array}{llll} \text{a domain of } \mathbb{R}^2 & \rightarrow S_{0c}^{(i)} & \rightarrow S_c^{(i)} & \\ \boldsymbol{\xi}^{(i)} & \mapsto \mathbf{X}^{(i)} = \boldsymbol{\theta}^{(i)}(\boldsymbol{\xi}^{(i)}) & \mapsto \mathbf{x}^{(i)} = \boldsymbol{\phi}^{(i)}(\mathbf{X}^{(i)}, t) = \boldsymbol{\psi}^{(i)}(\boldsymbol{\xi}^{(i)}, t) \end{array} \qquad [2.17]$$

The normal at the current point $\mathbf{x}^{(2)}$ to $S_c^{(2)}$ is denoted as $\mathbf{n}$.

– The symbol $\mathbf{x}$ denotes a *particular* point of $S_c^{(1)}$ and $\mathbf{y}$ the projection point of $\mathbf{x}$ onto $S_c^{(2)}$, defined by [2.12]. The inverse images of $\mathbf{x}$ are $\mathbf{X}$ and $\boldsymbol{\xi}$, and those of $\mathbf{y}$ are $\mathbf{Y}$ and $\boldsymbol{\eta}$, as can be seen in the following chains:

$$\begin{array}{llll} \text{a domain in } \mathbb{R}^2 & \rightarrow S_{0c}^{(1)} & \rightarrow S_c^{(1)} & \\ \boldsymbol{\xi} & \mapsto \mathbf{X} = \boldsymbol{\theta}^{(1)}(\boldsymbol{\xi}) & \mapsto \mathbf{x} = \boldsymbol{\phi}^{(1)}(\mathbf{X}, t) = \boldsymbol{\psi}^{(1)}(\boldsymbol{\xi}, t) \end{array} \qquad [2.18]$$

$$\begin{array}{llll} \text{a domain in } \mathbb{R}^2 & \rightarrow S_{0c}^{(2)} & \rightarrow S_c^{(2)} & \\ \boldsymbol{\eta} & \mapsto \mathbf{Y} = \boldsymbol{\theta}^{(2)}(\boldsymbol{\eta}) & \mapsto \mathbf{y} = \boldsymbol{\phi}^{(2)}(\mathbf{Y}, t) = \boldsymbol{\psi}^{(2)}(\boldsymbol{\eta}, t) \end{array} \qquad [2.19]$$

The normal at the point $\mathbf{y}$, which reappears constantly in contact calculations, is denoted by $\boldsymbol{\nu}$.

According to the typical linguistic contractions that arise in continuum mechanics, the particle with reference position $\mathbf{X}$ (respectively, $\mathbf{Y}$) is just said to be the particle $\mathbf{X}$ (respectively, $\mathbf{Y}$).

In general, if we fix the particle $\mathbf{X}$, then $\mathbf{Y}(\mathbf{X}, t)$ will not represent the same particle as time varies.

Calculation of the projection

Let t be a fixed given time, and consider a point $\mathbf{x}$ on the contactor surface $S_c^{(1)}$. We will show how the nearest point $\mathbf{y}$ on the target surface $S_c^{(2)}$ can be determined which, by definition [2.12], is the point realizing the minimum of $\|\mathbf{x} - \mathbf{x}^{(2)}\|$, $\mathbf{x}^{(2)} \in S_c^{(2)}$.

Since the target surface $S_c^{(2)}$ at time t is parameterized by [2.8] $\boldsymbol{\xi}^{(2)} \mapsto \mathbf{x}^{(2)} = \boldsymbol{\psi}^{(2)}(\boldsymbol{\xi}^{(2)}, t)$, knowledge of $\mathbf{y}$ amounts to knowing its inverse image $\boldsymbol{\eta}$. Hence, the question is reduced to the determination of $\boldsymbol{\eta}$.

THEOREM.–

Assumption: we know the parameterization [2.8] of the target surface $S_c^{(2)}$ at the time t: $\boldsymbol{\xi}^{(2)} \mapsto \mathbf{x}^{(2)} = \boldsymbol{\psi}^{(2)}(\boldsymbol{\xi}^{(2)}, t)$.

Then, the point $\boldsymbol{\eta}$ is the root of the algebraic equation

$$\forall \alpha \in \{1, 2\}, \quad \boxed{\left(\mathbf{x} - \boldsymbol{\psi}^{(2)}(\boldsymbol{\eta}, t)\right) \cdot \frac{\partial \boldsymbol{\psi}^{(2)}}{\partial \xi^{(2)\alpha}}(\boldsymbol{\eta}, t) = 0} \qquad [2.20]$$

As $\dfrac{\partial \mathbf{x}^{(2)}}{\partial \xi^{(2)\alpha}} = \mathbf{a}_\alpha$, the previous equation states: the vector $\mathbf{x} - \mathbf{y}$ is normal to the vectors $\mathbf{a}_1$ and $\mathbf{a}_2$ of the natural basis at $\mathbf{y}$, and hence to the surface $S_c^{(2)}$.

PROOF. The point $\boldsymbol{\eta}$ is determined by equation [2.16], which is to say:

$$\forall \alpha \in \{1, 2\}, \quad 0 = \frac{\partial}{\partial \xi^{(2)\alpha}} \|\mathbf{x} - \mathbf{x}^{(2)}\| \Big|_{\mathbf{x}^{(2)} = \mathbf{y}} = -\frac{\mathbf{x} - \mathbf{x}^{(2)}}{\|\mathbf{x} - \mathbf{x}^{(2)}\|} \frac{\partial \mathbf{x}^{(2)}}{\partial \xi^{(2)\alpha}} \Big|_{\boldsymbol{\xi}^{(2)} = \boldsymbol{\eta}} \qquad \blacksquare$$

In general, the parameterization $\boldsymbol{\psi}^{(2)}$ is not linear in $\boldsymbol{\eta}$, in which case equation [2.20] is nonlinear and must be solved iteratively. The point $\boldsymbol{\eta}$ can only obtained up to a specific predetermined precision.

Once we know $\boldsymbol{\eta}$, then the desired projection point $\mathbf{y}$ is obtained by $\mathbf{y} = \boldsymbol{\psi}^{(2)}(\boldsymbol{\eta}, t)$. We can then deduce the other quantities related to $\mathbf{y}$ such as the proximity g, the normal $\boldsymbol{\nu}$, etc.

2.4. Variation of kinematic quantities with respect to time

This section is dedicated to calculating the time derivatives of the various kinematic quantities. The problem posed here is the following:

- Consider a given particle on the contactor surface, that is, fix a point $\mathbf{X}$ and its inverse image $\boldsymbol{\xi}$.

- Consider a given time t.

- Consider a small change Δt in time and see how the following quantities vary: the point $\mathbf{x}$ (current position of $\mathbf{X}$), the point $\mathbf{y}$ (projection point of $\mathbf{x}$), the point $\boldsymbol{\eta}$ (inverse image of $\mathbf{Y}$), the vectors $\mathbf{a}_\alpha$ (natural basis vectors at $\mathbf{y}$), the vector $\boldsymbol{\nu}$ (normal at $\mathbf{y}$) and the proximity g (Figure 2.6).

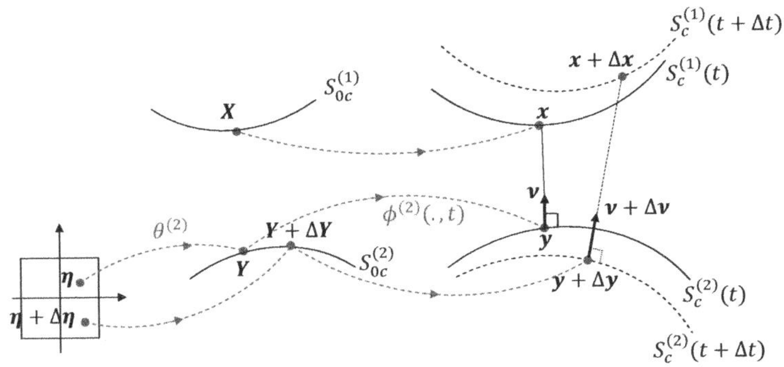

Figure 2.6. *Variations of the kinematic quantities with respect to time. For a color version of this figure, see www.iste.co.uk/levan/contact.zip*

The time derivative of $\mathbf{x}$ is straightforward:

$$\boxed{\dot{\mathbf{x}} = \frac{d\boldsymbol{\phi}^{(1)}(\mathbf{X},t)}{dt} = \mathbf{V}^{(1)}(\mathbf{X},t)} \qquad [2.21]$$

where $\mathbf{V}^{(1)}(\mathbf{X},t)$ is the velocity (in the reference frame R_0 at time t) of the particle $\mathbf{X}$. On the other hand, the derivatives of the other quantities are more difficult to calculate, since they depend on the two surfaces – the target surface and the contactor surface – which concurrently vary themselves.

In accordance with definition [2.9], the natural basis vectors at $\mathbf{y} \in S_c^{(2)}$ are written as

$$\boxed{\mathbf{a}_1 = \frac{\partial \psi^{(2)}}{\partial \xi^{(2)1}}(\eta(\mathbf{X},t),t) \qquad \mathbf{a}_2 = \frac{\partial \psi^{(2)}}{\partial \xi^{(2)2}}(\eta(\mathbf{X},t),t)} \qquad [2.22]$$

THEOREM.–

1. The time derivative of the projection point $\mathbf{y}$ is given by

$$\boxed{\dot{\mathbf{y}} = \frac{d\boldsymbol{\phi}^{(2)}(\mathbf{Y}(\mathbf{X},t),t)}{dt} = \mathbf{V}^{(2)}(\mathbf{Y}(\mathbf{X},t),t) + \dot{\eta}^\alpha \mathbf{a}_\alpha} \qquad [2.23]$$

where $\dot{\eta}^\alpha$ is given by [2.26], $\mathbf{V}^{(2)}(\mathbf{Y}(\mathbf{X},t),t)$ is the velocity (in reference frame R_0 at time t) of the particle $\mathbf{Y}(\mathbf{X},t)$:

$$\mathbf{V}^{(2)}(\mathbf{Y}(\mathbf{X},t),t) = \left.\frac{\partial \boldsymbol{\phi}^{(2)}(\mathbf{X}^{(2)},t)}{\partial t}\right|_{\mathbf{X}^{(2)}=\mathbf{Y}(\mathbf{X},t)}$$

Recall that $\mathbf{Y}(\mathbf{X},t)$ is *not* the same particle at each time t, since the particle $\mathbf{X}$ is fixed.

2. The time derivatives of the natural basis vectors $\mathbf{a}_\alpha$ at $\mathbf{y}$ are given by

$$\boxed{\dot{\mathbf{a}}_\alpha = \mathbf{V}^{(2)}_{,\alpha}(\boldsymbol{\eta},t) + \boldsymbol{\psi}^{(2)}_{,\alpha\beta}\,\dot{\eta}^\beta} \qquad [2.24]$$

where $\dot{\eta}^\beta$ is given by [2.26], and where we have denoted $\mathbf{V}^{(2)}(\boldsymbol{\eta},t) \equiv \mathbf{V}^{(2)}(\mathbf{Y}(\boldsymbol{\eta}),t)$, $(\bullet)_{,\alpha} \equiv \dfrac{\partial(\bullet)}{\partial\xi^{(2)\alpha}}$ as an abbreviation.

3. For the derivatives $\dot{\eta}^\alpha$, the notation $\bar{a}^{\alpha\beta}$ that we define below is required:

   – We denote by $b_{\alpha\beta}$ the covariant rank-2 components of the curvature tensor:

$$\boxed{b_{\alpha\beta} \equiv \boldsymbol{\nu}.\boldsymbol{\psi}^{(2)}_{,\alpha\beta}} \qquad [2.25]$$

   – We define $\boxed{\bar{a}_{\alpha\beta} \equiv a_{\alpha\beta} + g b_{\alpha\beta}}$ (there exists a symmetry $\bar{a}_{\alpha\beta} = \bar{a}_{\beta\alpha}$).

   – Then, we denote by $\bar{a}^{\alpha\beta}$ the components of the 2×2 matrix $[\bar{a}^{\cdot\cdot}]$, which is the inverse of $[\bar{a}_{\cdot\cdot}]$ (there exists a symmetry $\bar{a}^{\alpha\beta} = \bar{a}^{\beta\alpha}$).

With the notation $\bar{a}^{\alpha\beta}$, the derivatives $\dot{\eta}^\alpha$ can now be expressed as functions of the Lagrangian velocities of the particle:

$$\boxed{\dot{\eta}^\alpha = \bar{a}^{\alpha\beta}\mathbf{a}_\beta.\left[\mathbf{V}^{(1)}(\mathbf{X},t) - \mathbf{V}^{(2)}(\mathbf{Y},t)\right] - g\bar{a}^{\alpha\beta}\boldsymbol{\nu}.\mathbf{V}^{(2)}_{,\beta}(\boldsymbol{\eta},t)} \qquad [2.26]$$

PROOF.–

1. The time derivative of the projection point $\mathbf{y}$ is obtained by applying the chain rule to the function $\mathbf{y} = \boldsymbol{\phi}^{(2)}(\mathbf{Y}(\boldsymbol{\eta}),t) = \boldsymbol{\psi}^{(2)}(\boldsymbol{\eta},t)$:

$$\dot{\mathbf{y}} = \underbrace{\left.\frac{\partial\boldsymbol{\psi}^{(2)}(\boldsymbol{\xi}^{(2)},t)}{\partial t}\right|_{\boldsymbol{\xi}^{(2)}=\boldsymbol{\eta}}}_{\left.\frac{\partial\boldsymbol{\phi}^{(2)}(\mathbf{X}^{(2)},t)}{\partial t}\right|_{\mathbf{X}^{(2)}=\mathbf{Y}(\boldsymbol{\eta})}} + \underbrace{\frac{\partial\boldsymbol{\psi}^{(2)}}{\partial\xi^{(2)\alpha}}(\boldsymbol{\eta},t)}_{\mathbf{a}_\alpha}\,\dot{\eta}^\alpha$$

from which we obtain [2.23].

2. The derivatives of the natural basis vectors follow from definition [2.22]:

$$\frac{d\mathbf{a}_\alpha}{dt} = \left.\left(\frac{\partial}{\partial t}\frac{\partial\boldsymbol{\psi}^{(2)}(\boldsymbol{\xi}^{(2)},t)}{\partial\xi^{(2)\alpha}}\right)\right|_{\boldsymbol{\xi}^{(2)}=\boldsymbol{\eta}} + \left.\left(\frac{\partial}{\partial\xi^{(2)\beta}}\frac{\partial\boldsymbol{\psi}^{(2)}(\boldsymbol{\xi}^{(2)},t)}{\partial\xi^{(2)\alpha}}\right)\right|_{\boldsymbol{\xi}^{(2)}=\boldsymbol{\eta}}\dot{\eta}^\beta$$

However:

$$\frac{\partial}{\partial t}\frac{\partial\boldsymbol{\psi}^{(2)}(\boldsymbol{\xi}^{(2)},t)}{\partial\xi^{(2)\alpha}} = \frac{\partial}{\partial\xi^{(2)\alpha}}\frac{\partial\boldsymbol{\psi}^{(2)}(\boldsymbol{\xi}^{(2)},t)}{\partial t} = \frac{\partial\mathbf{V}^{(2)}(\boldsymbol{\xi}^{(2)},t)}{\partial\xi^{(2)\alpha}}$$

where we have denoted $\mathbf{V}^{(2)}(\boldsymbol{\xi}^{(2)},t) \equiv \mathbf{V}^{(2)}(\mathbf{X}^{(2)}(\boldsymbol{\xi}^{(2)}),t)$ as an abuse of notation, from which we obtain [2.24].

3. To determine the derivatives $\dot{\eta}^\alpha$, recall equation [2.20]:

$$\forall\alpha \in \{1,2\}, \quad \left(\mathbf{x} - \boldsymbol{\psi}^{(2)}(\boldsymbol{\eta},t)\right) \cdot \frac{\partial\boldsymbol{\psi}^{(2)}}{\partial\xi^{(2)\alpha}}(\boldsymbol{\eta},t) = 0 \qquad \text{that is,} \qquad (\mathbf{x}-\mathbf{y}).\mathbf{a}_\alpha = 0 \quad [2.27]$$

As the time t and the point $\mathbf{x}$ are given, this equation constitutes two scalar equations with two unknowns $\boldsymbol{\eta} = (\eta_1, \eta_2)$. Suppose that these equations have already been solved, and that the time t and the points $\mathbf{x}, \boldsymbol{\eta}$ are all known. Then we can differentiate [2.27] with respect to time to obtain two scalar equations that are related to the derivatives of $\dot{\eta}^\alpha$:

$$(\dot{\mathbf{x}} - \dot{\mathbf{y}}).\mathbf{a}_\alpha + \underbrace{(\mathbf{x}-\mathbf{y})}_{-g\boldsymbol{\nu}}.\dot{\mathbf{a}}_\alpha = 0$$

By using equations [2.21]–[2.24] for the derivatives $\dot{\mathbf{x}}$, $\dot{\mathbf{y}}$ and $\dot{\mathbf{a}}_\alpha$, we arrive at

$$\left[\mathbf{V}^{(1)}(\mathbf{X},t) - \mathbf{V}^{(2)}(\mathbf{Y},t)\right].\mathbf{a}_\alpha - \underbrace{\left(a_{\alpha\beta} + g\boldsymbol{\nu}.\boldsymbol{\psi}^{(2)}_{,\alpha\beta}\right)}_{\bar{a}_{\alpha\beta}}\dot{\eta}^\beta - g\boldsymbol{\nu}.\mathbf{V}^{(2)}_{,\alpha}(\boldsymbol{\eta},t) = 0$$

namely:

$$\bar{a}_{\alpha\beta}\,\dot{\eta}^\beta = \left[\mathbf{V}^{(1)}(\mathbf{X},t) - \mathbf{V}^{(2)}(\mathbf{Y},t)\right].\mathbf{a}_\alpha - g\boldsymbol{\nu}.\mathbf{V}^{(2)}_{,\alpha}(\boldsymbol{\eta},t)$$

Then, as desired, we get two scalar equations for the two unknowns $\dot{\eta}^\alpha$. Finally, we can invert this system to get [2.26]. ■

THEOREM.–

1. The time derivative of the proximity g is given by

$$\boxed{\dot{g} = -\left[\mathbf{V}^{(1)}(\mathbf{X},t) - \mathbf{V}^{(2)}(\mathbf{Y}(\mathbf{X},t),t)\right].\boldsymbol{\nu}} \quad [2.28]$$

2. The time derivative of the normal $\boldsymbol{\nu}$ at $\mathbf{y}$ is given by

$$\boxed{\dot{\boldsymbol{\nu}} = -\Big[\boldsymbol{\nu}.\mathbf{V}^{(2)}_{,\alpha}(\boldsymbol{\eta},t) + \underbrace{\boldsymbol{\nu}.\boldsymbol{\psi}^{(2)}_{,\alpha\beta}}_{b_{\alpha\beta}}\dot{\eta}^\beta\Big]\mathbf{a}^\alpha} \quad [2.29]$$

There is another expression for $\dot{\boldsymbol{\nu}}$, in which only the velocities $\mathbf{V}^{(1)}$ and $\mathbf{V}^{(2)}$ appear and the derivatives $\dot{\eta}^\alpha$ are absent:

$$\boxed{\dot{\boldsymbol{\nu}} = -\left(b_{\alpha\gamma}\bar{a}^{\gamma\beta}\mathbf{a}^\alpha \otimes \mathbf{a}_\beta\right).(\mathbf{V}^{(1)} - \mathbf{V}^{(2)}) - (\bar{a}^{\alpha\beta}\mathbf{a}_\beta \otimes \boldsymbol{\nu}).\mathbf{V}^{(2)}_{,\alpha}} \quad [2.30]$$

The term $b_{\alpha\gamma}\bar{a}^{\gamma\beta}\mathbf{a}^\alpha \otimes \mathbf{a}_\beta$ can be replaced by various other equivalent expressions:

$$\boxed{\begin{aligned} b_{\alpha\gamma}\bar{a}^{\gamma\beta}\mathbf{a}^\alpha \otimes \mathbf{a}_\beta &= \bar{a}^{\alpha\gamma}b_{\gamma\beta} \quad \mathbf{a}_\alpha \otimes \mathbf{a}^\beta \\ &= a^{\alpha\gamma}b_{\gamma\delta}\bar{a}^{\delta\beta} \quad \mathbf{a}_\alpha \otimes \mathbf{a}_\beta \\ &= \bar{a}^{\alpha\gamma}b_{\gamma\delta}a^{\delta\beta} \quad \mathbf{a}_\alpha \otimes \mathbf{a}_\beta \end{aligned}} \text{(both hands are } \textit{symmetric} \text{ tensors)} \quad [2.31]$$

PROOF.–

1. To calculate $\dot{g}$, we differentiate equation [2.13] with respect to time:

$$\begin{aligned} -\dot{g}\boldsymbol{\nu} - g\dot{\boldsymbol{\nu}} &= \dot{\mathbf{x}} - \dot{\mathbf{y}} \\ &= \mathbf{V}^{(1)}(\mathbf{X},t) - \mathbf{V}^{(2)}(\mathbf{Y},t) - \dot{\eta}^{\alpha}\mathbf{a}_{\alpha} \quad \text{from [2.21] and [2.23]} \end{aligned} \tag{2.32}$$

   Since $\|\boldsymbol{\nu}\|^2 = 1$, we have $\boldsymbol{\nu}.\dot{\boldsymbol{\nu}} = 0$, which is to say $\dot{\boldsymbol{\nu}} \perp \boldsymbol{\nu}$. Then projecting equation [2.32] onto $\boldsymbol{\nu}$ gives us [2.28].

2. Since $\dot{\boldsymbol{\nu}} \perp \boldsymbol{\nu}$, the vector $\dot{\boldsymbol{\nu}}$ lies in the tangent plane at $\mathbf{y}$ defined by $(\mathbf{a}_1, \mathbf{a}_2)$, so we can write

$$\begin{aligned} \dot{\boldsymbol{\nu}} &= (\dot{\boldsymbol{\nu}}.\mathbf{a}_{\alpha})\mathbf{a}^{\alpha} \\ &= -(\boldsymbol{\nu}.\dot{\mathbf{a}}_{\alpha})\mathbf{a}^{\alpha} \quad \text{since } 0 = \frac{d}{dt}(\boldsymbol{\nu}.\mathbf{a}_{\alpha}) = \dot{\boldsymbol{\nu}}.\mathbf{a}_{\alpha} + \boldsymbol{\nu}.\dot{\mathbf{a}}_{\alpha} \end{aligned}$$

   Then by taking [2.24] into account, we obtain [2.29].

3. To get [2.30], we replace $\dot{\eta}^{\beta}$ in [2.29] with its expression in [2.26]:

$$\dot{\boldsymbol{\nu}} = -\left(\boldsymbol{\nu}.\mathbf{V}^{(2)}_{,\alpha}\right)\mathbf{a}^{\alpha} - b_{\alpha\beta}\bar{a}^{\beta\gamma}\left(\mathbf{a}^{\alpha}\otimes\mathbf{a}_{\gamma}\right).\left(\mathbf{V}^{(1)} - \mathbf{V}^{(2)}\right) + \underbrace{gb_{\alpha\beta}}_{=\bar{a}_{\alpha\beta}-a_{\alpha\beta}}\bar{a}^{\beta\gamma}\left(\boldsymbol{\nu}.\mathbf{V}^{(2)}_{,\gamma}\right)\mathbf{a}^{\alpha}$$

   Then, since we know that $\bar{a}_{\alpha\beta}\bar{a}^{\beta\gamma} = \delta^{\gamma}_{\alpha}$ and $a_{\alpha\beta}\mathbf{a}^{\alpha} = \mathbf{a}_{\beta}$, we arrive at

$$\begin{aligned} \dot{\boldsymbol{\nu}} &= -\cancel{\left(\boldsymbol{\nu}.\mathbf{V}^{(2)}_{,\alpha}\right)\mathbf{a}^{\alpha}} - b_{\alpha\beta}\bar{a}^{\beta\gamma}\left(\mathbf{a}^{\alpha}\otimes\mathbf{a}_{\gamma}\right).\left(\mathbf{V}^{(1)} - \mathbf{V}^{(2)}\right) + \cancel{\left(\boldsymbol{\nu}.\mathbf{V}^{(2)}_{,\alpha}\right)\mathbf{a}^{\alpha}} - \bar{a}^{\beta\gamma}\left(\boldsymbol{\nu}.\mathbf{V}^{(2)}_{,\gamma}\right)\mathbf{a}_{\beta} \\ &= -\left(b_{\alpha\beta}\bar{a}^{\beta\gamma}\mathbf{a}^{\alpha}\otimes\mathbf{a}_{\gamma}\right).\left(\mathbf{V}^{(1)} - \mathbf{V}^{(2)}\right) - \left(\bar{a}^{\beta\gamma}\mathbf{a}_{\beta}\otimes\boldsymbol{\nu}\right).\mathbf{V}^{(2)}_{,\gamma} \end{aligned}$$

   Lastly, by rewriting $b_{\alpha\beta}\bar{a}^{\beta\gamma}\mathbf{a}^{\alpha}\otimes\mathbf{a}_{\gamma} = b_{\alpha\gamma}\bar{a}^{\gamma\beta}\mathbf{a}^{\alpha}\otimes\mathbf{a}_{\beta}$, we obtain [2.30].

4. Now, let us prove $[2.31]_1$. On the one hand:

$$\begin{aligned} b_{\alpha\gamma}\bar{a}^{\gamma\beta} &= \frac{1}{g}(\bar{a}_{\alpha\gamma} - a_{\alpha\gamma})\bar{a}^{\gamma\beta} \qquad \text{by the definition of } \bar{a}_{\alpha\gamma} \\ &= \frac{1}{g}(\bar{a}_{\alpha\gamma}\bar{a}^{\gamma\beta} - a_{\alpha\gamma}\bar{a}^{\gamma\beta}) = \frac{1}{g}(\delta^{\beta}_{\alpha} - a_{\alpha\gamma}\bar{a}^{\gamma\beta}) \end{aligned}$$

   from which we get $b_{\alpha\gamma}\bar{a}^{\gamma\beta}\mathbf{a}^{\alpha}\otimes\mathbf{a}_{\beta} = \frac{1}{g}(\mathbf{a}^{\alpha}\otimes\mathbf{a}_{\alpha} - \bar{a}^{\gamma\beta}\mathbf{a}_{\gamma}\otimes\mathbf{a}_{\beta})$. On the other hand:

$$\begin{aligned} \bar{a}^{\alpha\gamma}b_{\gamma\beta} &= \bar{a}^{\alpha\gamma}\frac{1}{g}(\bar{a}_{\gamma\beta} - a_{\gamma\beta}) \qquad \text{by the definition of } \bar{a}_{\gamma\beta} \\ &= \frac{1}{g}(\bar{a}^{\alpha\gamma}\bar{a}_{\gamma\beta} - \bar{a}^{\alpha\gamma}a_{\gamma\beta}) = \frac{1}{g}(\delta^{\alpha}_{\beta} - \bar{a}^{\alpha\gamma}a_{\gamma\beta}) \end{aligned}$$

   from which we get $\bar{a}^{\alpha\gamma}b_{\gamma\beta}\mathbf{a}_{\alpha}\otimes\mathbf{a}^{\beta} = \frac{1}{g}(\mathbf{a}_{\alpha}\otimes\mathbf{a}^{\alpha} - \bar{a}^{\alpha\gamma}\mathbf{a}_{\alpha}\otimes\mathbf{a}_{\gamma})$. Since $\mathbf{a}^{\alpha}\otimes\mathbf{a}_{\alpha} = \mathbf{a}_{\alpha}\otimes\mathbf{a}^{\alpha}$ (both are equal to $\mathbf{I} - \boldsymbol{\nu}\otimes\boldsymbol{\nu}$), we obtain $[2.31]_1$.

5. Let us show that $b_{\alpha\gamma}\bar{a}^{\gamma\beta}\mathbf{a}^{\alpha}\otimes\mathbf{a}_{\beta}$ is symmetric.

   (a) For this, first we will see that its transpose equals itself. We have

$$\begin{aligned} \left(b_{\alpha\gamma}\bar{a}^{\gamma\beta}\mathbf{a}^{\alpha}\otimes\mathbf{a}_{\beta}\right)^T &= b_{\alpha\gamma}\bar{a}^{\gamma\beta}\mathbf{a}_{\beta}\otimes\mathbf{a}^{\alpha} \\ &= \bar{a}^{\alpha\gamma}b_{\gamma\beta}\mathbf{a}_{\alpha}\otimes\mathbf{a}^{\beta} \quad \text{by permuting } \alpha \text{ and } \beta \\ &= b_{\alpha\gamma}\bar{a}^{\gamma\beta}\mathbf{a}^{\alpha}\otimes\mathbf{a}_{\beta} \quad \text{from equality } [2.31]_1 \text{ just proven} \end{aligned}$$

(b) Another way to prove this symmetry is based on the property that:

$$\text{the tensor } \mathbf{T} \text{ is symmetric } \Leftrightarrow \forall i, j, T_i^{\ j} = T^j_{\ i}$$

On the one hand, we have

$$\mathbf{T} = b_{\alpha\gamma}\bar{a}^{\gamma\beta}\mathbf{a}^{\alpha} \otimes \mathbf{a}_{\beta} \quad \Rightarrow \quad T_{\alpha}^{\ \beta} = b_{\alpha\gamma}\bar{a}^{\gamma\beta} \qquad [2.33]$$

On the other hand, we also have

$$\begin{aligned} \mathbf{T} = \bar{a}^{\alpha\gamma}b_{\gamma\beta}\mathbf{a}_{\alpha} \otimes \mathbf{a}^{\beta} \Rightarrow T^{\alpha}_{\ \beta} &= \bar{a}^{\alpha\gamma}b_{\gamma\beta} \\ \Rightarrow T^{\beta}_{\ \alpha} &= \bar{a}^{\beta\gamma}b_{\gamma\alpha} \quad \text{in permuting } \alpha \text{ and } \beta \\ &= b_{\alpha\gamma}\bar{a}^{\gamma\beta} \quad \text{since } b_{\gamma\alpha} = b_{\alpha\gamma}, \bar{a}^{\beta\gamma} = \bar{a}^{\gamma\beta} \end{aligned} \qquad [2.34]$$

After comparing [2.33] with [2.34], we indeed see that $T_{\alpha}^{\ \beta} = T^{\beta}_{\ \alpha}$.

6. Finally, $[2.31]_2$ is obtained from $[2.31]_1$ by setting $\mathbf{a}^{\alpha} = a^{\alpha\delta}\mathbf{a}_{\delta}$ and then relabeling the indices.

   $[2.31]_3$ is obtained from $[2.31]_2$ by setting $\mathbf{a}^{\beta} = a^{\beta\delta}\mathbf{a}_{\delta}$ and then relabeling the indices. ■

REMARK.–

1. Equation [2.21] was obtained from the usual parameterization: $\mathbf{X} \mapsto \mathbf{x} = \boldsymbol{\phi}^{(1)}(\mathbf{X}, t)$. It is nothing special.

2. Equations [2.23]–[2.24] and [2.29] giving $\dot{\mathbf{y}}, \dot{\mathbf{a}}_{\alpha}$ and the first equation for $\dot{\boldsymbol{\nu}}$ were obtained uniquely from the parameterization

   $$\boldsymbol{\eta} \mapsto \mathbf{Y} = \boldsymbol{\theta}^{(2)}(\boldsymbol{\eta}) \mapsto \mathbf{y} = \boldsymbol{\phi}^{(2)}(\mathbf{Y}, t) = \boldsymbol{\psi}^{(2)}(\boldsymbol{\eta}, t)$$

   and not from the fact that $\mathbf{y} = \text{proj}_{S_c^{(2)}}\mathbf{x}$. Thus, *these equations hold for any point* $\mathbf{y}$ *of* $S_c^{(2)}$. We equally could have also written $\boldsymbol{\xi}^{(2)}, \mathbf{X}^{(2)}, \mathbf{x}^{(2)}$ in the places of $\boldsymbol{\eta}, \mathbf{Y}, \mathbf{y}$.

3. On the other hand, equations [2.26]–[2.28] and [2.30] giving $\dot{\eta}^{\alpha}$ and $\dot{g}$ and the second equation for $\dot{\boldsymbol{\nu}}$ use the fact that $\mathbf{y} = \text{proj}_{S_c^{(2)}}\mathbf{x}$, which is to say $\mathbf{x} - \mathbf{y} = -g\boldsymbol{\nu}$. Hence, *these equations are specific to contact kinematics*.

2.5. Tangential contact kinematics – Relative velocity

An important variable that appears in tangential contact calculations with friction is relative velocity, whose definition must be stated properly. Intuitively, the first idea that comes to mind is to introduce the relative velocity between a particle $\mathbf{X} \in S_{0c}^{(1)}$ and its associated particle $\mathbf{Y} \in S_{0c}^{(2)}$ via equation [2.14]. This relative velocity, which is considered to be *standard*, is not objective as we will soon show. It does have some use however, since it will serve as a guideline when introducing a different velocity which will be *objective* (it is better to use an objective velocity when establishing an objective contact law).

DEFINITION.– Consider a particle on $S_c^{(1)}$ with reference position $\mathbf{X} \in S_{0c}^{(1)}$, and the particle opposite it with reference position $\mathbf{Y} \in S_{0c}^{(2)}$. The *standard relative velocity* of the point $\mathbf{X}$ at time t is defined to be:

$$\begin{aligned}\mathbf{V}_{standard} &= \mathbf{V}^{(1)}(\mathbf{X},t) - \mathbf{V}^{(2)}(\mathbf{Y}(\mathbf{X},t),t)\\ &= \frac{\partial \boldsymbol{\phi}^{(1)}}{\partial t}(\mathbf{X},t) - \frac{\partial \boldsymbol{\phi}^{(2)}(\mathbf{X}^{(2)},t)}{\partial t}\Big|_{\mathbf{X}^{(2)}=\mathbf{Y}(\mathbf{X},t)}\end{aligned} \qquad [2.35]$$

where $\mathbf{V}^{(1)}(\mathbf{X},t)$ is the velocity (in the reference frame R_0 at time t) of the particle $\mathbf{X}$, and $\mathbf{V}^{(2)}(\mathbf{Y},t)$ is the velocity (in the reference frame R_0 at time t) of the particle $\mathbf{Y}(\mathbf{X},t)$.

As it is well known, the velocity of a particle is not an objective quantity. The following theorem shows that $\mathbf{V}_{standard}$, which was defined to be the difference between the two velocities, is not an objective quantity either.

PROPOSITION.– In general, the standard relative velocity is *not* objective. It is objective if, and only if, $g = 0$, that is, when there is contact.

PROOF.– Consider the change of frame $R_0 \to R^*$ defined by

$$\mathbf{x}^* = \mathbf{Q}(t)\mathbf{x} + \mathbf{b}(t)$$

where $\mathbf{Q}(t)$ is a second-order orthogonal tensor and $\mathbf{b}(t)$ is a vector. The reference positions $\mathbf{X}$ and $\mathbf{X}^*$, corresponding, respectively, to $\mathbf{x}$ and $\mathbf{x}^*$, are related via $\mathbf{X}^* = \mathbf{Q}(0)\mathbf{X} + \mathbf{b}(0)$, hence $\mathbf{X}^*$ is a fixed vector since $\mathbf{X}$ is one too. We have

$$\begin{cases}\boldsymbol{\phi}^{(1)*}(\mathbf{X}^*,t) = \mathbf{Q}(t)\boldsymbol{\phi}^{(1)}(\mathbf{X},t) + \mathbf{b}(t)\\ \boldsymbol{\phi}^{(2)*}(\mathbf{Y}^*,t) = \mathbf{Q}(t)\boldsymbol{\phi}^{(2)}(\mathbf{Y},t) + \mathbf{b}(t)\end{cases}$$

Hence the standard relative velocity in the new R^* frame:

$$\begin{aligned}\mathbf{V}^*_{standard} &= \frac{\partial \boldsymbol{\phi}^{(1)*}}{\partial t}(\mathbf{X}^*,t) - \frac{\partial \boldsymbol{\phi}^{(2)*}(\mathbf{X}^{(2)*},t)}{\partial t}\Big|_{\mathbf{X}^{(2)*}=\mathbf{Y}^*(\mathbf{X}^*,t)}\\ &= \left(\mathbf{Q}\frac{\partial \boldsymbol{\phi}^{(1)}}{\partial t}(\mathbf{X},t) + \dot{\mathbf{Q}}\boldsymbol{\phi}^{(1)}(\mathbf{X},t) + \dot{\mathbf{b}}\right) - \left(\mathbf{Q}\frac{\partial \boldsymbol{\phi}^{(2)}}{\partial t}\Big|_{\mathbf{X}^{(2)}=\mathbf{Y}(\mathbf{X},t)} + \dot{\mathbf{Q}}\boldsymbol{\phi}^{(2)}(\mathbf{Y}(\mathbf{X},t),t) + \dot{\mathbf{b}}\right)\\ &= \mathbf{Q}\left(\frac{\partial \boldsymbol{\phi}^{(1)}}{\partial t}(\mathbf{X},t) - \frac{\partial \boldsymbol{\phi}^{(2)}}{\partial t}\Big|_{\mathbf{X}^{(2)}=\mathbf{Y}(\mathbf{X},t)}\right) + \dot{\mathbf{Q}}\left(\boldsymbol{\phi}^{(1)}(\mathbf{X},t) - \boldsymbol{\phi}^{(2)}(\mathbf{Y}(\mathbf{X},t),t)\right)\\ &= \mathbf{Q}.\mathbf{V}_{standard} + \dot{\mathbf{Q}}(-g\boldsymbol{\nu}) \neq \mathbf{Q}.\mathbf{V}_{standard} \qquad \text{QED} \blacksquare\end{aligned}$$

[2.36]

It is therefore necessary to provide a suitable alternative definition for the relative velocity, satisfying the following main requirements: (i) it must be objective and (ii) it must coincide with the standard relative velocity [2.35] when there is contact ($g = 0$). Klarbring et al. [30, 73] succeeded in extracting such a velocity from the standard relative velocity in the following way. They began by recasting definition [2.35] of $\mathbf{V}_{standard}$, with the help from equations [2.21]–[2.23]:

$$\mathbf{V}_{standard} = \frac{d}{dt}(\mathbf{x} - \mathbf{y}) + \dot{\eta}^\alpha \mathbf{a}_\alpha = -\dot{g}\boldsymbol{\nu} - g\dot{\boldsymbol{\nu}} + \dot{\eta}^\alpha \mathbf{a}_\alpha \qquad [2.37]$$

After examining the right-hand side in [2.37] and identifying the term responsible for the nonobjectivity of $\mathbf{V}_{standard}$, Klarbring et al. then had the idea to remove the term $g\dot{\boldsymbol{\nu}}$ in [2.37] to define the following new velocity:

Definition.– *The relative contact velocity* is defined by

$$\boxed{\mathbf{V}_c = \mathbf{V}^{(1)}(\mathbf{X},t) - \mathbf{V}^{(2)}(\mathbf{Y}(\mathbf{X},t),t) + g\dot{\nu} = -\dot{g}\nu + \dot{\eta}^\alpha \mathbf{a}_\alpha} \qquad [2.38]$$

By construction, the velocity $\mathbf{V}_c$ indeed does coincide with $\mathbf{V}_{standard}$ when contact occurs ($g = 0$). On the other hand, the same reasoning as in [2.36] allows us to prove the following proposition:

Proposition.– $\mathbf{V}_c$ is *objective.*

It is the relative contact velocity $\mathbf{V}_c$ that we will use throughout this book, and several new definitions are based on it:

Definition.–

(i) *The relative normal velocity* or the *proximity velocity* $\mathbf{V}_N$ is defined as the projection of $\mathbf{V}_c$ onto the normal vector ν.

(ii) *The relative tangential velocity* or the *slip velocity* $\mathbf{V}_T$ is the projection of $\mathbf{V}_c$ onto the tangent plane at $\mathbf{y}$ to $S_c^{(2)}$:

$$\boxed{\begin{aligned} \mathbf{V}_N &= -\dot{g}\nu \\ \mathbf{V}_T &= \dot{\eta}^\alpha \mathbf{a}_\alpha \end{aligned}} \qquad [2.39]$$

Note that the relative tangential velocity $\mathbf{V}_T$ is associated with the particle $\mathbf{X} \in S_{0c}^{(1)}$, but it is decomposed with respect to the Eulerian basis $(\mathbf{a}_1, \mathbf{a}_2)$ at the point $\mathbf{y} \in S_c^{(2)}$. As we will see in Chapter 6, by involving both the contactor and target surfaces simultaneously, we can reduce the asymmetry that arises when distinguishing between the contactor and target surfaces.

Remark.– Some authors – such as [138], page 53 – introduce the notion of the slip arc length (or simply the slip) between two times t_1 and t_2, defined by

$$g_T = \int_{t_1}^{t_2} \|\dot{\eta}^\alpha \mathbf{a}_\alpha\| dt = \int_{t_1}^{t_2} \sqrt{\dot{\eta}^\alpha \dot{\eta}^\beta a_{\alpha\beta}}\, dt$$

The differential dg_T of g_T corresponds to what we will call the *slip* $\mathbf{V}_T \Delta t$ in section 8.11. ■

3

Sthenics of Contact

Sthenics (from Greek *sthenos* force) is the area of mechanics that focuses on forces and stresses: the external forces exerted on each body in the system under consideration, the internal forces within each body called stresses and in particular the interactive forces between each body, which appear in contact problems.

The keys notions found within this chapter are the contact stress vector, the normal contact stress or the contact pressure t_N, the frictional stress $\mathbf{t}_T$ and also their nominal versions T_N and $\mathbf{T}_T$.

3.1. Stresses in bodies

Consider a system made up of two bodies numbered 1 and 2, which occupy the domains $\Omega^{(1)}$ and $\Omega^{(2)}$ at time t, respectively. The stresses in body number i can be represented in a few different albeit equivalent ways:

1. As the Cauchy stress tensor $\boldsymbol{\sigma}^{(i)}(\mathbf{x}^{(i)}, t)$ (symmetric), which depends on the point $\mathbf{x}^{(i)} \in \Omega^{(i)}$ and time t.

2. As the Piola–Kirchhoff stress tensor of the first kind $\boldsymbol{\Pi}^{(i)}(\mathbf{X}^{(i)}, t)$ (PK1 or nominal stress tensor, asymmetric), which depends on the point $\mathbf{X}^{(i)} \in \Omega_0^{(i)}$, the inverse image of $\mathbf{x}^{(i)}$ and the time t.

3. As the Piola–Kirchhoff stress tensor of the second kind $\boldsymbol{\Sigma}^{(i)}(\mathbf{X}^{(i)}, t)$ (PK2 or the material stress tensor, symmetric).

These three stress tensors are related by

$$\boldsymbol{\Pi}^{(i)} = \mathbf{F}^{(i)}.\boldsymbol{\Sigma}^{(i)} = J^{(i)}\,\boldsymbol{\sigma}^{(i)}.[\mathbf{F}^{(i)}]^{-T} \qquad (J^{(i)} \equiv \det \mathbf{F}^{(i)}) \tag{3.1}$$

The local dynamic equation for body i can be written in terms of the Lagrangian variables as

$$\mathbf{Div}\boldsymbol{\Pi}^{(i)} + \rho_0^{(i)}\mathbf{f}^{(i)} = \rho_0^{(i)}\ddot{\mathbf{U}}^{(i)} \tag{3.2}$$

where $\rho_0^{(i)}$ is the reference density of body i and $\mathbf{f}^{(i)}$ is the force per unit mass. The symbol **Div** denotes the divergence vector relative to Lagrangian variables, whose derivatives are taken with respect to the reference coordinates $\mathbf{X}^{(i)}$.

3.2. Contact stress vector

Let $\mathbf{t}(\mathbf{x}, t)$ be the Cauchy stress vector at the point $\mathbf{x} \in S_c^{(1)}$ associated with the outwards normal to $\Omega^{(1)}$ (this normal, $-\boldsymbol{\nu}$, is not shown in Figure 3.1). This vector can be decomposed with respect to the natural basis at the projection point $\mathbf{y} \in S_c^{(2)}$ as follows:

$$\boxed{\mathbf{t} = t_N \boldsymbol{\nu} - \mathbf{t}_T} \quad \text{where } \mathbf{t}_T = t_{T\alpha}\mathbf{a}^\alpha \qquad [3.3]$$

where $\mathbf{a}^\alpha$ is a dual basis vector in the tangent plane to the target surface, defined by [2.11] (see Figure 3.1).

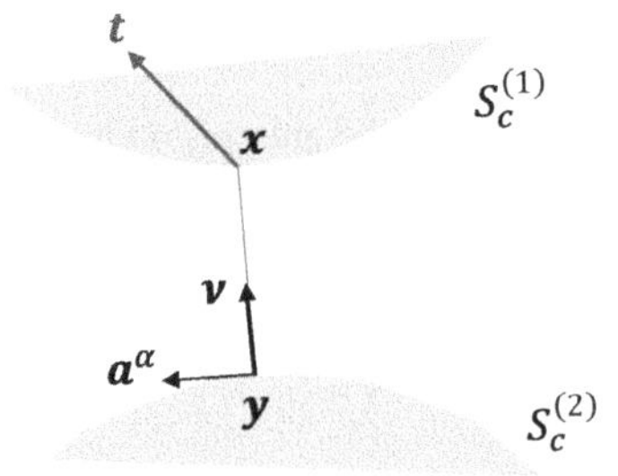

Figure 3.1. *Contact stress vector. For a color version of this figure, see www.iste.co.uk/levan/contact.zip*

Decomposing the stress vector $\mathbf{t}$ onto $\boldsymbol{\nu}$ means that its normal component t_N is *positive* when compressed. This component is called the *contact pressure*. The minus sign adopted in definition [3.3] is not natural, but allows us to obtain convenient expressions for the contact law in the sequel. The $\mathbf{t}_T$ component is *the opposite of the frictional stress.*

We can easily check the following property:

PROPOSITION.– The contact pressure t_N and the frictional stress $\mathbf{t}_T$ are *objective* quantities.

Let $\mathbf{X}$ be a point of $S_{0c}^{(1)}$ and $\mathbf{N}$ is the outwards normal at this point. We denote by $\mathbf{T}(\mathbf{X}, t) = \mathbf{\Pi}^{(1)}.\mathbf{N}$ the nominal stress vector at the point $\mathbf{X} \in S_{0c}^{(1)}$. As with the Cauchy stress vector, we can decompose $\mathbf{T}$ relative to the natural basis at $\mathbf{y} \in S_c^{(2)}$ as follows:

$$\boxed{\mathbf{T}(\mathbf{X}, t) = T_N(\mathbf{X}, t)\,\boldsymbol{\nu} - \mathbf{T}_T(\mathbf{X}, t)} \quad \text{where } \mathbf{T}_T = T_{T\alpha}\mathbf{a}^\alpha \qquad [3.4]$$

The normal component T_N is called *nominal contact pressure*. The component $\mathbf{T}_T$ is *the opposite of the nominal frictional stress*. Note that the nominal contact stress vector at the point $\mathbf{X} \in S_{0c}^{(1)}$ is decomposed on the *current* surface $S_c^{(2)}$.

From the equality $\mathbf{t}dS^{(1)} = \mathbf{T}dS_o^{(1)}$, where $dS_o^{(1)}$ is the reference surface area element and $dS^{(1)}$ is the deformed area, we obtain the following equations relating the (normal and tangential) Eulerian and nominal components:

$$\begin{aligned} t_N dS^{(1)} &= T_N dS_o^{(1)} \\ \mathbf{t}_T dS^{(1)} &= \mathbf{T}_T dS_o^{(1)} \end{aligned} \qquad [3.5]$$

4

The Constitutive Law

The bodies studied, which appear in the mechanical systems, can be made up of various types of materials, and the constitutive laws of these materials will of course dictate how the system responds. On the other hand, the contact formulation presented in this book will be the same regardless of the constituent materials involved. Hence, to fix ideas, the contactor and the target will be considered from now on to be made up of two typical materials: compressible hyperelastic materials or isotropic hardening elastoplastic materials. Other more sophisticated materials may be considered by just changing the constitutive laws in the problem equations.

4.1. Hyperelastic materials

A compressible hyperelastic material is completely described by the strain-energy density function in the reference configuration, denoted w, which is a function of the Green–Lagrange strain tensor $\mathbf{E}$ or of the right Cauchy–Green stretch tensor $\mathbf{C}$ (in this section, the indices $i \in \{1, 2\}$ for the bodies are omitted to simplify the notation). The second Piola–Kirchhoff stress tensor $\mathbf{\Sigma}$ is obtained via the state law:

$$\mathbf{\Sigma} = 2\,\mathrm{SYM}\frac{\partial w}{\partial \mathbf{C}} = \mathrm{SYM}\frac{\partial w}{\partial \mathbf{E}} \qquad [4.1]$$

where the symbol SYM denotes the symmetric part of a second-order tensor.

In numerical calculations, the tangent matrix that arises from the internal forces will need to be known. This matrix is constructed from the nominal tangent tensor $\dfrac{\partial \mathbf{\Pi}}{\partial \mathbf{F}}$, which itself is related to the material tangent tensor $\dfrac{\partial \mathbf{\Sigma}}{\partial \mathbf{E}}$ (elasticity tensor) by

$$\frac{\partial \mathbf{\Pi}}{\partial \mathbf{F}} = \mathbf{F}.\frac{\partial \mathbf{\Sigma}}{\partial \mathbf{E}}.\mathbf{F}^T + (\mathbb{I}.\mathbf{\Sigma})^T \qquad [4.2]$$

where $\mathbb{I}$ is the fourth-order unit tensor defined by $\forall i, j, k, \ell \in \{1, 2, 3\}$, $\mathbb{I}_{ijk\ell} = \delta_{i\ell}\delta_{kj}$ (δ_{ij} is the Kronecker symbol), and the superscript T denotes the major transposition of a fourth-order tensor such that, for any fourth-order tensor $\mathbb{D}$, $\left(\mathbb{D}^T\right)_{ijkl} = \mathbb{D}_{klij}$.

We will discuss how the stress tensor $\mathbf{\Sigma}$ and the material tensor $\dfrac{\partial \mathbf{\Sigma}}{\partial \mathbf{E}}$ can be expressed in two representative models of a hyperelastic material.

Saint Venant–Kirchhoff material

The Saint Venant–Kirchhoff material is the simplest model used in hyperelasticity and is valid in the cases involving moderate strains. The elastic strain energy is a quadratic function of the strain:

$$w(\mathbf{E}) = \bar{\mu} tr(\mathbf{E}^2) + \frac{\lambda}{2}(tr\mathbf{E})^2 \qquad [4.3]$$

where $tr\mathbf{E} = E_{ii}$ is the trace of $\mathbf{E}$, and where $\bar{\mu}$ and λ are the Lamé constants, which are related to Young's modulus E and to Poisson's ratio ν in the same way as in linear elasticity:

$$\lambda = \frac{E\nu}{(1+\nu)(1-2\nu)} \qquad \bar{\mu} = \frac{E}{2(1+\nu)} \qquad [4.4]$$

We have used the notation $\bar{\mu}$ and reserved the symbol μ for the coefficient of friction, which appears frequently in contact theory and will feature in Chapter 5. We can also write the energy w in the form:

$$w(\mathbf{E}) = \frac{1}{2}\mathbf{E} : \mathbb{D} : \mathbf{E} = \frac{1}{2}E^e_{ji} D_{ijkl} E_{lk}$$

where $\mathbb{D}$ is a constant fourth-order tensor whose components relative to an orthonormal Cartesian basis are $D_{ijkl} = \lambda\delta_{ij}\delta_{kl} + \bar{\mu}\left(\delta_{ik}\delta_{jl} + \delta_{il}\delta_{kj}\right)$.

The second Piola–Kirchhoff stress tensor is given by [4.1]:

$$\mathbf{\Sigma} = \text{SYM}\frac{\partial w}{\partial \mathbf{E}} = 2\bar{\mu}\mathbf{E} + \lambda\, tr\mathbf{E}\,\mathbf{I} = \mathbb{D} : \mathbf{E} \qquad [4.5]$$

The material tangent tensor is $\dfrac{\partial \mathbf{\Sigma}}{\partial \mathbf{E}} = \mathbb{D} = \text{constant}$.

Compressible neo-Hookean material

In problems involving large compressive stresses, the Saint Venant–Kirchhoff material is no longer suitable, since it becomes unstable and calculations begin to diverge. In this case, another more sophisticated material model must be used, for example, the neo-Hookean compressible model, defined by the elastic deformation energy [58]:

$$w(\mathbf{C}) = \frac{\bar{\mu}}{2}(tr\mathbf{C} - 3 - \ln\det\mathbf{C}) + \frac{\lambda}{8}(\ln\det\mathbf{C})^2 \qquad [4.6]$$

where $\bar{\mu}$ and λ are the given material parameters (we can also write everything in terms of E and ν, which are related to $\bar{\mu}$ and λ by equations [4.4]). From [4.1], we obtain the stress:

$$\mathbf{\Sigma} = \bar{\mu}(\mathbf{I} - \mathbf{C}^{-1}) + \frac{\lambda}{2}\ln\det\mathbf{C}.\mathbf{C}^{-1} \qquad [4.7]$$

The material tangent tensor is given by:

$$\left(\frac{\partial \mathbf{\Sigma}}{\partial \mathbf{E}}\right)_{ijk\ell} = \frac{\partial \Sigma_{ij}}{\partial E_{\ell k}} = \lambda(C^{-1})_{ij}(C^{-1})_{k\ell} + (\mu - \frac{\lambda}{2}\ln\det\mathbf{C})\left[(C^{-1})_{i\ell}(C^{-1})_{kj} + (C^{-1})_{ik}(C^{-1})_{\ell j}\right] \qquad [4.8]$$

4.2. Elastoplastic materials with isotropic hardening

In the presence of an elastoplastic material undergoing finite deformations, we will adopt the additive Green–Nagdhi decomposition [47, 48], which states that the total strain $\mathbf{E}$ decomposes into a sum of the elastic strain, denoted by $\mathbf{E}^e$, and the plastic strain, denoted by $\mathbf{E}^p$:

$$\mathbf{E} = \mathbf{E}^e + \mathbf{E}^p \tag{4.9}$$

In this approach, the total strain $\mathbf{E}$ and the plastic strain $\mathbf{E}^p$ are considered to be two independent variables, while the elastic strain $\mathbf{E}^e$ is just the difference between $\mathbf{E}$ and $\mathbf{E}^p$ [99, 19, 98]. The plastic strain $\mathbf{E}^p$ is assumed to be a symmetric and objective tensor, so that $\mathbf{E}^e$ is too.

In the literature, there are several models that describe the kinematics of plasticity in finite deformations, with Lee's multiplicative decomposition [87, 86] being the most widely used today. This book uses the additive Green–Nagdhi decomposition, as it allows for a system of equations similar to those in small deformations and simplifies programming. The focus of this book is the study of the contact problem, not plasticity in large strains. It is important to ensure the proposed contact formulation can be applied to other material responses, including plasticity. However, the choice of kinematic decomposition is secondary in the context of this book.

The following description is based on the theory of generalized standard materials [51, 42, 65]. Let us assume that the strain-energy density w in the reference configuration can be written as the sum of the elastic energy w^e, which depends on the elastic strain $\mathbf{E}^e$, and the hardening energy w^α, which depends on the hardening variable α. By assuming that the elastic energy takes on the same form as in the Saint Venant–Kirchhoff model [4.3], that is, as a quadratic function of the elastic strain, the strain–energy density can be written as

$$\begin{aligned} w\left(\mathbf{E}^e, \alpha\right) = w^e\left(\mathbf{E}^e\right) + w^\alpha(\alpha) &= \frac{1}{2}\mathbf{E}^e : \mathbb{D} : \mathbf{E}^e + w^\alpha(\alpha) \\ &= \frac{1}{2}E^e_{ji} D_{ijkl} E^e_{lk} + w^\alpha(\alpha) \end{aligned} \tag{4.10}$$

where $\mathbb{D}$ is a fourth-order constant tensor.

With the additive decomposition [4.9], the state and evolution laws have the same form as in the case of small strains. The state laws state that the second Piola–Kirchhoff stress tensor $\mathbf{\Sigma}$ and the thermodynamic force A (conjugate of the work hardening variable α) are the partial derivatives of the elastic energy with respect to the strain and the work hardening variable, respectively:

$$\mathbf{\Sigma} = \frac{\partial w}{\partial \mathbf{E}^e} = \mathbb{D} : \mathbf{E}^e \qquad A = \frac{\partial w}{\partial \alpha} \tag{4.11}$$

It is assumed that the elastic domain is defined using the von Mises criterion, with an isotropic work hardening:

$$f(\mathbf{\Sigma}, A) = \sqrt{\frac{3}{2}\mathbf{S} : \mathbf{S}} - A - \sigma_0 \tag{4.12}$$

where $\mathbf{S} = \mathbf{\Sigma} - \frac{1}{3}(tr\mathbf{\Sigma})\mathbf{I}$ is the deviatoric part of the stress $\mathbf{\Sigma}$ and σ_0 the initial yield stress.

The evolution laws are normality laws, which state that the rate of the plastic strain $\mathbf{E}^p$ and the rate of the hardening variable α are derived from a dissipation potential, which itself is directly linked to the yield function f:

$$\dot{\mathbf{E}}^p = \dot{\Lambda}\frac{\partial f}{\partial \Sigma} \qquad \dot{\alpha} = -\dot{\Lambda}\frac{\partial f}{\partial A}$$

where $\dot{\Lambda} \geq 0$ is the plastic multiplier subject to plasticity conditions $\dot{\Lambda} = 0$ if $f < 0$ or $f = 0$, and $\frac{\partial f}{\partial \Sigma} : \dot{\Sigma} < 0$, $\dot{\Lambda} \geq 0$ if $f = 0$ and $\frac{\partial f}{\partial \Sigma} : \dot{\Sigma} \geq 0$. The rates of the Lagrangian variables $\mathbf{E}^p$ and α are just the usual time derivatives. According to equation [4.12] of the yield function f, we obtain:

$$\dot{\mathbf{E}}^p = \dot{\Lambda}\sqrt{\frac{3}{2}}\frac{\mathbf{S}}{\|\mathbf{S}\|} \qquad \dot{\alpha} = -\dot{\Lambda} \tag{4.13}$$

We may define an equivalent plastic strain P, which is a scalar representative of the plastic strain tensor $\mathbf{E}^p$:

$$\dot{P} \equiv \sqrt{\frac{2}{3}\dot{\mathbf{E}}^p : \dot{\mathbf{E}}^p} \tag{4.14}$$

By substituting equation $[4.13]_1$ into the previous definition, we see that $\dot{P} = \dot{\Lambda}$ which, when combined with $[4.13]_2$, gives us the following simple equation that relates the hardening variable α and the equivalent plastic strain P (assuming the initial values are zero):

$$\alpha = P$$

When simulating elastoplasticity numerically, we will utilize the following expression for the work hardening energy:

$$w^\alpha(\alpha) = c_0\alpha + \frac{1}{2}c_2\alpha^2 + (c_1 - c_0)\left(\alpha + \frac{\exp(-c_3\alpha)}{c_3}\right)$$

where c_0, c_1, c_2 and c_3 are the given material coefficients. Using $[4.11]_2$, the thermodynamic force A can then be written as:

$$A = c_0 + c_2\alpha + (c_1 - c_0)\left(1 - \exp(-c_3\alpha)\right) \tag{4.15}$$

The case $c_0 = c_1 = 0$ corresponds to linear isotropic work hardening.

In the elastoplastic case, the material tangent tensor equals:

$$\frac{\partial \Sigma}{\partial \mathbf{E}} = \mathbb{D} - \frac{\mathbb{D} : \frac{\partial f}{\partial \Sigma} \otimes \frac{\partial f}{\partial \Sigma} : \mathbb{D}}{H + \frac{\partial f}{\partial \Sigma} : \mathbb{D} : \frac{\partial f}{\partial \Sigma}} \tag{4.16}$$

where $H = \left(\frac{\partial f}{\partial A}\right)^2 \frac{\partial^2 w}{\partial \alpha^2}$ is called the strain hardening modulus, and where the tensor product $\otimes$ of two second-order tensors $\mathbf{S}$ and $\mathbf{T}$ is given by $(\mathbf{S} \otimes \mathbf{T})_{ijkl} = S_{ij}T_{kl}$.

In the numerical computations, a local resolution must be performed at each iterative step of the calculation and at every integration point. This involves solving the system of equations [4.9] and [4.11]–[4.13] given the current displacement field $\mathbf{U}$ and the information that comes from the previous increment. The rates in [4.13] are discretized with respect to time using the implicit Euler scheme, and the solution is obtained using the radial return scheme [123, 124]. The resulting tangent tensor $\frac{\partial \Sigma}{\partial \mathbf{E}}$ differs slightly from its theoretical counterpart [4.16], and is essential in construction of the coherent tangent matrix since it ensures convergence.

The Saint Venant–Kirchhoff model, which comes into play in the elastic part of the energy [4.10], is simple yet flawed in the sense that it becomes unstable under strong compressive stresses [58, 65]. This is demonstrated in the fitting example discussed in Chapter 10, where the calculation fails to converge due to the instability under large compressive stresses. In such cases, the Saint Venant–Kirchhoff model is replaced by the Mooney model, which is slightly more complex but can handle large compression. As used in [111], this model involves an intermediate configuration referred to as the Mooney configuration, in which the strain $\mathbf{E}$ is replaced by the Mooney strain denoted $\mathbf{E}^M$ and the stress $\boldsymbol{\Sigma}$ is replaced by the Mooney stress denoted $\boldsymbol{\Sigma}^M$. The Mooney strain is defined by

$$\mathbf{E}^M = \frac{1}{4}(\mathbf{C} - \mathbf{C}^{-1}) \quad [4.17]$$

The Mooney stress $\boldsymbol{\Sigma}^M$ is defined by the equation:

$$\boldsymbol{\Sigma} : \dot{\mathbf{E}} = \boldsymbol{\Sigma}^M : \dot{\mathbf{E}}^M \quad [4.18]$$

Equations [4.9]–[4.13] can then be rewritten in the Mooney configuration which, for example, gives

$$\mathbf{E}^M = \mathbf{E}^{e,M} + \mathbf{E}^{p,M} \quad [4.19]$$

$$\boldsymbol{\Sigma}^M = 2\bar{\mu}\mathbf{E}^{e,M} + \lambda tr(\mathbf{E}^{e,M})\mathbf{I} \quad [4.20]$$

Once the local integration has been performed in the Mooney configuration, we obtain the stress $\boldsymbol{\Sigma}$ via

$$\boldsymbol{\Sigma} = \frac{1}{2}(\boldsymbol{\Sigma}^M + \mathbf{C}^{-1}.\boldsymbol{\Sigma}^M.\mathbf{C}^{-1}) \quad [4.21]$$

The calculation scheme is presented in Table 4.1.

1. Knowing $\mathbf{U}$, compute the strains $\mathbf{E}, \mathbf{C}$.
2. Switch to the Mooney configuration.

   Calculate the strain $\mathbf{E}^M$ using [4.17].

   Perform the local integration to obtain the Mooney variables: $\boldsymbol{\Sigma}^M, \mathbf{E}^{e,M}, \mathbf{E}^{p,M}$, etc.
3. Return to Lagrangian configuration.

   Calculate $\boldsymbol{\Sigma}$ using [4.21].

Table 4.1. *Mooney model calculation scheme*

5

Contact Laws

Contact laws are equations that describe what happens at the contact interface when two solids in the considered system touch. They are special boundary conditions that are added in addition to the standard ones (displacements and/or stresses constraints), and which are essential when trying to solve a contact problem. Contact laws are studied in tribology, which is namely the science that studies the interactions between solids and which provides a wide variety of more or less complex contact models, which may take into account adhesion, lubrication, friction and wear. An interested reader can find detailed descriptions of the different contact models and laws in [41] and [119].

The simplest contact laws in the literature that are frequently used are the adhesion-free normal contact law and Coulomb's dry friction tangential contact law. Less frequently used examples include more sophisticated friction models, which come into play when Coulomb's model no longer suffices. See some previous studies [102, 142, 138, 77] for examples of friction laws that take microscopic effects into account as well as examples of nonlocal friction laws. For the numerical implementation of the orthotropic contact law that includes both adhesion and friction, the reader can also consult the work of [75].

In this work, we only consider the *nonadhesive* normal contact law and *Coulomb's dry friction* tangential contact law. Despite their simple physical nature, these laws are mathematically complicated: they are not regular as they are multivalued and are given by non-differentiable equations.

The laws of contact are presented as relations between the contact kinematic quantities (studied in Chapter 2) and contact stresses (discussed in Chapter 3) in a way analogous to the constitutive laws that connect the strains to the stresses within the bodies.

We will first state the contact laws in the Eulerian form, which is natural. Then, thanks to the relation [3.5], we will transform them into a mixed form where the Lagrangian sliding velocity and the nominal contact stresses intervene, knowing that the mixed form lends itself better to numerical resolution in total Lagrangian formulation.

5.1. Normal contact law

The normal contact law translates the simple facts:

- A contactor particle cannot penetrate inside the target: this is equivalent to saying that the proximity g must be negative or zero.

- When the two bodies are separated ($g < 0$), the normal contact stress is zero.

– When there is contact ($g = 0$), the normal stress at the point of contact must be compressive.

The law of normal contact is expressed in the following form, linking the proximity g and the contact pressure t_N:

$$\forall t, \quad \forall \mathbf{X} \in S_{0c}^{(1)}, \quad \begin{array}{l} \text{(a) } g(\mathbf{X},t) \leq 0 \text{ (non-penetrability condition)} \\ \text{(b) if } g(\mathbf{X},t) < 0, \text{then } t_N(\mathbf{X},t) = 0 \\ \text{(c) if } g(\mathbf{X},t) = 0, \text{then } t_N(\mathbf{X},t) \geq 0 \text{ (non-adhesion condition)} \end{array} \qquad [5.1]$$

The previous equations, which are called the *Hertz-Signorini conditions*, are summarized in Figure 5.1. The normal contact law is a relation involving g and t_N, which is non-differentiable and multivalued at $g = 0$.

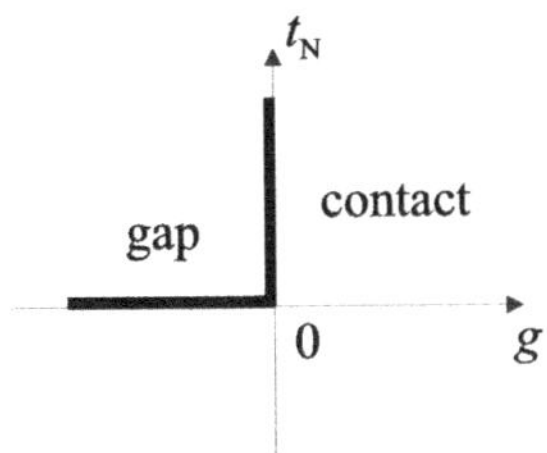

Figure 5.1. *Normal contact law*

From [3.5], we know that the real stress t_N and the nominal stress T_N have the same sign. We can therefore rewrite the normal contact law [5.1] in terms of T_N:

$$\forall t, \quad \forall \mathbf{X} \in S_{0c}^{(1)}, \quad \begin{array}{l} \text{(a) } g(\mathbf{X},t) \leq 0 \text{ (non-penetrability condition)} \\ \text{(b) if } g(\mathbf{X},t) < 0, \text{then } T_N(\mathbf{X},t) = 0 \\ \text{(c) if } g(\mathbf{X},t) = 0, \text{then } T_N(\mathbf{X},t) \geq 0 \text{ (non-adhesion condition)} \end{array} \qquad [5.2]$$

We are used to expressing the normal contact law in the form of *the Kuhn–Tucker optimality conditions* when there are no more cases, $g < 0$ or $g = 0$, to distinguish, but only equality or inequality constraint equations:

$$\forall t, \quad \forall \mathbf{X} \in S_{0c}^{(1)}, \quad \begin{array}{l} \text{(a) } g(\mathbf{X},t) \leq 0 \text{ (non-penetrability condition)} \\ \text{(b) } T_N(\mathbf{X},t) \geq 0 \text{ (non-adhesion condition)} \\ \text{(c) } T_N(\mathbf{X},t)g(\mathbf{X},t) = 0 \text{ (complementarity condition)} \end{array} \qquad [5.3]$$

We can check that the contact law (5.3) is *objective*.

By using convex analysis, we can further express the multivalued normal contact law $t_N = t_N(g)$ in terms of the subgradient of a non-differentiable convex potential [95]: the normal stress t_N is a subdifferential of the indicator function $I_{\mathbb{R}^-}$ on the half-line $\mathbb{R}^-$:

$$t_N \in \partial I_{\mathbb{R}^-}(g) \qquad [5.4]$$

where the indicator function of a convex set C is defined by

$$I_C(x) = \begin{cases} 0 & \text{if } x \in C \\ +\infty & \text{if } x \notin C \end{cases} \qquad [5.5]$$

Since the Legendre–Fenchel transform $I^*_{\mathbb{R}^-}$ of $I_{\mathbb{R}^-}$ is equal to $I_{\mathbb{R}^+}$, we obtain the inverse equation to [5.4]: the proximity g is a subdifferential of the indicator function $I_{\mathbb{R}^+}$ on the ray $\mathbb{R}^+$:

$$g \in \partial I^*_{\mathbb{R}^-}(t_N) = \partial I_{\mathbb{R}^+}(t_N) \quad [5.6]$$

5.2. Tangential contact law

In addition to the normal contact law, one needs a tangential contact law to take into account the friction at the interface of the two bodies. In this book, we consider *Coulomb's friction law*, which is frequently used in practice, and which consists of taking the same law found in rigid-body mechanics and adapting it to continuum mechanics by writing it at each point of the contact surface.

The Coulomb model involves a material coefficient μ, called *the coefficient of friction*, which is assumed to be constant throughout time and space. Coulomb's friction law can be stated as follows:

$$\begin{array}{l}\forall t, \quad \forall \mathbf{X} \in S^{(1)}_{0c},\\ \text{a) if } g(\mathbf{X},t) < 0, \text{ then } \mathbf{t}_T(\mathbf{X},t) = \mathbf{0} \text{ (with no condition on } \mathbf{V}_T(\mathbf{X},t))\\ \text{b) if } g(\mathbf{X},t) = 0, \text{ then } \|\mathbf{t}_T(\mathbf{X},t)\| \le \mu t_N(\mathbf{X},t) \text{ (Coulomb's criterion)}\\ \qquad \text{and } \mathbf{V}_T \text{ is given by } \left\{\begin{array}{l}\text{if } \|\mathbf{t}_T\| < \mu t_N, \text{ then } \mathbf{V}_T = \mathbf{0} \ (\textit{stick})\\ \text{if } \|\mathbf{t}_T\| = \mu t_N, \text{ then } \mathbf{V}_T \times \mathbf{t}_T = \mathbf{0}, \mathbf{V}_T.\mathbf{t}_T \ge 0\\ \qquad \text{(i.e., } \mathbf{V}_T \upuparrows \mathbf{t}_T) \ (\textit{slip})\end{array}\right.\end{array} \quad [5.7]$$

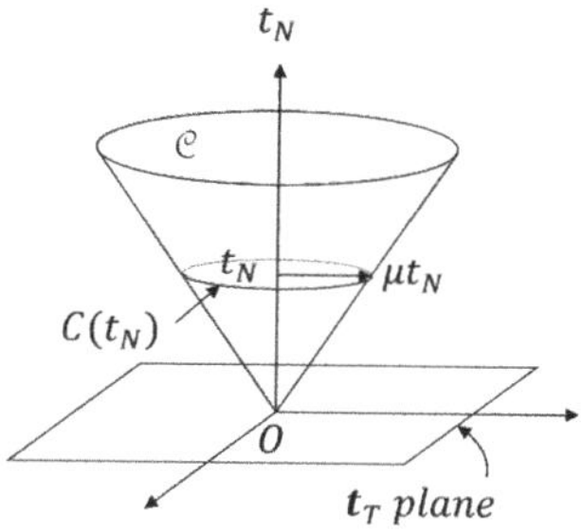

Figure 5.2. *Coulomb's friction cone in the stress space* $(t_N, \mathbf{t}_T)$

Thus, in constraint space $(t_N, \mathbf{t}_T)$, the point $(t_N, \mathbf{t}_T)$ never leaves *Coulomb's friction cone* $\mathcal{C} = \{(t_N, \mathbf{t}_T) \mid \|\mathbf{t}_T\| \le \mu t_N\}$ (see Figures 5.2 and 5.3).

- When we are inside the cone ($\|\mathbf{t}_T\| < \mu t_N$), no slipping occurs (*stick*).
- When we are on the boundary of the cone ($\|\mathbf{t}_T\| = \mu t_N$), *slip* can occur. Any slip that takes place must be collinear to the frictional stress and in the same direction as it.

The friction law [5.1] has $\mathbf{V}_T$ and $\mathbf{t}_T$ as the main variables, and *contact pressure* t_N *as the parameter*.

The stress $\mathbf{t}_T$ is related to the sliding velocity $\mathbf{V}_T$ by a multivalued function at $\mathbf{V}_T = \mathbf{0}$ and non-differentiable. The value $\mathbf{V}_T = \mathbf{0}$ possesses a corresponding set of values of $\mathbf{t}_T$ satisfying $\|\mathbf{t}_T\| < \mu t_N$.

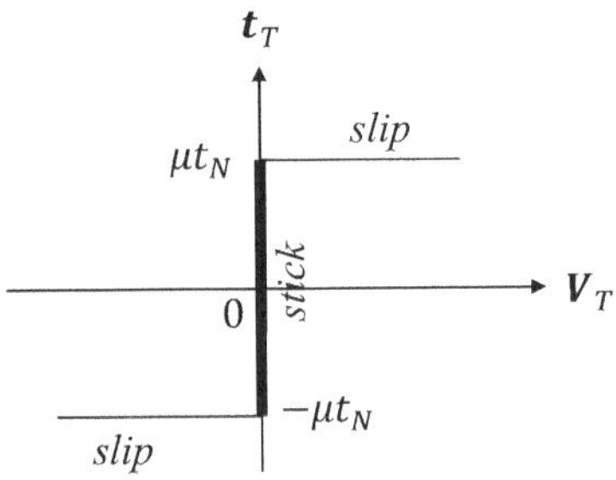

Figure 5.3. *Coulomb's friction law for tangential contact, many-to-many equation between $\mathbf{t}_T$ and $\mathbf{V}_T$*

From [3.5], we know that the real stresses $(t_N, \mathbf{t}_T)$ and nominal stresses $(T_N, \mathbf{t}_T)$ are proportional. Hence, the statement in [5.1] remains valid even if we replace $(t_N, \mathbf{t}_T)$ by $(T_N, \mathbf{t}_T)$:

$$\begin{array}{l} \forall t, \quad \forall \mathbf{X} \in S_{0c}^{(1)}, \\ \text{(a) if } g(\mathbf{X},t) < 0, \text{ then } \mathbf{T}_T(\mathbf{X},t) = \mathbf{0} \text{ (and no relation on } \mathbf{V}_T(\mathbf{X},t)) \\ \text{(b) if } g(\mathbf{X},t) = 0, \text{ then } \|\mathbf{T}_T(\mathbf{X},t)\| \le \mu T_N(\mathbf{X},t) \text{ (Coulomb's criterion)} \\ \qquad \text{and } \mathbf{V}_T \text{ is given by } \begin{cases} \text{if } \|\mathbf{T}_T\| < \mu T_N, \text{ then } \mathbf{V}_T = \mathbf{0} \; (\textit{stick}) \\ \text{if } \|\mathbf{T}_T\| = \mu T_N, \text{ then } \mathbf{V}_T \times \mathbf{T}_T = \mathbf{0}, \mathbf{V}_T.\mathbf{T}_T \ge 0 \\ \qquad \text{(i.e., } \mathbf{V}_T \upuparrows \mathbf{T}_T) \; (\textit{slip}) \end{cases} \end{array} \qquad [5.8]$$

Thus, in constraint space $(T_N, \mathbf{T}_T)$, the point $(T_N, \mathbf{T}_T)$ never leaves the Coulomb cone $\mathcal{C} = \{(T_N, \mathbf{T}_T) \mid \|\mathbf{T}_T\| \le \mu T_N\}$ (see Figures 5.4 and 5.5).

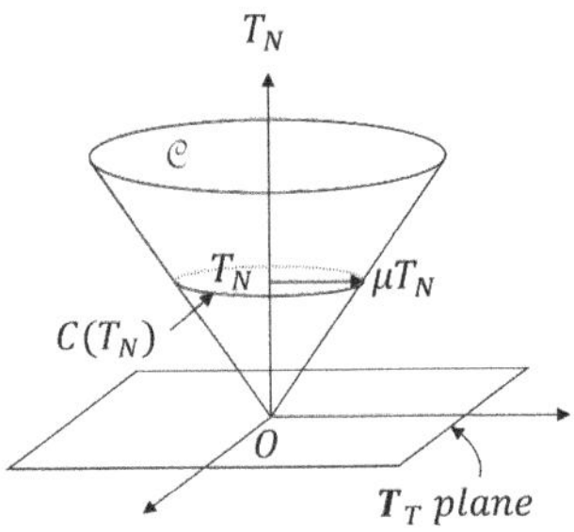

Figure 5.4. *Coulomb cone in stress space $(T_N, \mathbf{T}_T)$*

- When we are inside the cone $(\|\mathbf{T}_T\| < \mu T_N)$, no slip can occur (*stick*).
- When we are on the border of the cone $(\|\mathbf{T}_T\| = \mu T_N)$, then *slip* can occur. Any slip that takes place must be collinear to the frictional stress and in the same direction as it.

Just as we wrote the law of normal contact in the form of Kuhn–Tucker [5.3], we will write the law of tangential contact in the following form, where there are no more case distinctions,

$||T_T|| < \mu T_N$ or $||T_T|| = \mu T_N$, but where there are only equality or inequality constraint equations:

$$\forall t, \forall \mathbf{X} \in S^{(1)}_{0c},\ \text{(a) if } g < 0, \text{then } \mathbf{T}_T = \mathbf{0} \text{ (and no relation on } \mathbf{V}_T)$$
$$\text{(b) if } g = 0, \text{then } \begin{cases} f(\mathbf{T}_T, T_N) = ||\mathbf{T}_T|| - \mu T_N \le 0 \text{ (Coulomb's criterion)} \\ \gamma \ge 0 \\ \gamma f = 0 \text{ (complementarity condition)} \end{cases}$$
$$\text{and } \mathbf{V}_T \text{ is given by } \mathbf{V}_T = \gamma \mathbf{T}_T \qquad [5.9]$$

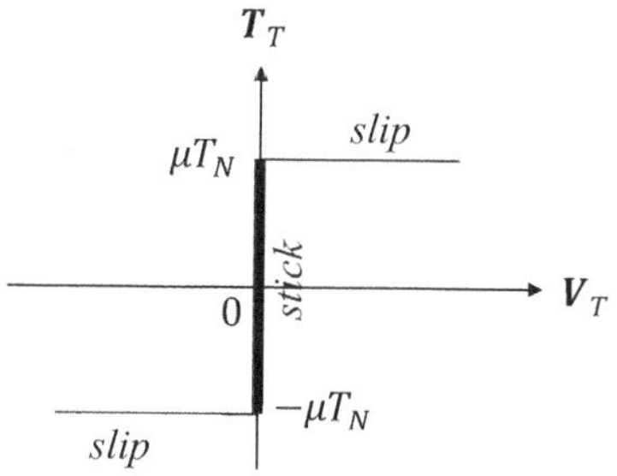

Figure 5.5. *Coulomb's tangential contact law, multivalued equation between T_T and V_T*

We can check that the tangential contact law [5.9] is *objective*.

We can also rewrite the tangential contact law in a more condensed form using convex analysis. For a fixed normal stress t_N, let us denote (see Figure 5.2):

$$C(t_N) = \{\mathbf{t}_T \mid \ ||\mathbf{t}_T|| \le \mu t_N\} = \text{the } \textit{closed} \text{ disk of radius } k \equiv \mu t_N \qquad [5.10]$$

The tangential contact is written then as:

$$\mathbf{V}_T \in \partial I_{C(t_N)}(\mathbf{t}_T) \qquad [5.11]$$

or conversely:

$$\mathbf{t}_T \in \partial I^*_{C(t_N)}(\mathbf{V}_T) \qquad [5.12]$$

where $I^*_{C(t_N)}(\mathbf{V}_T) = k||\mathbf{V}_T||$ is the conjugate of the indicator function $I_{C(t_N)}(\mathbf{t}_T)$.

REMARK.– In plasticity, if at any given moment we are on the elastic boundary so that $f(\sigma, a) = 0$ (where f is the load function and a is a hardening variable), then there exist two possible subsequent events:

– We can remain on this boundary (f remains zero, which is the plastic loading case).

– We can go back into the elastic domain (f becomes negative, which is the unloading case).

Numerically, it is necessary to write the plasticity test in the form $f \ge 0$ rather than as $f = 0$, since there is a small chance of having $f = 0$ exactly.

Similarly, in contact mechanics, at a given moment, if *slip* occurs, $f(\mathbf{T}_T, T_N) = ||\mathbf{T}_T|| - \mu T_N = 0$, then subsequently we can either:

- stay in the *slipping* regime (f stays zero);
- or move into the *sticking* regime (f becomes negative).

Numerically, the *slip* test would have to be written in the form $||\mathbf{T}_T|| \geq \mu T_N$ instead of $||\mathbf{T}_T|| = \mu T_N$, since there is a small chance of having the equality again exactly. In fact, we will see in Chapter 8 that the test is more about the neighboring quantities denoted $||\hat{\boldsymbol{\lambda}}_T||$ and $\hat{\lambda}_N$, so that the test can then be written as $||\hat{\boldsymbol{\lambda}}_T|| \geq \mu\hat{\lambda}_N$. ■

6

Strong Formulation of the Contact Problem

The various equations seen in previous chapters that are necessary to solve finite deformation contact problems are collected together in this chapter. These form a system of partial differential equations with respect to both time and space, coupled with inequalities specific to the contact problem.

6.1. Field equations

The field equations consisted of the following equations and are to be satisfied at any time t in the time interval $[0, t_{max}]$ in question, and at any point $\mathbf{X}^{(i)}$ of the reference domains $\Omega_0^{(i)}$, $i \in \{1, 2\}$.

• The dynamical equation of the two bodies:

$$\forall i \in \{1,2\}, \qquad \boxed{\mathbf{Div}\boldsymbol{\Pi}^{(i)} + \rho_0^{(i)}\mathbf{f}^{(i)} = \rho_0^{(i)}\ddot{\mathbf{U}}^{(i)}} \qquad [6.1]$$

where we recall that the **Div** symbol denotes the Lagrangian divergence of a vector, which involves derivatives taken with respect to the reference coordinates $\mathbf{X}^{(i)}$.

• The relation between the displacements and strains:

$$\forall i \in \{1,2\}, \qquad \boxed{\mathbf{F}^{(i)} = \mathbf{I} + \mathbf{Grad}\mathbf{U}^{(i)}} \qquad \boxed{\mathbf{C}^{(i)} = \mathbf{F}^{(i)T}\mathbf{F}^{(i)}} \qquad [6.2]$$

where the **Grad** symbol denotes the Lagrangian gradient of a vector, which involves derivatives taken with respect to the reference coordinates $\mathbf{X}^{(i)}$.

• The hyperelastic constitutive laws of the bodies:

$$\forall i \in \{1,2\}, \qquad \boxed{\boldsymbol{\Sigma}^{(i)} = 2\,\mathrm{SYM}\frac{\partial w^{(i)}}{\partial \mathbf{C}^{(i)}}} \qquad \text{where } w^{(i)} = w^{(i)}(\mathbf{C}^{(i)}) \text{ is given} \qquad [6.3]$$

In elastoplasticity, the constitutive laws are more complicated. In brief, the constitutive laws are represented symbolically using the operators $\mathcal{L}^{(i)}$:

$$\forall i \in \{1,2\}, \qquad \boxed{\boldsymbol{\Sigma}^{(i)}(t) = \mathcal{L}^{(i)}\left(\{\mathbf{C}^{(i)}(\tau)\}_{0\le\tau\le t}\right)} \qquad [6.4]$$

6.2. Boundary conditions

We recall that the boundary of the reference body $S_0^{(i)}$ is partitioned into three disjoint sets $S_{0U}^{(i)}$, $S_{0T}^{(i)}$ and $S_{0c}^{(i)}$, where $S_{0U}^{(i)}$ and $S_{0T}^{(i)}$ are boundary portions where the displacement and the stress vector are prescribed, respectively, and where $S_{0c}^{(i)}$ is the boundary portion where contact is likely to occur.

The boundary conditions are composed of the classical ones on the surfaces $S_{0U}^{(i)}$, $S_{0T}^{(i)}$ and the contact conditions on the contact surface $S_{0c}^{(1)}$.

- The classical boundary conditions on $S_{0U}^{(i)}$ and $S_{0T}^{(i)}$, $i \in \{1,2\}$ are: $\forall t \in [0, t_{max}]$:

$$\begin{array}{ll} \forall \mathbf{X}^{(i)} \in S_{0U}^{(i)}, & \boxed{\mathbf{U}^{(i)} = \bar{\mathbf{U}}^{(i)}} \\ \forall \mathbf{X}^{(i)} \in S_{0T}^{(i)}, & \boxed{\mathbf{\Pi}^{(i)}.\mathbf{N}^{(i)} = \bar{\mathbf{T}}^{(i)}} \end{array} \qquad [6.5]$$

where $\bar{\mathbf{U}}^{(i)}, \bar{\mathbf{T}}^{(i)}$ represent, respectively, the prescribed displacement and the prescribed nominal stress vector, and where $\mathbf{N}^{(i)}$ represents the unit normal vector to the point $\mathbf{X}^{(i)}$ on $\Omega_0^{(i)}$.

- Boundary conditions due to contact: As seen in Chapters 2 and 3, $\forall t, \forall \mathbf{X} \in S_{0c}^{(1)}$, the relative contact velocity $\mathbf{V}_c(\mathbf{X}, t)$ and the nominal stress vector at a point $\mathbf{X}$, $\mathbf{T}(\mathbf{X}, t) = \mathbf{\Pi}^{(1)}(\mathbf{X}, t).\mathbf{N}(\mathbf{X})$, can be decomposed as follows:

$$\boxed{\mathbf{V}_c = V_N \boldsymbol{\nu} + \mathbf{V}_T} \qquad [6.6]$$

$$\boxed{\mathbf{T}(\mathbf{X}, t) = T_N \boldsymbol{\nu} - \mathbf{T}_T} \qquad [6.7]$$

where $\boldsymbol{\nu}$ is the outer normal to $S_c^{(2)}$ at the point $\mathbf{y} = \text{proj}_{S_c^{(2)}} \mathbf{x}$.

The contact laws can then be written (see Chapter 5) as follows:

(i) The normal contact law:

$$\boxed{\begin{array}{lll} \forall t, & \forall \mathbf{X} \in S_{0c}^{(1)}, & \text{(a) } g(\mathbf{X}, t) \leq 0 \text{ (non-penetrability condition)} \\ & & \text{(b) if } g(\mathbf{X}, t) < 0, \text{then } T_N(\mathbf{X}, t) = 0 \\ & & \text{(c) if } g(\mathbf{X}, t) = 0, \text{then } T_N(\mathbf{X}, t) \geq 0 \text{ (non-adhesion condition)} \end{array}} \qquad [6.8]$$

(ii) The tangential contact law:

$$\boxed{\begin{array}{l} \forall t, \quad \forall \mathbf{X} \in S_{0c}^{(1)}, \\ \text{(a) if } g(\mathbf{X}, t) < 0, \text{then } \mathbf{T}_T(\mathbf{X}, t) = \mathbf{0} \text{ (no condition on } \mathbf{V}_T(\mathbf{X}, t)) \\ \text{(b) if } g(\mathbf{X}, t) = 0, \text{then } \|\mathbf{T}_T(\mathbf{X}, t)\| \leq \mu T_N(\mathbf{X}, t) \text{ (Coulomb criterion)} \\ \qquad \text{and } \mathbf{V}_T \text{ is given by} \left\{ \begin{array}{l} \text{if } \|\mathbf{T}_T\| < \mu T_N, \text{ then } \mathbf{V}_T = \mathbf{0} \text{ (\textit{stick})} \\ \text{if } \|\mathbf{T}_T\| = \mu T_N, \text{ then } \mathbf{V}_T \times \mathbf{T}_T = \mathbf{0}, \mathbf{V}_T.\mathbf{T}_T \geq 0 \\ \qquad \text{(i.e. } \mathbf{V}_T \upuparrows \mathbf{T}_T) \text{ (\textit{slip})} \end{array} \right. \end{array}}$$

[6.9]

(iii) We must also satisfy the equilibrium of the contact stresses at the contact interface, that is, the momentum balance between the normal stresses (or the tangential tractions) on the contact interface:

$$\forall t, \quad \forall \mathbf{X} \in S_{0c}^{(1)}, \text{ if } g = 0 : \boxed{\underbrace{\mathbf{T}^{(1)}}_{\text{this is } \mathbf{T}} (\mathbf{X}, t) dS_o^{(1)} = -\mathbf{T}^{(2)}(\mathbf{Y}(\mathbf{X}, t), t) dS_o^{(2)}} \qquad [6.10]$$

where $dS_o^{(1)}$ (respectively, $dS_o^{(2)}$) is the surface element of $S_{0c}^{(1)}$ (respectively, $S_{0c}^{(2)}$), such that the two surface elements coincide at the current time instant: $dS^{(1)} = dS^{(2)}$.

Equation [6.10] can be extended to the case $g < 0$. Indeed, when there is no contact, $\mathbf{T}^{(1)} = \mathbf{T} = \mathbf{0}$ and $\mathbf{T}^{(2)} = \mathbf{0}$, so that [6.10] is trivially satisfied.

6.3. Initial conditions

For a dynamical problem, we need to specify the initial conditions at the initial time $t = 0$:

$$\forall i \in \{1, 2\}, \forall \mathbf{X}^{(i)} \in \Omega_0^{(i)}, \quad \boxed{\mathbf{U}^{(i)}(\mathbf{X}^{(i)}, 0) = \mathbf{0}} \quad \boxed{\dot{\mathbf{U}}^{(i)}(\mathbf{X}^{(i)}, 0) = \mathbf{V}_0^{(i)}(\mathbf{X}^{(i)})} \qquad [6.11]$$

where $\mathbf{V}_0^{(i)}$ is the prescribed initial velocity field for body i.

6.4. Remarks

Strong problem

We therefore have to solve the following problem:

Look for the displacement fields $\mathbf{U}^{(i)}$ and the stress fields $\mathbf{\Sigma}^{(i)}$, $i \in \{1, 2\}$, in particular, the stress vector fields T_N and $\mathbf{T}_T$ along the contact surface, that satisfy each one of the previous equations, $\forall t \in [0, t_{max}]$, $\forall \mathbf{X}^{(i)} \in \Omega_0^{(i)}$, $i \in \{1, 2\}$, which is a system of coupled partial differential equations with respect to both time and space; this is accompanied by the inequalities specific to the contact problem.

This is an initial-boundary value problem that is subject to inequality constraints.

Case of statics

To treat a static problem, we take $\ddot{\mathbf{U}}^{(i)} = \mathbf{0}$ in [6.1] and remove the initial condition $[6.11]_2$.

Even in the static case, the strong problem is complex and difficult to solve analytically. In the sequel, we will seek to replace it by a weak formulation in order to solve the contact problem numerically.

Choice of the target surface

The choice of the contactor surface (and therefore of the target surface) is arbitrary and gives rise to an artificial asymmetry in the system of equations. There is therefore an asymmetry arising in the solution of the problem: equations [6.8] and [6.9] are written relative to any point of the contactor surface $S_{0c}^{(1)}$, and not the target surface $S_{0c}^{(2)}$.

In theory, the solution to the problem should be independent of the choice of the contactor surface. When solving numerically, this choice will inevitably have repercussions on the results, but typically its influence is not too significant.

One way to correct this asymmetry is to solve the problem in two cycles by interchanging the roles of the contactor and the target for the second cycle [132]. The contact variables are determined by a linear combination of the results obtained from each cycle. This method increases the computational cost and requires carefully considering the weighting of each cycle.

Another, more simple, method for mitigating this asymmetry consists of decomposing specific vector quantities that are related to the contactor surface relative to a basis attached to the target surface that is opposite. This is what we did in equations [6.6] and [6.7].

7

Weak Formulation of the Contact Problem

As it is difficult to analytically solve a contact problem, we instead turn to a weak formulation of the problem, which is more suitable for a numerical solution using the finite element method.

To solve a contactless mechanical problem by the finite element method, the weak form that is used the most frequently is the principle of virtual power (PVP). In this chapter, we will propose a weak form to generalize the PVP, and which can be applied to contact problems. Like the PVP, this weak form is constituted by a weighted residual relationship. On the other hand, it is a mixed form in the sense that it involves both the displacement field of the bodies and the multiplier field on the contact surfaces, with virtual fields for the virtual velocity fields and the virtual multiplier fields.

We will prove that the proposed weak form is equivalent to the strong form that we saw in Chapter 6. In particular, it restores the contact laws on contact surfaces.

7.1. Transforming the contact laws into equalities

In Chapter 5, we saw, on the one hand, that g and T_N must satisfy the normal contact law [5.2] and, on the other hand, that g, $\mathbf{V}_T$, T_N and $\mathbf{T}_T$ must satisfy Coulomb's tangential contact law [5.8]. These laws are difficult to handle numerically since they involve inequalities. Thus, we are going to establish two theorems that allow us to transform the contact laws into other equivalent expressions that only contain *equalities*.

In fact, these theorems are general in scope, as they can be applied to any functions g, $\mathbf{V}_T$, T_N or $\mathbf{T}_T$, which are *not* necessarily the proximity, slip velocity, normal contact stress and tangential contact stress functions. To emphasize this fact, we are going to write λ_N and $\boldsymbol{\lambda}_T$ in the place of T_N and $\mathbf{T}_T$ (however, we will keep the notation for g and $\mathbf{V}_T$). The symbols λ_N and $\boldsymbol{\lambda}_T$ also appear in the weak form [7.21], which will be stated later.

THEOREM.– [7.1]

Consider any two scalar functions g and λ_N. Then the following two statements are *equivalent*:

(i)

$$g \leq 0, \quad \begin{cases} \text{if } g < 0, \text{then } \lambda_N = 0 \\ \text{if } g = 0, \text{then } \lambda_N \geq 0 \end{cases} \qquad [7.2]$$

on a specified spatial domain (in contact mechanics, it is the contacting surface $S_{0c}^{(1)}$) and at a given time;

(ii)

$$\boxed{\lambda_N = \langle \lambda_N + \epsilon_N g \rangle} \qquad [7.3]$$

within the same spatial domain and at the same given time, where ϵ_N is an arbitrary fixed strictly positive scalar, and $\langle a \rangle$ is the Macaulay bracket that denotes the positive part of a: $\langle a \rangle = \dfrac{|a| + a}{2}$.

PROOF.

($\Rightarrow$) Suppose that g, λ_N satisfy the assertion [7.2]. Then, it is easy to check that the equality [7.3] is satisfied.

($\Leftarrow$) Conversely, suppose that g and λ_N both satisfy [7.3]. We will show that they both satisfy [7.2].

First of all, note that [7.3] implies $\lambda_N \geq 0$. Then, there are two exclusive cases which can arise:

$$\text{. if } \lambda_N + \epsilon_N g < 0, \text{ then [7.3] gives } \lambda_N = 0, \text{ from which } g < 0 \text{ since } \lambda_N + \epsilon_N g < 0; \qquad [7.4]$$

$$\text{. if } \lambda_N + \epsilon_N g \geq 0, \text{ then [7.3] gives } g = 0, \text{ from which } \lambda_N \geq 0 \text{ since } \lambda_N + \epsilon_N g \geq 0. \qquad [7.5]$$

(The inequality $\lambda_N \geq 0$ was previously found above).

We can collect the assertions [7.4]–[7.6] into the form [7.2]. ■

The proof of the converse ($\Leftarrow$) shows that, if we have $\lambda_N = \langle \lambda_N + \epsilon_N g \rangle$, then we have the following two equivalences:

$$\begin{cases} g < 0 \Leftrightarrow \lambda_N + \epsilon_N g < 0 \\ g = 0 \Leftrightarrow \lambda_N + \epsilon_N g \geq 0 \end{cases}$$

The second theorem takes longer to prove.

THEOREM.– [7.6]

Consider two arbitrary scalar functions, g and λ_N, as well as two arbitrary vector functions, λ_T and $\mathbf{V}_T$ (here, we do not need to assume that $\mathbf{V}_T \perp \boldsymbol{\nu}$ and $\lambda_T \perp \boldsymbol{\nu}$; however, it is in the statement of the weak form [7.21] that it will become necessary to assume $\lambda_T \perp \boldsymbol{\nu}$).

Hypothesis: Relations [7.2] or [7.3] are satisfied, that is to say, g and λ_N satisfy:

$$\lambda_N = \langle \lambda_N + \epsilon_N g \rangle, \text{ where } \epsilon_N > 0 \text{ is a fixed scalar.} \qquad [7.7]$$

Then the following two statements are *equivalent*:

(i)

$$\boxed{\begin{array}{l}\text{if } g < 0, \text{then } \lambda_T = \mathbf{0} \\ \text{if } g = 0, \text{then } \|\lambda_T\| \leq \mu\lambda_N, \text{ and } \left\{\begin{array}{l}\text{if } \|\lambda_T\| < \mu\lambda_N, \text{then } \mathbf{V}_T = \mathbf{0} \\ \text{if } \|\lambda_T\| = \mu\lambda_N, \text{then } \mathbf{V}_T \times \lambda_T = \mathbf{0}, \mathbf{V}_T.\lambda_T \geq 0 \\ \text{(that is to say } \mathbf{V}_T \upuparrows \lambda_T)\end{array}\right.\end{array}} \quad [7.8]$$

in a specific spatial domain and at a given instant in time;

(ii)

$$\boxed{\lambda_T = \left(1 - \langle 1 - \frac{\mu\langle\lambda_N + \epsilon_N g\rangle}{\|\lambda_T + \epsilon_T \mathbf{V}_T\|}\rangle\right)(\lambda_T + \epsilon_T \mathbf{V}_T)} \quad [7.9]$$

in the same spatial domain and at the same instant in time, where ϵ_T is a strictly positive arbitrary fixed scalar, just like ϵ_N.

PROOF.

($\Rightarrow$) Suppose that $g, \mathbf{V}_T, \lambda_N, \lambda_T$ satisfy [7.8]. We can then check that the equality [7.9] is satisfied – its verification is trivial but it takes longer than that of [7.3] in the previous theorem. Thus, let us do it only in the case when $g < 0$.

If $g < 0$, then, on the one hand, [7.8] implies that $\lambda_T = \mathbf{0}$, whereas, on the other hand, the hypothesis [7.7] applied to theorem [7.1] leads to $\lambda_N = 0$. The equality [7.9] is then just the trivial equality $\mathbf{0} = \mathbf{0}$.

($\Leftarrow$) Conversely, suppose that $g, \mathbf{V}_T, \lambda_N$ and λ_T satisfy [7.9], and let us show that we obtain [7.8].

The first consequence of [7.9] is that $\mathbf{V}_T$ is parallel to λ_T. We will show a more precise result, namely that the direction that $\mathbf{V}_T$ points in is the same as that for λ_T. As an abbreviation, let us provisionally denote $\alpha = \dfrac{\mu\langle\lambda_N + \epsilon_N g\rangle}{\|\lambda_T + \epsilon_T \mathbf{V}_T\|} \geq 0$, so that the relationship in [7.9] can then be written as

$$\langle 1 - \alpha\rangle\lambda_T = (1 - \langle 1 - \alpha\rangle)\epsilon_T \mathbf{V}_T \quad [7.10]$$

where $\langle 1 - \alpha\rangle \geq 0$ and $1 - \langle 1 - \alpha\rangle \geq 0, \forall\alpha$ (see Figure 7.1).

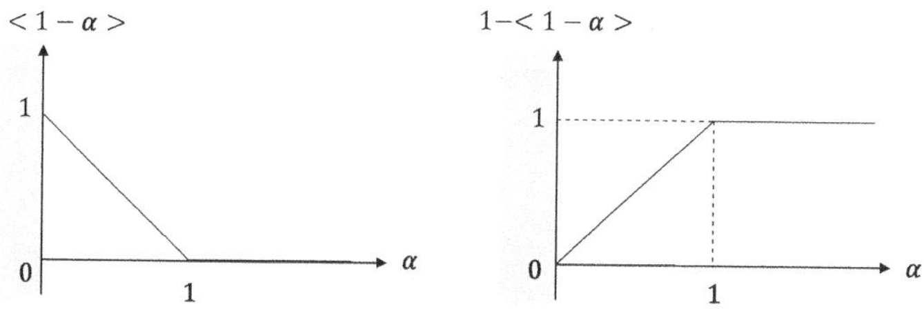

Figure 7.1. *Graphs of two specific functions of* α

Let us distinguish between the following exclusive cases which depend on the value of α:

- If $\alpha = 0$, then [7.10] gives $\boldsymbol{\lambda}_T = \mathbf{0}$.
- If $0 < \alpha < 1$, then [7.10] implies that $(1-\alpha)\boldsymbol{\lambda}_T = \alpha\epsilon_T\mathbf{V}_T$, with $1-\alpha > 0$ and $\alpha > 0$.
- If $\alpha \geq 1$, then [7.10] gives $\mathbf{V}_T = \mathbf{0}$.

In each of the three cases, the vector $\mathbf{V}_T$ does indeed point in the same direction as $\boldsymbol{\lambda}_T$:

$$\mathbf{V}_T \upuparrows \boldsymbol{\lambda}_T \qquad [7.11]$$

REMARK.– Even in the case when $\alpha = 0$ with $\boldsymbol{\lambda}_T = \mathbf{0}$, we always have that $\mathbf{V}_T \upuparrows \boldsymbol{\lambda}_T$. But in fact we will see in [7.25] later on that $\boldsymbol{\lambda}_T = \mathbf{0}$ means that $\mathbf{T}_T = \mathbf{0}$, which is to say that no contact occurs (or, in an exceptional case, that $\mathbf{V}_T = \mathbf{0}$ at a point of symmetrical contact), but then $\mathbf{V}_T$ becomes not very interesting. ■

Let us now distinguish between two exclusive cases:

(i) If $\lambda_N + \epsilon_N g < 0$, then:

– on the one hand, [7.4] implies $g < 0$; on the other hand, [7.9] implies $\boldsymbol{\lambda}_T = \mathbf{0}$. [7.12]

(ii) If $\lambda_N + \epsilon_N g \geq 0$, then:

– on the one hand, [7.6] implies $g = 0$;

– on the other hand, we have: $1 - \langle 1 - \dfrac{\mu\langle\lambda_N + \epsilon_N g\rangle}{\|\boldsymbol{\lambda}_T + \epsilon_T\mathbf{V}_T\|}\rangle = 1 - \langle 1 - \dfrac{\mu(\lambda_N + \epsilon_N g)}{\|\boldsymbol{\lambda}_T + \epsilon_T\mathbf{V}_T\|}\rangle$. [7.13]

There are a further two exclusive sub-cases that present themselves here:

(i) If $\|\boldsymbol{\lambda}_T + \epsilon_T\mathbf{V}_T\| < \mu(\lambda_N + \epsilon_N g)$, then $1 - \langle 1 - \dfrac{\mu\langle\lambda_N + \epsilon_N g\rangle}{\|\boldsymbol{\lambda}_T + \epsilon_T\mathbf{V}_T\|}\rangle = 1$ due to [7.13], which implies that $\mathbf{V}_T = \mathbf{0}$ due to [7.9]. As we have just shown that $g = 0$, the inequality $\|\boldsymbol{\lambda}_T + \epsilon_T\mathbf{V}_T\| < \mu(\lambda_N + \epsilon_N g)$ becomes

$$\|\boldsymbol{\lambda}_T\| < \mu\lambda_N \qquad [7.14]$$

(ii) If $\|\boldsymbol{\lambda}_T + \epsilon_T\mathbf{V}_T\| \geq \mu(\lambda_N + \epsilon_N g)$, then $1 - \langle 1 - \dfrac{\mu\langle\lambda_N + \epsilon_N g\rangle}{\|\boldsymbol{\lambda}_T + \epsilon_T\mathbf{V}_T\|}\rangle = \dfrac{\mu(\lambda_N + \epsilon_N g)}{\|\boldsymbol{\lambda}_T + \epsilon_T\mathbf{V}_T\|}$ due to [7.13], from which $\boldsymbol{\lambda}_T = \mu(\lambda_N + \epsilon_N g)\dfrac{\boldsymbol{\lambda}_T + \epsilon_T\mathbf{V}_T}{\|\boldsymbol{\lambda}_T + \epsilon_T\mathbf{V}_T\|}$ due to [7.9]. By taking the norm of the last equality and knowing that $\lambda_N + \epsilon_N g \geq 0$ and $g = 0$ here, we obtain:

$$\|\boldsymbol{\lambda}_T\| = \mu\lambda_N \qquad [7.15]$$

Note that by substituting [7.15] into the original inequality $\|\boldsymbol{\lambda}_T + \epsilon_T\mathbf{V}_T\| \geq \mu(\lambda_N + \epsilon_N g)$ and then taking into account $g = 0$, we find that $\|\boldsymbol{\lambda}_T + \epsilon_T\mathbf{V}_T\| \geq \|\boldsymbol{\lambda}_T\|$, which of course is true since we have just shown that $\mathbf{V}_T \upuparrows \boldsymbol{\lambda}_T$.

We can collect [7.11]–[7.15] together into the form [7.8]. ■

Definition.– The positive scalars ϵ_N and ϵ_T which appear in equations [7.3] and [7.9] are called *the penalty parameters*.

• In the proof of the previous theorem, it is clear why we wrote $\boldsymbol{\lambda}_T + \epsilon_T \mathbf{V}_T$ in the numerator and $||\boldsymbol{\lambda}_T + \epsilon_T \mathbf{V}_T||$ in the denominator on the right-hand side of [7.9]:

- On the one hand, the act of writing $\boldsymbol{\lambda}_T + \epsilon_T \mathbf{V}_T$ in the numerator allows us to have $\mathbf{V}_T = \mathbf{0}$ when $\lambda_N + \epsilon_N g \geq 0$ and $||\boldsymbol{\lambda}_T + \epsilon_T \mathbf{V}_T|| < \mu(\lambda_N + \epsilon_N g)$.
- On the other hand, the act of writing $||\boldsymbol{\lambda}_T + \epsilon_T \mathbf{V}_T||$ in the denominator allows us to have $||\boldsymbol{\lambda}_T|| = \mu \lambda_N$ when $\lambda_N + \epsilon_N g \geq 0$ and $||\boldsymbol{\lambda}_T + \epsilon_T \mathbf{V}_T|| \geq \mu(\lambda_N + \epsilon_N g)$ (when taking the norm of the equation $\boldsymbol{\lambda}_T = \mu(\lambda_N + \epsilon_N g)\dfrac{\boldsymbol{\lambda}_T + \epsilon_T \mathbf{V}_T}{||\boldsymbol{\lambda}_T + \epsilon_T \mathbf{V}_T||}$, we use the fact that the fraction $\dfrac{\boldsymbol{\lambda}_T + \epsilon_T \mathbf{V}_T}{||\boldsymbol{\lambda}_T + \epsilon_T \mathbf{V}_T||}$ which appears in the equation is a unit vector).

• According to the proof of the theorem, notice that the theorem remains valid even if we replace $\mathbf{V}_T$ with $\alpha \mathbf{V}_T$, with $\alpha > 0$ arbitrary. This property allows us to replace $\mathbf{V}_T$ everywhere by $\mathbf{V}_T \Delta t$, where $\Delta t > 0$ is some arbitrary finite interval in time, without changing anything in the form. This is what we will do in section 8.11 when we consider numerical solutions of the formulation.

Similarly, we will write $\mathbf{V}_T \Delta t = \gamma \dfrac{T_T}{||T_T||}$ in the place of $\mathbf{V}_T = \dot{\gamma} \dfrac{T_T}{||T_T||}$ (and γ in the place of $\dot{\gamma}$).

The advantage of writing $\mathbf{V}_T \Delta t$ is that ϵ_T *has the same physical dimensions* as ϵ_N. The penalty parameter ϵ_N has the dimensions of body force (N/m^3 in SI units). At the current stage, the penalty parameter ϵ_T has Ns/m^3 for its dimensions, but when we replace $\mathbf{V}_T$ by $\mathbf{V}_T \Delta t$, the new parameter ϵ_T will then have the same dimensions as ϵ_N.

• In fact, the equalities [7.3] and [7.9] contain the contact *inequalities*. These equalities, which are easier to deal with than inequalities, will be used in what follows to construct the weak form of the contact problem.

7.2. Preliminary ideas for the weak form

It is not easy to immediately find a weak form of the contact problem, which is equivalent to the strong one. Before we propose a weak form in the following section, let us first expose the ideas which led to an expression of this form.

1. Consider the PVPs for each of the two bodies 1 and 2 before taking the sum of the two respective expressions: $\forall t \in [0, t_{max}], \forall \mathbf{U}^{(1)*}, \forall \mathbf{U}^{(2)*}$:

$$\sum_{i=1}^{2} \Big\{ \underbrace{- \int_{\Omega_o^{(i)}} \boldsymbol{\Pi}^{(i)T} : \mathbf{Grad}_{\mathbf{X}^{(i)}} \mathbf{U}^{(i)*} d\Omega_o}_{(a)} + \underbrace{\int_{\Omega_o^{(i)}} \rho_0^{(i)} \mathbf{f}^{(i)} \mathbf{U}^{(i)*} d\Omega_o + \int_{S_{0U}^{(i)} \cup S_{0T}^{(i)}} \mathbf{T}^{(i)} \mathbf{U}^{(i)*} dS_0}_{(b)} \Big\}$$
$$+ \underbrace{\sum_{i=1}^{2} \int_{S_{0c}^{(i)}} \mathbf{T}^{(i)} \mathbf{U}^{(i)*} dS_0}_{(c)} = \underbrace{\sum_{i=1}^{2} \int_{\Omega_o^{(i)}} \rho_0^{(i)} \ddot{\mathbf{U}}^{(i)} \mathbf{U}^{(i)*} d\Omega_0}_{(d)} \quad [7.16]$$

In the previous equation, only term (c) deserves our attention, with the other terms being classical:

$$\begin{cases} (a) \equiv -\int_{\Omega_o^{(i)}} \mathbf{\Pi}^{(i)T} : \mathbf{Grad}_{\mathbf{X}^{(i)}}\mathbf{U}^{(i)*} d\Omega_o \text{ is the virtual power (VP)} \\ \qquad\qquad \text{of the internal stresses in body } i \\ (b) \equiv \int_{\Omega_o^{(i)}} \rho_0^{(i)}\mathbf{f}^{(i)}\mathbf{U}^{(i)*} d\Omega_o + \int_{S_{0U}^{(i)} \cup S_{0T}^{(i)}} \mathbf{T}^{(i)}\mathbf{U}^{(i)*} dS_0 \text{ is the VP of the exterior} \\ \qquad\qquad \text{forces in } \Omega_o^{(i)} \text{ and on } S_0^{(i)} \backslash S_{0c}^{(i)} \\ (d) \equiv \int_{\Omega_o^{(i)}} \rho_0^{(i)}\ddot{\mathbf{U}}^{(i)}\mathbf{U}^{(i)*} d\Omega_0 \text{ is the VP of the inertia forces in body } i \end{cases} \qquad [7.17]$$

2. Assume that the contact stresses balance one another out along the contact interface: $\mathbf{T}^{(1)}dS_{0c}^{(1)} = -\mathbf{T}^{(2)}dS_{0c}^{(2)}$. Then the sum (c) in [7.16] (VP of the contact forces on $S_{0c}^{(i)}$) can be transformed into

$$\sum_{i=1}^{2} \int_{S_{0c}^{(i)}} \mathbf{T}^{(i)}\mathbf{U}^{(i)*} dS_0 = \int_{S_{0c}^{(1)}} \mathbf{T}^{(1)}(\mathbf{U}^{(1)*} - \mathbf{U}^{(2)*}) dS_0 = \int_{S_{0c}^{(1)}} (T_N\boldsymbol{\nu} - \mathbf{T}_T)(\mathbf{U}^{(1)*} - \mathbf{U}^{(2)*}) dS_0 \qquad [7.18]$$

3. The contact laws also need to be satisfied. To do so, these laws can be transformed into equalities thanks to theorems [7.1] and [7.6], before being transformed into the weak form as follows: $\forall t, \forall \lambda^*(\mathbf{X})$:

$$\int_{S_{0c}^{(1)}} \left\{ (T_N - \langle T_N + \epsilon_N g \rangle)\, \boldsymbol{\nu} + \left[\mathbf{T}_T - \left(1 - \langle 1 - \frac{\mu \langle T_N + \epsilon_N g \rangle}{\|\mathbf{T}_T + \epsilon_T \mathbf{V}_T\|} \rangle \right) (\mathbf{T}_T + \epsilon_T \mathbf{V}_T) \right] \right\} \lambda^* dS_0 = 0 \qquad [7.19]$$

where we have used the fact that $\mathbf{T}_T$ and $\mathbf{V}_T$ are $\perp \boldsymbol{\nu}$.

4. The system to be solved consists of the PVPs [7.16], modified by [7.18], and the weak form [7.19]. The unknowns involved here are the displacement fields $\mathbf{U}^{(1)}$ and $\mathbf{U}^{(2)}$ of the bodies, knowing that the stress tensors $\mathbf{\Pi}^{(i)}$ are functions of the displacements $\mathbf{U}^{(i)}$ via the constitutive laws in $\Omega_o^{(i)}$, and that the stress vector $\mathbf{T}^{(1)}$ on $S_{0c}^{(1)}$ is given by $\mathbf{T}^{(1)} \equiv \mathbf{T} \equiv T_N\boldsymbol{\nu} - \mathbf{T}_T \equiv \mathbf{\Pi}^{(1)}\mathbf{N}^{(1)}$.

   This system with the unknowns $\mathbf{U}^{(1)}$ and $\mathbf{U}^{(2)}$ is correct, but it contains more equations than unknowns! The idea is to rewrite it as a *mixed* weak form that involves two types of *independent* unknowns: (i) the displacement fields, $\mathbf{U}^{(1)}$ and $\mathbf{U}^{(2)}$, and (ii) the fields λ_N and λ_T (with $\lambda_T \perp \boldsymbol{\nu}$ like $\mathbf{V}_T$), called *the multipliers*. It remains for us to prove that λ_N and λ_T are equal to the contact stresses, T_N and $\mathbf{T}_T$, respectively.

These preliminary ideas lead us to the statement of the mixed weak form in the following section.

7.3. Weak form of the contact problem

In this section, we propose a weak form [84, 85], which generalizes the PVPs and allows us to solve the contact problems numerically. Like the PVPs, this weak form is a weighted residual relation. On the other hand, it is a *mixed* form, since it involves both the displacement fields, $\mathbf{U}^{(1)}$ and $\mathbf{U}^{(2)}$, of the bodies and *the multiplier fields*, λ_N and λ_T, on the contacting surface.

The proposed weak form resembles that of Pietrzak and Curnier [111, 112], who constructed their weak formulation according to the augmented Lagrangian method. Here, we immediately state the weak form as a postulate and will prove later that it is *equivalent* to the strong form of the contact problem seen in Chapter 6. In particular, we will prove that the contact laws on the contact surfaces can be recovered using the weak form.

The weak form is designed for problems where there is frictionless or *Coulomb* frictional contact *uniquely*. It does *not* apply when the tangential contact is governed by a law different to that of Coulomb's.

Weak form statement of the contact problem.

Consider the following fields:

1. the vector fields $\mathbf{U}^{(1)}(\mathbf{X}^{(1)}, t)$ and $\mathbf{U}^{(2)}(\mathbf{X}^{(2)}, t)$, defined on $\Omega_0^{(1)}\times[0, t_{max}]$ and $\Omega_0^{(2)}\times[0, t_{max}]$, respectively. As it stands, they are not necessarily the displacement fields of bodies 1 and 2. Regardless, from these fields we can define the following ones on $S_{0c}^{(1)}$:
   - the fields $\boldsymbol{\nu}$ and g using [2.13];
   - the field $\mathbf{V}_T$ using [2.38] or [2.39];
2. the *multiplier* fields $\lambda_N(\mathbf{X}, t)$ and $\lambda_T(\mathbf{X}, t)$ defined on $S_{0c}^{(1)} \times [0, t_{max}]$.
   - Hypothesis: Suppose that $\boxed{\lambda_T \perp \boldsymbol{\nu}}$ [7.20]

We choose the two fixed positive penalty parameters ϵ_N and ϵ_T beforehand.

Then we can state the weak form as follows:

$\forall t \in [0, t_{max}]$, $\forall$ vector field $\mathbf{m}$ defined on $S_c^{(2)} \times [0, t_{max}]$ *arbitrarily* such that, at every point of $S_c^{(2)}$, $\mathbf{m}$ is *not* orthogonal to the normal $\boldsymbol{\nu}$, $\forall$ *virtual velocity* fields $\mathbf{U}^{(1)*}$ and $\mathbf{U}^{(2)*}$, $\forall$ *virtual multiplier* field λ^*:

$$\boxed{\begin{aligned}&\sum_{i=1}^{2}\left\{-\int_{\Omega_o^{(i)}} \mathbf{\Pi}^{(i)T} : \mathbf{Grad}_{\mathbf{X}^{(i)}}\mathbf{U}^{(i)*}d\Omega_o + \int_{\Omega_o^{(i)}} \rho_0^{(i)}\mathbf{f}^{(i)}.\mathbf{U}^{(i)*}d\Omega_o + \int_{S_{0U}^{(i)}\cup S_{0T}^{(i)}} \mathbf{T}^{(i)}.\mathbf{U}^{(i)*}dS_0\right\}\\
&+\int_{S_{0c}^{(1)}}\left[\langle\lambda_N + \epsilon_N g\rangle\boldsymbol{\nu} - \left(1 - \langle 1 - \frac{\mu\langle\lambda_N + \epsilon_N g\rangle}{\|\lambda_T + \epsilon_T\mathbf{V}_T\|}\rangle\right)(\lambda_T + \epsilon_T\mathbf{V}_T)\right].\left(\mathbf{U}^{(1)*}(\mathbf{X}) - \mathbf{U}^{(2)*}(\mathbf{Y}(\mathbf{X}))\right)dS_0\\
&+\int_{S_{0c}^{(1)}}\left\{\frac{1}{\epsilon_N}(\lambda_N - \langle\lambda_N + \epsilon_N g\rangle)\,\mathbf{m} - \frac{1}{\epsilon_T}\left[\lambda_T - \left(1 - \langle 1 - \frac{\mu\langle\lambda_N + \epsilon_N g\rangle}{\|\lambda_T + \epsilon_T\mathbf{V}_T\|}\rangle\right)(\lambda_T + \epsilon_T\mathbf{V}_T)\right]\right\}.\lambda^* dS_0\\
&= \sum_{i=1}^{2}\int_{\Omega_o^{(i)}} \rho_0^{(i)}\ddot{\mathbf{U}}^{(i)}.\mathbf{U}^{(i)*}d\Omega_0\end{aligned}}$$

[7.21]

In the previous equation, each virtual field was chosen arbitrarily, with the only requirement being that they are sufficiently regular so that they can be integrated.

• Compared to equation [7.19], we see that there are the additional $1/\epsilon_N$ and $1/\epsilon_T$ factors in the integrand of the last $\int_{S_{0c}^{(1)}}$ integral of [7.21]. This does not change anything in the local equations deduced from the weak form. However, as we will see later on, these additional features provide the following two advantages:

. They allow for the dimensions of the different integrals to be identical.

. In the case of frictionless contact, the $1/\epsilon_N$ factor allows for the tangent matrix to be symmetric.

In theory, the values of the penalty parameters ϵ_N, ϵ_T can be chosen arbitrarily without changing anything, though, on the other hand, we should notice that they must be chosen suitably in numerical applications so that the computation converges.

• The vector field $\mathbf{m}$ in [7.21] is chosen at any time t. As noted in the statement above, it is any field that satisfies the condition $\mathbf{m} \not\perp \boldsymbol{\nu}$ at any point. It is *independent* of the actual fields $\mathbf{U}^{(1)}, \mathbf{U}^{(2)}, \lambda_N, \boldsymbol{\lambda}_T$ and also of the virtual fields $\mathbf{U}^{(1)*}, \mathbf{U}^{(2)*}, \boldsymbol{\lambda}^*$. As we will see in equation [9.67] in Chapter 9, it is in the final formulae that we will fix $\mathbf{m} = \boldsymbol{\nu}$ (the condition $\mathbf{m} \not\perp \boldsymbol{\nu}$ is automatically satisfied with this choice).

• Several terms in [7.21] are classical, and they have already been identified in [7.17]. Only the integrals over $S_{0c}^{(1)}$ are new and deserve to be given a name:

Definition.– *The contact virtual power* is, by definition, the following expression contained in the weak form equation [7.21]:

$$
\begin{aligned}
&\mathcal{P}^*_{\text{contact}} \equiv \\
&\int_{S_{0c}^{(1)}} \left[\langle \lambda_N + \epsilon_N g \rangle \boldsymbol{\nu} - \left(1 - \langle 1 - \frac{\mu \langle \lambda_N + \epsilon_N g \rangle}{\| \boldsymbol{\lambda}_T + \epsilon_T \mathbf{V}_T \|} \rangle \right) (\boldsymbol{\lambda}_T + \epsilon_T \mathbf{V}_T) \right] . \left(\mathbf{U}^{(1)*}(\mathbf{X}) - \mathbf{U}^{(2)*}(\mathbf{Y}(\mathbf{X})) \right) dS_0 \\
&+ \int_{S_{0c}^{(1)}} \left\{ \frac{1}{\epsilon_N} (\lambda_N - \langle \lambda_N + \epsilon_N g \rangle) \, \mathbf{m} - \frac{1}{\epsilon_T} \left[\boldsymbol{\lambda}_T - \left(1 - \langle 1 - \frac{\mu \langle \lambda_N + \epsilon_N g \rangle}{\| \boldsymbol{\lambda}_T + \epsilon_T \mathbf{V}_T \|} \rangle \right) (\boldsymbol{\lambda}_T + \epsilon_T \mathbf{V}_T) \right] \right\} . \boldsymbol{\lambda}^* dS_0
\end{aligned}
\tag{7.22}
$$

The classical PVP considers the unknowns $\mathbf{U}^{(1)}$, $\mathbf{U}^{(2)}$ (given that the constraints $\boldsymbol{\Pi}^{(i)}$ are functions of the $\mathbf{U}^{(i)}$) and also involves the virtual fields $\mathbf{U}^{(1)*}(\mathbf{X}^{(1)})$, $\mathbf{U}^{(2)*}(\mathbf{X}^{(2)})$. Here:

- the weak form [7.21] is a *mixed* principle, as it considers the displacement fields $\mathbf{U}^{(1)}$, $\mathbf{U}^{(2)}$ and the multiplier fields $\lambda_N, \boldsymbol{\lambda}_T$ defined on $S_{0c}^{(1)}$;
- it involves the virtual velocity fields $\mathbf{U}^{(1)*}$, $\mathbf{U}^{(2)*}$ as well as the virtual mutiplier fields $\boldsymbol{\lambda}^*$ that depend on the point $\mathbf{X} \in S_{0c}^{(1)}$ and is *independent* of $\mathbf{U}^{(1)*}$, $\mathbf{U}^{(2)*}$.

7.4. Equivalence between the strong and the weak forms

We will show that the proposed weak form [7.21] is equivalent to the strong form of the contact problem seen in Chapter 6. First of all, we can easily check (although it is a bit long) that the strong form implies the weak form:

Theorem.– Provided that we set $\lambda_N = T_N$ and $\boldsymbol{\lambda}_T = \mathbf{T}_T$ in the statement [7.21], and that the solution fields of the strong problem – namely, $\mathbf{U}^{(1)}, \mathbf{U}^{(2)}$ in $\Omega_0^{(1)}, \Omega_0^{(2)}$ and $T_N, \mathbf{T}_T, \mathbf{V}_T$ on $S_{0c}^{(1)}$ – satisfy the weak form [7.21].

Conversely, the following proposition shows that the weak form implies the strong form (except for the material constitutive laws).

CONVERSE THEOREM.–

Hypothesis: Let the surfaces $S_c^{(1)}$ and $S_c^{(2)}$ be parameterized by

$$\begin{array}{ccc} \text{a domain of } \mathbb{R}^2 & \to & S_c^{(1)} \\ \boldsymbol{\xi} & \mapsto & \mathbf{x} \end{array} \qquad \begin{array}{ccc} \text{a domain of } \mathbb{R}^2 & \to & S_c^{(2)} \\ \boldsymbol{\eta} & \mapsto & \mathbf{y} \end{array}$$

Suppose that there exists *a bijection* (eventually piecewise) between the two contact surfaces:

$$\begin{array}{ccc} S_c^{(1)} & \to & S_c^{(2)} \\ \mathbf{x} & \mapsto & \mathbf{y} \end{array}$$

(this bijection can be the projection mapping, see Figure 7.2, or something similar).

Then, at any time t, the weighted residual equation [7.21] implies the following local equations:

1. The local dynamical equation [6.1] of the two bodies 1 and 2:

$$\forall i \in \{1,2\}, \forall \mathbf{X}^{(i)} \in \Omega_0^{(i)}, \quad \mathbf{Div}\boldsymbol{\Pi}^{(i)} + \rho_0^{(i)}\mathbf{f}^{(i)} = \rho_0^{(i)}\ddot{\mathbf{U}}^{(i)} \qquad [7.23]$$

2. The following equation on the boundary sections $S_0^{(i)} \backslash S_{0c}^{(i)} = S_{0T}^{(i)} \cup S_{0U}^{(i)}$:

$$\forall i \in \{1,2\}, \forall \mathbf{X}^{(i)} \in S_{0T}^{(i)} \cup S_{0U}^{(i)}, \quad \boldsymbol{\Pi}^{(i)}\mathbf{N}^{(i)} = \mathbf{T}^{(i)} \qquad [7.24]$$

On $S_{0T}^{(i)}$, the stress vector $\mathbf{T}^{(i)}$ is given and equation [7.24] is a stress boundary condition, while on $S_{0U}^{(i)}$, $\mathbf{T}^{(i)}$ is an unknown stress vector that equation [7.24] allows us to precisely calculate.

3. The following equations can be found on the contacting surface $S_{0c}^{(1)}$:

3.1. the equalities between the nominal stresses and the multipliers on $S_{0c}^{(1)}$:

$$\forall \mathbf{X} \in S_{0c}^{(1)}, \quad \boxed{\mathbf{T} \equiv \boldsymbol{\Pi}^{(1)}.\mathbf{N}^{(1)} = \lambda_N \boldsymbol{\nu} - \boldsymbol{\lambda}_T}$$

Noting that in [7.21] we assumed that $\boldsymbol{\lambda}_T \perp \boldsymbol{\nu}$, then the previous equation becomes equivalent to

$$\boxed{T_N \equiv \mathbf{T}.\boldsymbol{\nu} = \lambda_N \geq 0 \qquad \text{and} \qquad \mathbf{T}_T = \boldsymbol{\lambda}_T} \qquad [7.25]$$

3.2. the normal [5.2] and the tangential [5.8] contact laws, that is to say, [6.8] and [6.9];

3.3. the following equation, which is [6.10], and that translates the contact stress balance onto the contact interface:

$$\forall \mathbf{X} \in S_{0c}^{(1)}, \boxed{\mathbf{T}^{(2)}(\mathbf{Y}(\mathbf{X},t))dS_o^{(2)} = -(\lambda_N \boldsymbol{\nu} - \boldsymbol{\lambda}_T)\frac{\|\mathbf{X}_{,\xi^1} \times \mathbf{X}_{,\xi^2}\|}{\|\mathbf{Y}_{,\eta^1} \times \mathbf{Y}_{,\eta^2}\|}\frac{D\xi}{D\eta}dS_o^{(2)} = -\mathbf{T}^{(1)}(\mathbf{X})dS_o^{(1)}}$$

[7.26]

where $dS_o^{(1)}$ is the area element of $S_{0c}^{(1)}$, $dS_o^{(2)}$ is the area element of $S_{0c}^{(2)}$ corresponding to $dS_o^{(1)}$ via the bijection mentioned in the hypothesis above, and lastly the Jacobian $\frac{D\xi}{D\eta}$ is $\begin{vmatrix} \frac{\partial \eta^1}{\partial \xi^1} & \frac{\partial \eta^1}{\partial \xi^2} \\ \frac{\partial \eta^2}{\partial \xi^1} & \frac{\partial \eta^2}{\partial \xi^2} \end{vmatrix}$.

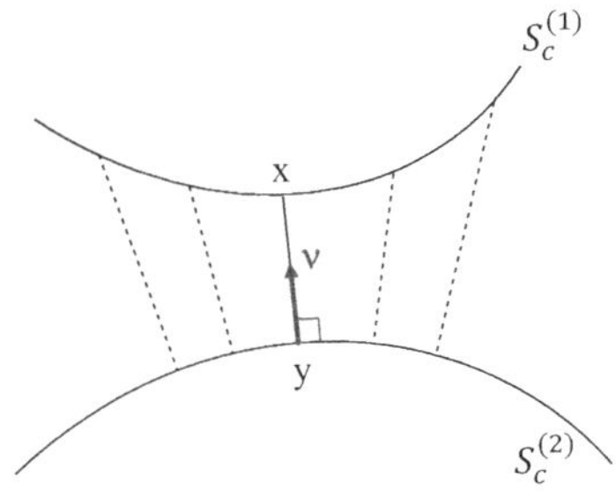

Figure 7.2. *Local bijection between the two contact surfaces. For a color version of this figure, see www.iste.co.uk/levan/contact.zip*

PROOF. We argue this with a fixed time instant t in mind.

1. Since [7.21] holds independently of the virtual fields $\mathbf{U}^{(1)*}, \mathbf{U}^{(2)*}$ and $\boldsymbol{\lambda}^*$, let us take $\mathbf{U}^{(1)*} = \mathbf{U}^{(2)*} = \mathbf{0}$ while keeping $\boldsymbol{\lambda}^*$ arbitrary. The weak form [7.21] then becomes: $\forall \boldsymbol{\lambda}^*$:

$$\int_{S_{0c}^{(1)}} \left\{ \frac{1}{\epsilon_N}(\lambda_N - \langle \lambda_N + \epsilon_N g \rangle)\,\mathbf{m} - \frac{1}{\epsilon_T}\left[\boldsymbol{\lambda}_T - \left(1 - \langle 1 - \frac{\mu\langle \lambda_N + \epsilon_N g\rangle}{\|\boldsymbol{\lambda}_T + \epsilon_T \mathbf{V}_T\|}\rangle\right)(\boldsymbol{\lambda}_T + \epsilon_T \mathbf{V}_T)\right]\right\} .\boldsymbol{\lambda}^* dS_0$$

The localization theorem, which can be applied since the coefficient of $\boldsymbol{\lambda}^*$ is continuous, leads to

$$\frac{1}{\epsilon_N}(\lambda_N - \langle \lambda_N + \epsilon_N g \rangle)\,\mathbf{m} - \frac{1}{\epsilon_T}\left[\boldsymbol{\lambda}_T - \left(1 - \langle 1 - \frac{\mu\langle \lambda_N + \epsilon_N g\rangle}{\|\boldsymbol{\lambda}_T + \epsilon_T \mathbf{V}_T\|}\rangle\right)(\boldsymbol{\lambda}_T + \epsilon_T \mathbf{V}_T)\right] = \mathbf{0}$$

This is an equality of the form $a\,\mathbf{m} + \mathbf{b} = \mathbf{0}$, where a is a scalar, $\mathbf{m}$ is a vector such that $\mathbf{m} \not\perp \boldsymbol{\nu}$ and $\mathbf{b}$ is a vector such that $\mathbf{b} \perp \boldsymbol{\nu}$ (since $\boldsymbol{\lambda}_T \perp \boldsymbol{\nu}$ according to the hypothesis [7.20] and since $\mathbf{V}_T \perp \boldsymbol{\nu}$). By taking the inner product of this equality with $\boldsymbol{\nu}$, we see that $a\,\mathbf{m}.\boldsymbol{\nu} = 0$, and therefore $a = 0$. From which $\mathbf{b} = \mathbf{0}$, and thus:

$$\begin{aligned} \lambda_N &= \langle \lambda_N + \epsilon_N g \rangle && \text{on } S_{0c}^{(1)} \\ \boldsymbol{\lambda}_T &= \left(1 - \langle 1 - \frac{\mu\langle \lambda_N + \epsilon_N g\rangle}{\|\boldsymbol{\lambda}_T + \epsilon_T \mathbf{V}_T\|}\rangle\right)(\boldsymbol{\lambda}_T + \epsilon_T \mathbf{V}_T) && \text{on } S_{0c}^{(1)} \end{aligned} \quad [7.27]$$

According to the equivalence in [7.1], equation $[7.27]_1$ is equivalent to

$$g \le 0, \quad \begin{cases} \text{if } g < 0, \text{then } \lambda_N = 0 \\ \text{if } g = 0, \text{then } \lambda_N \ge 0 \end{cases} \quad [7.28]$$

and, according to the equivalence [7.6], equation $[7.27]_2$ is then equivalent to

$$\begin{cases} \text{if } g < 0, \text{ then } \boldsymbol{\lambda}_T = \mathbf{0} \\ \text{if } g = 0, \text{ then } \|\boldsymbol{\lambda}_T\| \le \mu\lambda_N, \text{ and } \begin{cases} \text{if } \|\boldsymbol{\lambda}_T\| < \mu\lambda_N, \text{ then } \mathbf{V}_T = \mathbf{0} \\ \text{if } \|\boldsymbol{\lambda}_T\| = \mu\lambda_N, \text{ then } \mathbf{V}_T \times \boldsymbol{\lambda}_T = \mathbf{0}, \mathbf{V}_T.\boldsymbol{\lambda}_T \ge 0 \\ \text{(i.e., } \mathbf{V}_T \upuparrows \boldsymbol{\lambda}_T) \end{cases} \end{cases}$$

[7.29]

2. Let us now write [7.21] here by setting $\boldsymbol{\lambda}^* = \mathbf{0}$ and letting $\mathbf{U}^{(1)*}$ and $\mathbf{U}^{(2)*}$ be arbitrary. The weak form [7.21] then becomes: $\forall \mathbf{U}^{(1)*}, \forall \mathbf{U}^{(2)*}$:

$$\sum_{i=1}^{2}\left\{-\int_{\Omega_o^{(i)}} \boldsymbol{\Pi}^{(i)T} : \mathbf{Grad}_{\mathbf{X}^{(i)}}\mathbf{U}^{(i)*}d\Omega_o + \int_{\Omega_o^{(i)}} \rho_0^{(i)}\mathbf{f}^{(i)}.\mathbf{U}^{(i)*}d\Omega_o + \int_{S_{0U}^{(i)}\cup S_{0T}^{(i)}} \mathbf{T}^{(i)}.\mathbf{U}^{(i)*}dS_0\right\}$$
$$+\int_{S_{0c}^{(1)}}\left[\langle\lambda_N + \epsilon_N g\rangle\boldsymbol{\nu} - \left(1-\langle 1-\frac{\mu\langle\lambda_N+\epsilon_N g\rangle}{\|\boldsymbol{\lambda}_T+\epsilon_T\mathbf{V}_T\|}\rangle\right)(\boldsymbol{\lambda}_T+\epsilon_T\mathbf{V}_T)\right].\left(\mathbf{U}^{(1)*}(\mathbf{X})-\mathbf{U}^{(2)*}(\mathbf{Y}(\mathbf{X}))\right)dS_0$$
$$=\sum_{i=1}^{2}\int_{\Omega_o^{(i)}}\rho_0^{(i)}\ddot{\mathbf{U}}^{(i)}.\mathbf{U}^{(i)*}d\Omega_0$$

By applying the divergence theorem, $\forall i \in \{1,2\}$:

$$-\int_{\Omega_0^{(i)}} \mathbf{Div}\boldsymbol{\Pi}^{(i)}.\mathbf{U}^{(i)*}d\Omega_0 = \int_{\Omega_0^{(i)}} \boldsymbol{\Pi}^{(i)T} : \mathbf{Grad}\mathbf{U}^{(i)*}d\Omega_0 - \int_{S_0^{(i)}=S_{0U}^{(i)}\cup S_{0T}^{(i)}\cup S_{0c}^{(i)}} \mathbf{U}^{(i)*}.\boldsymbol{\Pi}^{(i)}.\mathbf{N}^{(i)}dS_0$$

we obtain the following equality:

$$\begin{aligned}
&\forall \mathbf{U}^{(1)*}, \forall \mathbf{U}^{(2)*}, \sum_{i=1}^{2}\left\{\int_{\Omega_o^{(i)}} \mathbf{U}^{(i)*}.(\mathbf{div}\boldsymbol{\Pi}^{(i)} + \rho_0^{(i)}\mathbf{f}^{(i)})d\Omega_o - \int_{S_{0U}^{(i)}\cup S_{0T}^{(i)}} \mathbf{U}^{(i)*}.(\boldsymbol{\Pi}^{(i)}.\mathbf{N}^{(i)} - \mathbf{T}^{(i)})dS_0\right\}\\
&-\int_{S_{0c}^{(1)}} \mathbf{U}^{(1)*}.\boldsymbol{\Pi}^{(1)}.\mathbf{N}^{(1)}dS_0\\
&+\int_{S_{0c}^{(1)}} \mathbf{U}^{(1)*}.\left[\langle\lambda_N + \epsilon_N g\rangle\boldsymbol{\nu} - \left(1-\langle 1-\frac{\mu\langle\lambda_N+\epsilon_N g\rangle}{\|\boldsymbol{\lambda}_T+\epsilon_T\mathbf{V}_T\|}\rangle\right)(\boldsymbol{\lambda}_T+\epsilon_T\mathbf{V}_T)\right]dS_0\\
&-\int_{S_{0c}^{(2)}} \mathbf{U}^{(2)*}.\boldsymbol{\Pi}^{(2)}.\mathbf{N}^{(2)}dS_0\\
&-\int_{S_{0c}^{(1)}} \mathbf{U}^{(2)*}.\left[\langle\lambda_N + \epsilon_N g\rangle\boldsymbol{\nu} - \left(1-\langle 1-\frac{\mu\langle\lambda_N+\epsilon_N g\rangle}{\|\boldsymbol{\lambda}_T+\epsilon_T\mathbf{V}_T\|}\rangle\right)(\boldsymbol{\lambda}_T+\epsilon_T\mathbf{V}_T)\right]dS_0\\
&=\sum_{i=1}^{2}\int_{\Omega_o^{(i)}}\rho_0^{(i)}\ddot{\mathbf{U}}^{(i)}\mathbf{U}^{(i)*}d\Omega_0
\end{aligned} \qquad [7.30]$$

Note that the last integral on the left-hand side extends over $S_{0c}^{(1)}$. We will transform this into an integral over $S_{0c}^{(2)}$ using the sequence of bijections $\boldsymbol{\xi} \mapsto \mathbf{X} \mapsto \mathbf{Y} \mapsto \boldsymbol{\eta}$, which allows us to perform a change of integration variables:

$$\begin{aligned}
dS_0^{(1)} &= \|\mathbf{X}_{,\xi^1} \times \mathbf{X}_{,\xi^2}\|d\xi^1 d\xi^2 = \|\mathbf{X}_{,\xi^1} \times \mathbf{X}_{,\xi^2}\|\frac{D\boldsymbol{\xi}}{D\boldsymbol{\eta}}d\eta^1 d\eta^2\\
dS_0^{(2)} &= \|\mathbf{Y}_{,\eta^1} \times \mathbf{Y}_{,\eta^2}\|d\eta^1 d\eta^2
\end{aligned}$$

where the Jacobian $\frac{D\boldsymbol{\xi}}{D\boldsymbol{\eta}}$ is defined after [7.26]. Hence

$$dS_0^{(1)} = \frac{\|\mathbf{X}_{,\xi^1} \times \mathbf{X}_{,\xi^2}\|}{\|\mathbf{Y}_{,\eta^1} \times \mathbf{Y}_{,\eta^2}\|}\frac{D\boldsymbol{\xi}}{D\boldsymbol{\eta}}dS_0^{(2)} \qquad [7.31]$$

If we recall [7.31] in [7.30], we get equations [7.23], [7.24] and

$$\begin{aligned}
&\mathbf{T}^{(1)} = \boldsymbol{\Pi}^{(1)}.\mathbf{N}^{(1)} = \langle\lambda_N + \epsilon_N g\rangle\boldsymbol{\nu} - \left(1-\langle 1-\frac{\mu\langle\lambda_N+\epsilon_N g\rangle}{\|\boldsymbol{\lambda}_T+\epsilon_T\mathbf{V}_T\|}\rangle\right)(\boldsymbol{\lambda}_T+\epsilon_T\mathbf{V}_T) \qquad \text{on } S_{0c}^{(1)}\\
&\mathbf{T}^{(2)} = \boldsymbol{\Pi}^{(2)}.\mathbf{N}^{(2)}\\
&= -\left[\langle\lambda_N + \epsilon_N g\rangle\boldsymbol{\nu} - \left(1-\langle 1-\frac{\mu\langle\lambda_N+\epsilon_N g\rangle}{\|\boldsymbol{\lambda}_T+\epsilon_T\mathbf{V}_T\|}\rangle\right)(\boldsymbol{\lambda}_T+\epsilon_T\mathbf{V}_T)\right]\frac{\|\mathbf{X}_{,\xi^1} \times \mathbf{X}_{,\xi^2}\|}{\|\mathbf{Y}_{,\eta^1} \times \mathbf{Y}_{,\eta^2}\|}\frac{D\boldsymbol{\xi}}{D\boldsymbol{\eta}} \qquad \text{on } S_{0c}^{(2)}
\end{aligned}$$

When using equations [7.27], the last two relations become

$$\begin{aligned} \mathbf{T}^{(1)} = \mathbf{\Pi}^{(1)}.\mathbf{N}^{(1)} &= \lambda_N \boldsymbol{\nu} - \boldsymbol{\lambda}_T & \text{on } S^{(1)}_{0c} \\ \mathbf{T}^{(2)} = \mathbf{\Pi}^{(2)}.\mathbf{N}^{(2)} &= -(\lambda_N \boldsymbol{\nu} - \boldsymbol{\lambda}_T)\frac{\|\mathbf{X}_{,\xi^1} \times \mathbf{X}_{,\xi^2}\| \, D\xi}{\|\mathbf{Y}_{,\eta^1} \times \mathbf{Y}_{,\eta^2}\| \, D\eta} & \text{on } S^{(2)}_{0c} \end{aligned} \quad [7.32]$$

Comparing $[7.32]_1$ with definition [3.4], namely $\mathbf{T}^{(1)} = T_N \boldsymbol{\nu} - \mathbf{T}_T$, we obtain

$$\lambda_N \boldsymbol{\nu} - \boldsymbol{\lambda}_T = T_N \boldsymbol{\nu} - \mathbf{T}_T \quad [7.33]$$

To proceed further, we have to distinguish between several exclusive cases and then exploit [7.29], which was obtained above:

$$\begin{cases} \text{if } g < 0, \text{ then } \boldsymbol{\lambda}_T = \mathbf{0}. \text{ We then deduce from [7.33] that } T_N = \lambda_N \text{ and } \mathbf{T}_T = \mathbf{0} \\ \text{if } g = 0 \text{ and } \begin{cases} . \text{ if } \|\boldsymbol{\lambda}_T\| = \mu\lambda_N, \text{then } \boldsymbol{\lambda}_T \,/\!/\, \mathbf{V}_T, \text{ that is to say } \boldsymbol{\lambda}_T \perp \boldsymbol{\nu}. \\ \quad \text{We then deduce from [7.33] that } T_N = \lambda_N \text{ and } \mathbf{T}_T = \boldsymbol{\lambda}_T \\ . \text{ if } \|\boldsymbol{\lambda}_T\| < \mu\lambda_N, \text{then there is no information on the direction of } \boldsymbol{\lambda}_T. \\ \quad \text{We must invoke hypothesis [7.20], that is, } \boldsymbol{\lambda}_T \perp \boldsymbol{\nu}, \text{ to deduce from [7.33]} \\ \quad \text{that } T_N = \lambda_N \text{ and } \mathbf{T}_T = \boldsymbol{\lambda}_T \end{cases} \end{cases}$$

In either case, we obtain [7.25].

Once [7.25] has been obtained, we get the normal [5.2] and the tangential [5.8] contact laws thanks to [7.28] and [7.29].

On the other hand, if we compare $[7.32]_1$ and $[7.32]_2$, and by taking [7.31] into account, we obtain [7.26]. ■

7.5. Final remarks

1. To summarize, we have shown that the weak form [7.21] is equivalent to the following set of relations:

   - the local dynamical equation defined inside the bodies;
   - the stress vector equation defined on the boundaries of the bodies;
   - *additionally*, the contact laws defined on the contacting surface.

   Note that the form in [7.21] is an *equality* that also includes the *inequalities* of the contact laws.

   Another interesting property is that equation [7.26], which expresses the balance of the contact stresses at the contact interface, is obtained as a *consequence* of the weak form [7.21].

2. In some formulations, such as those in [69, 43, 142, 121, 83, 81, 77], when there is contact with friction, we aim to solve, at the contact points, the contact relations of the form [5.9]:

$$f(\mathbf{T}_T, T_N) = \|\mathbf{T}_T\| - \mu T_N \leq 0 \qquad \dot{\gamma} \geq 0 \qquad \dot{\gamma} f = 0 \qquad \mathbf{V}_T = \dot{\gamma}\frac{\mathbf{T}_T}{\|\mathbf{T}_T\|}$$

   The solution of these frictional rate equations are obtained using a local integration scheme, analogous to the radial-return mapping in plasticity theory.

In our formulation, we satisfy the contact laws by solving the weak form, but we do not perform any local integration of the friction law. Moreover, we do not use the relation $\mathbf{V}_T = \dot{\gamma}\dfrac{\mathbf{T}_T}{||\mathbf{T}_T||}$ to obtain the slip velocity, since the factor $\dot{\gamma}$ is not calculated through the solution process. To obtain $\mathbf{V}_T$, we simply use the kinematic equation $[2.39]_b$ (or its discrete form [8.56]), which yields $\mathbf{V}_T$ as a function of the displacement fields $\mathbf{U}^{(1)}, \mathbf{U}^{(2)}$.

3. Returning to the hypothesis $\boldsymbol{\lambda}_T \perp \boldsymbol{\nu}$, that is [7.20], we saw in the proof above that sometimes we automatically have $\boldsymbol{\lambda}_T \perp \boldsymbol{\nu}$, although at other times this has to be assumed. The following theorem gives yet another case where we automatically obtain $\boldsymbol{\lambda}_T \perp \boldsymbol{\nu}$:

THEOREM.– [7.34]

Hypothesis: Assume equation [7.9], namely: $\boldsymbol{\lambda}_T = \left(1 - \langle 1 - \dfrac{\mu\langle\hat{\lambda}_N\rangle}{||\hat{\boldsymbol{\lambda}}_T||}\rangle\right)\hat{\boldsymbol{\lambda}}_T$, with $\dfrac{\mu\langle\hat{\lambda}_N\rangle}{||\hat{\boldsymbol{\lambda}}_T||} < 1$.

Then: $\boldsymbol{\lambda}_T \perp \boldsymbol{\nu}$.

PROOF. Multiplying [7.9] by $\boldsymbol{\nu}$, we get $\langle 1 - \dfrac{\mu\langle\hat{\lambda}_N\rangle}{||\hat{\boldsymbol{\lambda}}_T||}\rangle \boldsymbol{\lambda}_T.\boldsymbol{\nu} = 0$. ■

4. The current formulation involves spatial derivatives *that are at most of second order.* Second-order derivatives appear, for instance, in definition [2.25] of the curvatures $b_{\alpha\beta}$.

We will discretize the weak form [7.21] in Chapter 8, so that we can obtain the matrix equations that describe the contact problem.

8

Matrix Equations of the Contact Problem

The strong problem is difficult to solve, since the inequalities that reflect the one-sided contact condition and the Coulomb friction law must be satisfied. In the previous chapter, we saw that this difficulty no longer exists when posed in the weak form, since each contact equality and inequality can then be incorporated into the same equality, which is the weak form [7.21]. In this chapter, we will discretize the weak formulation [7.21] using the finite element method, and we will arrive at a system of coupled nonlinear differential equations that are time-dependent, and whose unknowns are the nodal displacements and the nodal multipliers.

This presentation will be established in the 3D framework, though we should keep in mind that any two-dimensional problem can be treated in an analogous way.

8.1. Introduction

In the weak form [7.21], the terms relating to the virtual powers of the acceleration quantities and to those of the internal and external forces can be discretized in a classical way. For this reason, we will be interested only in the discretization of the virtual contact power $\mathcal{P}^*_{\text{contact}}$ posed in the weak form [7.21], which was defined by [7.22]. We will henceforth adopt the following notational abbreviations:

$$\boxed{\begin{array}{l} \hat{\lambda}_N \equiv \lambda_N + \epsilon_N g \\ \hat{\boldsymbol{\lambda}}_T \equiv \boldsymbol{\lambda}_T + \epsilon_T \mathbf{V}_T \end{array}}$$

and we will rewrite the virtual contact power [7.22] as follows:

$$\boxed{\begin{aligned} \mathcal{P}^*_{\text{contact}} \equiv & \int_{S^{(1)}_{0c}} \left[\langle \hat{\lambda}_N \rangle \boldsymbol{\nu} - \left(1 - \langle 1 - \frac{\mu \langle \hat{\lambda}_N \rangle}{\|\hat{\boldsymbol{\lambda}}_T\|} \rangle \right) \hat{\boldsymbol{\lambda}}_T \right] . \left(\mathbf{U}^{(1)*}(\mathbf{X}) - \mathbf{U}^{(2)*}(\mathbf{Y}(\mathbf{X},t)) \right] dS_0 \\ & + \int_{S^{(1)}_{0c}} \left\{ \frac{1}{\epsilon_N} \left(\lambda_N - \langle \hat{\lambda}_N \rangle \right) \mathbf{m} - \frac{1}{\epsilon_T} \left[\boldsymbol{\lambda}_T - \left(1 - \langle 1 - \frac{\mu \langle \hat{\lambda}_N \rangle}{\|\hat{\boldsymbol{\lambda}}_T\|} \rangle \right) \hat{\boldsymbol{\lambda}}_T \right] \right\} . \boldsymbol{\Lambda}^* dS_0 \end{aligned}} \quad [8.1]$$

Recall that the unknowns of the problem are:

1. the displacement fields $\mathbf{U}^{(1)}$ and $\mathbf{U}^{(2)}$ in $\Omega_0^{(1)}$ and $\Omega_0^{(2)}$, respectively. These fields include the displacement fields on the contact surfaces or, which amounts to the same thing, the current positions of the contact surfaces;
2. the multiplier fields λ_N and $\boldsymbol{\lambda}_T$ on $S^{(1)}_{0c}$.

The discretization of the virtual contact power $\mathcal{P}^*_{\text{contact}}$ and the calculation of the contact tangent matrix are complicated tasks, since $\mathcal{P}^*_{\text{contact}}$ is of the form:

$$\mathcal{P}^*_{\text{contact}} \equiv \int_{S^{(1)}_{0c}} f(\mathbf{U}^{(1)}, \mathbf{U}^{(2)}, \lambda_N, \boldsymbol{\lambda}_T; \mathbf{U}^{(1)*}, \mathbf{U}^{(2)*}, \boldsymbol{\Lambda}^*) dS_0$$

where f is a complicated functional that involves the various quantities defined either on $S^{(1)}_{0c}$ or $S^{(1)}_{c}$, and either on $S^{(2)}_{0c}$ or $S^{(2)}_{c}$ (we do not consider the vector field $\mathbf{m}$ to be among the arguments, since it is not an important variable).

8.2. Meshes

The reference volumes $\Omega^{(1)}_0$ and $\Omega^{(2)}_0$ of the two bodies in question, as well as the surfaces $S^{(i)}_{0U}$ and $S^{(i)}_{0T}$, $i \in \{1,2\}$, where the boundary conditions are prescribed by [6.5], are meshed using finite elements in the usual way. In a contact problem, we must also mesh the reference contactor surface $S^{(1)}_{0c}$ and the reference target surface $S^{(2)}_{0c}$ using finite elements as in Figure 8.1 (for example, by using triangles with either three or six nodes, or quadrilaterals with either four or eight nodes). Systematically, we will use *isoparametric* elements for the contact surfaces. A current element of $S^{(1)}_{0c}$ will be denoted by $e^{(1)}$.

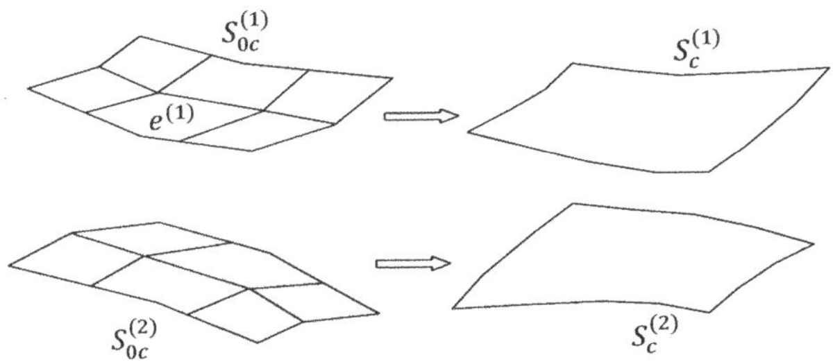

Figure 8.1. *Mesh of the reference contact surfaces $S^{(1)}_{0c}$ and $S^{(2)}_{0c}$. Only reference contact surfaces are meshed, and not the current surfaces*

The virtual contact power $\mathcal{P}^*_{\text{contact}}$ in relation [8.1] can then be written as the sum:

$$\mathcal{P}^*_{\text{contact}} = \sum_{e^{(1)} \subset S^{(1)}_{0c}} (\mathcal{P}^*_{\text{contact}})^{e^{(1)}} \qquad [8.2]$$

where $(\mathcal{P}^*_{\text{contact}})^{e^{(1)}}$ denotes *the element virtual contact power*, which is relative to the element $e^{(1)}$ and defined by

$$\boxed{\begin{aligned}(\mathcal{P}^*_{\text{contact}})^{e^{(1)}} \equiv & \int_{e^{(1)}} \left[\langle \hat{\lambda}_N \rangle \boldsymbol{\nu} - \left(1 - \langle 1 - \frac{\mu \langle \hat{\lambda}_N \rangle}{\|\hat{\boldsymbol{\lambda}}_T\|} \rangle \right) \hat{\boldsymbol{\lambda}}_T \right] . \left(\mathbf{U}^{(1)*}(\mathbf{X}) - \mathbf{U}^{(2)*}(\mathbf{Y}(\mathbf{X},t)) \right] dS_0 + \\ & \int_{e^{(1)}} \left\{ \frac{1}{\epsilon_N} \left(\lambda_N - \langle \hat{\lambda}_N \rangle \right) \mathbf{m} - \frac{1}{\epsilon_T} \left[\boldsymbol{\lambda}_T - \left(1 - \langle 1 - \frac{\mu \langle \hat{\lambda}_N \rangle}{\|\hat{\boldsymbol{\lambda}}_T\|} \rangle \right) \hat{\boldsymbol{\lambda}}_T \right] \right\} . \boldsymbol{\Lambda}^* dS_0\end{aligned}}$$

[8.3]

The integration variable in the integrand of [8.3] is the point $\mathbf{X}$ which runs over the surface $S_{0c}^{(1)}$. The other points appearing in this integrand are related to the point $\mathbf{X}$ via the chains [2.18] and [2.19], which we will recall here. Consider a particle on body 1 with reference position $\mathbf{X}$ that belongs to a given element $e^{(1)} \subset S_{0c}^{(1)}$, as in Figure 8.2. In section 2.3, we defined the following points, which are related to $\mathbf{X}$:

- the point $\boldsymbol{\xi} = (\xi^1, \xi^2) \in \mathbb{R}^2$ of the parent element of $e^{(1)}$, and which is the pre-image of $\mathbf{X}$;
- the point $\mathbf{x}$, which is the current position of the particle $\mathbf{X}$;
- the point $\mathbf{y} = \mathrm{proj}_{S_c^{(2)}}\mathbf{x}$, which is the projection of $\mathbf{x}$ onto the current target surface $S_c^{(2)}$;
- the point $\mathbf{Y}$, which is the pre-image of $\mathbf{y}$. More precisely, it is the reference position of the particle of body 2 situated at $\mathbf{y}$ at the current time instant. We denote by $e^{(2)}$ an element of $S_{0c}^{(2)}$ that contains $\mathbf{Y}$, of which there may be multiple. The element $e^{(2)}$ varies depending on the point $\mathbf{x}$;
- the point $\boldsymbol{\eta} = (\eta^2, \eta^2) \in \mathbb{R}^2$ of the parent element of $e^{(2)}$, which is the pre-image of $\mathbf{Y}$.

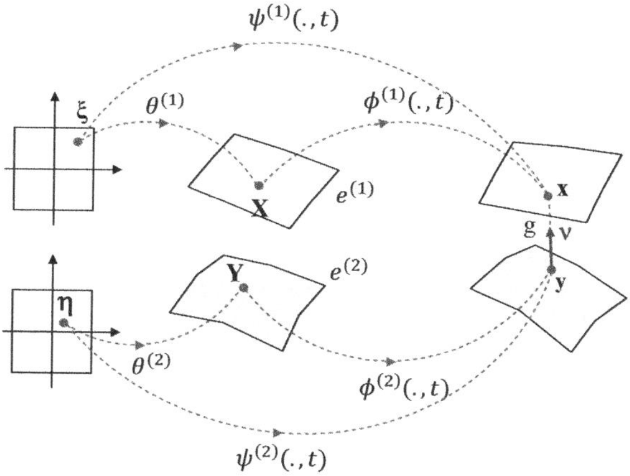

Figure 8.2. *The chain $\boldsymbol{\xi} \mapsto (\mathbf{X}, \mathbf{x}), \mapsto (\mathbf{y}, \mathbf{Y}, \boldsymbol{\eta})$. For a color version of this figure, see www.iste.co.uk/levan/contact.zip*

Thus, at each point $\mathbf{X}$ of an element $e^{(1)} \subset S_{0c}^{(1)}$, the points $(\mathbf{y}, \mathbf{Y}, \boldsymbol{\eta})$ are related to the target surface, as well as an element $e^{(2)} \subset S_{0c}^{(2)}$. Figure 8.2, which is a copy of Figure 2.5, shows how these different points are related to one another, except that here we have restricted our attention to the finite elements, $e^{(1)}$ and $e^{(2)}$, rather than to the entirety of the surfaces, $S_{0c}^{(1)}$ and $S_{0c}^{(2)}$. In Figure 8.2, the elements containing the points $\boldsymbol{\xi}$ and $\boldsymbol{\eta}$ are the parent elements (or the reference elements) of $e^{(1)}$ and $e^{(2)}$, respectively.

We will also denote:

- $nne^{(1)}$: the number of nodes of element $e^{(1)}$. For example, we have $nne^{(1)} = 6$ for a triangle with six nodes;
- $nne^{(2)}$: the number of nodes of element $e^{(2)}$.

8.3. Matrix notation in finite elements

Column-vectors, row-vectors and matrices

In the finite element method, vectors (respectively, tensors) can be represented using either column- or row-vectors (respectively, using matrices). A column-vector is just a single-column matrix, whereas a row-vector is just a single-row matrix. We will use the following three symbols:

- the braces $\{\bullet\}$ to designate a column-vector;
- the brackets $< \bullet >$ to designate a row-vector (there will be no possibility of confusion with the Macaulay bracket $\langle \bullet \rangle$, given that it is a scalar);
- the square brackets $[\bullet]$ to designate a matrix which is neither a column- nor a row-vector.

Moreover, in this chapter, when we speak of the components of a vector, a tensor, or the coordinates of a point, we will use the following convention:

- the components of any vector or tensor are to be considered relative to a fixed orthonormal Cartesian basis $(\mathbf{e}_1, \mathbf{e}_2, \mathbf{e}_3)$ of the Galilean frame R_0;
- the coordinates of any point are to be considered relative to a fixed orthonormal Cartesian coordinate frame $(\mathbf{O}; \mathbf{e}_1, \mathbf{e}_2, \mathbf{e}_3)$.

Matrix representations of vectors

A vector can be represented using either a column-vector or a row-vector with three components. Thus, for the normal vector $\boldsymbol{\nu}$ at the point $\mathbf{y}$, for example, we will denote the column-vector (respectively, the row-vector) containing the three components of $\boldsymbol{\nu}$ relative to the fixed orthonormal Cartesian basis $(\mathbf{e}_1, \mathbf{e}_2, \mathbf{e}_3)$ by $\{\boldsymbol{\nu}\}$ (respectively, by $< \boldsymbol{\nu} >$):

$$\{\boldsymbol{\nu}\} \equiv \begin{Bmatrix} \nu_1 \\ \nu_2 \\ \nu_3 \end{Bmatrix} \qquad \text{(respectively, } < \boldsymbol{\nu} > \;\equiv\; < \nu_1 \; \nu_2 \; \nu_3 > \text{)}$$

Positions, displacements (and virtual velocities) and multipliers will be represented using the following three-component column-vectors:

NOTATION.– [8.4]

1. POSITION VECTORS.
   - $\{\mathbf{X}\}$ (respectively, $\{\mathbf{x}\}$): the column-vector containing the three Cartesian coordinates of the point $\mathbf{X}$ (respectively, of $\mathbf{x}$);
   - $\{\mathbf{Y}\}$ (respectively, $\{\mathbf{y}\}$): the column-vector containing the three Cartesian coordinates of the point $\mathbf{Y}$ (respectively, of $\mathbf{y}$).
2. DISPLACEMENT AND VIRTUAL VELOCITY VECTORS.
   - $\{\mathbf{U}^{(1)}\}$ (respectively, $\{\mathbf{U}^{(2)}\}$): the column-vector containing the three components of the displacement vector $\mathbf{U}^{(1)}$ (respectively, of $\mathbf{U}^{(2)}$);

- Similarly, for virtual vectors: $\{\mathbf{U}^{(1)*}\}$ (respectively $\{\mathbf{U}^{(2)*}\}$) denotes the column-vector containing the three components of the virtual velocity vector $\mathbf{U}^{(1)*}$ (respectively, of $\mathbf{U}^{(2)*}$).

3. MULTIPLIER VECTORS. Aside from the virtual multiplier λ^* presented in [8.3], we will define a new vector, called the multiplier vector, in [8.14] later on; this will be denoted by λ. For these vectors λ and λ^*, we denote by:

   - $\{\lambda\}$ (respectively, $\{\lambda^*\}$): the column-vector containing the three components of the multiplier vector λ (respectively, of λ^*).

Matrix representation of a second-order tensor

The representative matrix of a second-order tensor $\mathbf{T}$ relative to the basis $(\mathbf{e}_1, \mathbf{e}_2, \mathbf{e}_3)$ is a 3×3 matrix that we will denote by $[\mathbf{T}]$. This matrix contains the components of the tensor $\mathbf{T}$ relative to the same basis:

$$\mathbf{T} = T_{ij}\mathbf{e}_i \otimes \mathbf{e}_j \quad \Rightarrow \quad [\mathbf{T}] = \begin{bmatrix} T_{11} & T_{12} & T_{13} \\ T_{21} & T_{22} & T_{23} \\ T_{31} & T_{32} & T_{33} \end{bmatrix} \quad (T_{ij} = \mathbf{e}_i.\mathbf{T}.\mathbf{e}_j)$$

Operations on column- and row-vectors

The product of a row-vector with a column-vector with the same dimension produces a scalar. For example, consider two vectors $\mathbf{a}$ and $\mathbf{b}$ with the components (a_1, a_2, a_3) and (b_1, b_2, b_3), respectively. Then we have:

$$< \mathbf{a} > \{\mathbf{b}\} = \{\mathbf{a}\}^T\{\mathbf{b}\} =< \mathbf{b} > \{\mathbf{a}\} = \{\mathbf{b}\}^T\{\mathbf{a}\} = a_1b_1 + a_2b_2 + a_3b_3 = \mathbf{a}.\mathbf{b}$$

The product of a column-vector with a row-vector produces a square matrix, provided that the column-vector has the same dimension as the row-vector. For example, using the column-vector $\{\mathbf{a}\}$ and the row-vector $< \mathbf{b} >$ from the previous example, we have:

$$\{\mathbf{a}\} < \mathbf{b} >= \begin{bmatrix} a_1b_1 & a_1b_2 & a_1b_3 \\ a_2b_1 & a_2b_2 & a_2b_3 \\ a_3b_1 & a_3b_2 & a_3b_3 \end{bmatrix} = [\mathbf{a} \otimes \mathbf{b}]$$

where $[\mathbf{a} \otimes \mathbf{b}]$ denotes the representative matrix of the tensor $\mathbf{a} \otimes \mathbf{b}$.

8.4. The element nodal vectors

To prepare for the discretization of [8.3], we will describe the interpolation of different fields on a current element of $S_{0c}^{(1)}$ or $S_{0c}^{(2)}$. While the interpolation of the position, displacement, and virtual velocity fields is a classical operation, interpolation of the multiplier fields is new.

The interpolation process consists of calculating a quantity at a current point using the quantities defined at the different nodes of an element. These *element nodal* quantities are then grouped together into column-vectors, which we say are the *element nodal vectors*.

Notation and definitions.– [8.5]

Nodal vectors relative to the element $e^{(1)} \subset S_{0c}^{(1)}$ (respectively, to $e^{(2)} \subset S_{0c}^{(2)}$) will be denoted with the superscript $e^{(1)}$ (respectively, with $e^{(2)}$). Each has dimension $3nne^{(1)}$ (respectively, $3nne^{(2)}$).

1. **Nodal coordinates.** The coordinates of the nodes of an element are represented using the following column-vectors, called *the element nodal coordinate vectors*:

   – $\{\mathbf{X}^{e^{(1)}}\}$ (respectively, $\{\mathbf{x}^{e^{(1)}}\}$): the column-vector formed from the three reference coordinates (respectively, from the current coordinates) for each node of the element $e^{(1)}$;

   – $\{\mathbf{Y}^{e^{(2)}}\}$, $\{\mathbf{y}^{e^{(2)}}\}$: the column-vector formed from the three reference coordinates (respectively, from the current coordinates) for each node of the element $e^{(2)}$.

2. **Nodal displacements and nodal virtual velocities.** The displacements and virtual velocities of the nodes of an element are represented using the following column-vectors:

   – $\{\mathbf{U}^{(1),e^{(1)}}\}$: the column-vector formed from the three components of the displacement vectors for each node of the element $e^{(1)}$;

   – $\{\mathbf{U}^{(1)*,e^{(1)}}\}$: the column-vector formed from the three components of the virtual velocity vectors for each node of the element $e^{(1)}$;

   – analogous definitions exist for the set of element nodal vectors $\{\mathbf{U}^{(2),e^{(2)}}\}$ and $\{\mathbf{U}^{(2)*,e^{(2)}}\}$ of the element $e^{(2)}$.

   The column-vectors $\{\mathbf{U}^{(1),e^{(1)}}\}$ and $\{\mathbf{U}^{(2),e^{(2)}}\}$ are said to be *the element nodal displacement vectors*, whereas the column-vectors $\{\mathbf{U}^{(1)*,e^{(1)}}\}$ and $\{\mathbf{U}^{(2)*,e^{(2)}}\}$ are called *the element nodal virtual velocity vectors*.

3. **Nodal multipliers.** The multipliers at the nodes of an element are represented using the following column-vectors:

   – $\{\mathbf{\Lambda}^{e^{(1)}}\}$: the column-vector formed from the three components of the vector λ for each node of the element $e^{(1)}$. This column-vector is called *the element nodal multiplier vector*. It is an unknown quantity in the same way that the element nodal displacements vectors $\{\mathbf{U}^{(1),e^{(1)}}\}$ and $\{\mathbf{U}^{(2),e^{(2)}}\}$ are;

   – $\{\mathbf{\Lambda}^{*,e^{(1)}}\}$: the column-vector formed from the three components of the vector λ^* for each node of the element $e^{(1)}$.

Again, according to the convention detailed in section 8.3, recall that each component in question is considered to be defined relative to the fixed orthonormal basis $(\mathbf{e}_1, \mathbf{e}_2, \mathbf{e}_3)$.

8.5. Interpolation of positions, displacements and virtual velocities

8.5.1. *Interpolation on the contactor surface*

• Consider an element $e^{(1)} \subset S^{(1)}_{0c}$ that has $nne^{(1)}$ nodes. Its position is obtained by interpolating the positions of its nodes:

$$\underset{}{\boldsymbol{\xi} = (\xi^1, \xi^2)} \mapsto \underset{(3,1)}{\{\mathbf{X}\}} = \underset{(3,\,3nne^{(1)})}{[\mathcal{N}^{e^{(1)}}(\boldsymbol{\xi})]} \quad \underset{(3nne^{(1)},\,1)}{\{\mathbf{X}^{e^{(1)}}\}} \tag{8.6}$$

In the previous expression:

1. the point $\boldsymbol{\xi} = (\xi^1, \xi^2) \in \mathbb{R}^2$ on the parent element of $e^{(1)}$ is the pre-image of $\mathbf{X}$ (see Figure 8.2);

2. the matrix $[\mathcal{N}^{e^{(1)}}(\boldsymbol{\xi})]$ is constructed using the interpolation functions $N_i, i \in \{1, 2, \dots, nne^{(1)}\}$ of the element $e^{(1)}$:

$$[\mathcal{N}^{e^{(1)}}(\boldsymbol{\xi})] = \left[\begin{array}{ccc|ccc|c} N_1 & & & N_2 & & & \\ & N_1 & & & N_2 & & \cdots \\ & & N_1 & & & N_2 & \end{array}\right] = [\; N_1\mathbf{I} \;|\; N_2\mathbf{I} \;|\; \cdots \;]$$

where $\mathbf{I}$ denotes the (3×3)-dimensional identity matrix.

The numbers beneath expression [8.6] indicate the sizes of the different terms: $\{\mathbf{X}^{e^{(1)}}\}$ has $3nne^{(1)}$ components, whereas the matrix $[\mathcal{N}^{e^{(1)}}(\boldsymbol{\xi})]$ has $(3 \times 3nne^{(1)})$ components.

Expression [8.6] is the finite element translation of the parameterization $\boldsymbol{\xi} \mapsto \mathbf{X} = \boldsymbol{\theta}^{(1)}(\boldsymbol{\xi})$ in [2.18], and it is a piecewise parameterization of the contactor surface.

• Consider a particle of body 1 with reference position $\mathbf{X} \in e^{(1)} \subset S^{(1)}_{0c}$ and whose pre-image is $\boldsymbol{\xi}$. The displacement $\mathbf{U}^{(1)}(\mathbf{X}, t)$ of this particle is obtained by interpolating the displacements of the nodes of the element $e^{(1)}$:

$$\boldsymbol{\xi} \mapsto \underset{(3,1)}{\{\mathbf{U}^{(1)}\}} = \underset{(3,\,3nne^{(1)})}{[\mathcal{N}^{e^{(1)}}(\boldsymbol{\xi})]} \quad \underset{(3nne^{(1)},\,1)}{\{\mathbf{U}^{(1),e^{(1)}}\}} \tag{8.7}$$

• Let $\mathbf{x}$ be the current position of a particle on body 1 that is located at $\mathbf{X}$ at the initial time. Equations [8.6]–[8.7] imply that $\mathbf{x}$ is obtained via interpolating in a similar way to what can be done for the reference position $\mathbf{X}$:

$$\boldsymbol{\xi} \quad \mapsto \quad \{\mathbf{x}\} = \{\mathbf{X}\} + \{\mathbf{U}^{(1)}\} = [\mathcal{N}^{e^{(1)}}(\boldsymbol{\xi})]\{\mathbf{x}^{e^{(1)}}\} \tag{8.8}$$

• The virtual velocity $\mathbf{U}^{(1)*}(\mathbf{X})$ of the contactor surface can be interpolated similarly to what can be done for the real displacement:

$$\boldsymbol{\xi} \quad \mapsto \quad \{\mathbf{U}^{(1)*}\} = [\mathcal{N}^{e^{(1)}}(\boldsymbol{\xi})]\{\mathbf{U}^{(1)*,e^{(1)}}\} \tag{8.9}$$

8.5.2. *Interpolation on the target surface*

Given a fixed point $\mathbf{X}$ of element $e^{(1)}$, we can define the corresponding points $\boldsymbol{\eta}$, $\mathbf{y}$ and $\mathbf{Y}$, as well as element $e^{(2)}$, as was described previously in section 8.2 and Figure 8.2.

• The position of element $e^{(2)}$ is obtained by interpolating the positions of each of its nodes:

$$\underset{}{\eta} \mapsto \underset{(3,1)}{\{\mathbf{Y}\}} = \underset{(3,3nne^{(2)})}{[\mathcal{N}^{e^{(2)}}(\eta)]} \underset{(3nne^{(2)},1)}{\{\mathbf{Y}^{e^{(2)}}\}} \quad [8.10]$$

where the matrix $[\mathcal{N}^{e^{(2)}}(\eta)]$ is constructed via the interpolation functions N_i, $i \in \{1, 2, \ldots, nne^{(2)}\}$ for element $e^{(2)}$:

$$[\mathcal{N}^{e^{(2)}}(\eta)] = \left[\begin{array}{ccc|ccc|c} N_1 & & & N_2 & & & \\ & N_1 & & & N_2 & & \cdots \\ & & N_1 & & & N_2 & \end{array}\right] = [\ N_1\mathbf{I} \ \mid \ N_2\mathbf{I} \ \mid \ \cdots \ ]$$

where, in general, the interpolation functions N_i for element $e^{(2)}$ differ from those for element $e^{(1)}$.

The numbers beneath expression [8.10] indicate that $\{\mathbf{U}^{(2),e^{(2)}}\}$ has $3nne^{(2)}$ components, whereas $[\mathcal{N}^{e^{(2)}}(\eta)]$ has $(3 \times 3nne^{(2)})$ components.

Equation [8.10] is the finite element translation of the parameterization $\boldsymbol{\eta} \mapsto \mathbf{Y} = \boldsymbol{\theta}^{(2)}(\boldsymbol{\eta})$ in [2.19], and it is a piecewise parameterization of the target surface.

• The displacement $\mathbf{U}^{(2)}(\mathbf{Y}, t)$ is obtained by interpolating on the element $e^{(2)} \subset S^{(2)}_{0c}$:

$$\eta \mapsto \underset{(3,1)}{\{\mathbf{U}^{(2)}\}} = \underset{(3,3nne^{(2)})}{[\mathcal{N}^{e^{(2)}}(\eta)]} \underset{(3nne^{(2)},1)}{\{\mathbf{U}^{(2),e^{(2)}}\}} \quad [8.11]$$

• The expressions [8.10]–[8.11] imply that the current position $\mathbf{y}$ can be interpolated in a similar way to what can be done for the reference position $\mathbf{Y}$:

$$\eta \quad \mapsto \quad \{\mathbf{y}\} = \{\mathbf{Y}\} + \{\mathbf{U}^{(2)}\} = [\mathcal{N}^{e^{(2)}}(\eta)]\{\mathbf{y}^{e^{(2)}}\} \quad [8.12]$$

• Lastly, the virtual velocity $\mathbf{U}^{(2)*}(\mathbf{Y})$ of the target surface can be interpolated in a similar way to what can be done for the real displacement, as:

$$\eta \quad \mapsto \quad \{\mathbf{U}^{(2)*}\} = [\mathcal{N}^{e^{(2)}}(\eta)]\{\mathbf{U}^{(2)*,e^{(2)}}\} \quad [8.13]$$

8.6. Interpolation of multipliers

8.6.1. *Definition of the vector* λ

• The situation is straightforward if we only consider frictionless contact; there is only the unknown scalar λ_N in the weak form, no vector $\boldsymbol{\lambda}_T$. We interpolate λ_N using the nodal values of λ_N : $\lambda_N =< N > \{\lambda^e_N\}$. Then we will work with a column-vector $\{\boldsymbol{\Lambda}_N\}$ that contains every nodal value λ_N on $S^{(1)}_{0c}$.

• When considering contact in general – with or without friction – the situation becomes more complicated, since then both λ_N and $\boldsymbol{\lambda}_T \bot \boldsymbol{\nu}$ feature. We could think of interpolating λ_N and $\boldsymbol{\lambda}_T \bot \boldsymbol{\nu}$ independently, as:

$$\lambda_N =< N^{e^{(1)}}(\boldsymbol{\xi}) > \{\lambda_N^{e^{(1)}}\}$$
$$\boldsymbol{\lambda}_T = [\mathcal{N}^{e^{(1)}}(\boldsymbol{\xi})]\{\boldsymbol{\lambda}_T^{e^{(1)}}\}$$

where the column-vector $\{\boldsymbol{\lambda}_T^{e^{(1)}}\}$ contains the three components of $\boldsymbol{\lambda}_T$ relative to the fixed orthonormal basis $(\mathbf{e}_1, \mathbf{e}_2, \mathbf{e}_3)$. However, this method of interpolation has a small drawback:

there are four unknown scalars for the nodal multiplier: one coming from λ_N, and the other three coming from $\boldsymbol{\lambda}_T$. Moreover, this method does not work since the second expression does not guarantee the condition $\boldsymbol{\lambda}_T \perp \boldsymbol{\nu}$.

We can also think of interpolating λ_N and the covariant components $\lambda_{T\alpha}$ relative to the dual basis ($\boldsymbol{\lambda}_T = \lambda_{T\alpha}\mathbf{a}^\alpha$, where $\mathbf{a}^\alpha$ is defined by [2.11]) in an independent way:

$$\begin{aligned} \lambda_N &= < N^{e^{(1)}}(\boldsymbol{\xi}) > \{\lambda_N^{e^{(1)}}\} \\ \forall\alpha \in \{1,2\},\ \lambda_{T\alpha} &= [\mathcal{N}^{e^{(1)}}(\boldsymbol{\xi})]\{\lambda_{T\alpha}^{e^{(1)}}\} \end{aligned}$$

This method of interpolation assumes that the nodal values of the normal and tangential contact stresses are unique, or in other words, that we know that a unique natural basis $(\mathbf{a}^1, \mathbf{a}^2)$ for each node of $S_c^{(2)}$ exists. However, in the finite element method, we do not have a unique parameterization $\boldsymbol{\psi}^{(2)}$ for the entire target surface, as we only have a piecewise parameterization, finite element-by-finite element, with no unique natural basis $(\mathbf{a}^1, \mathbf{a}^2)$ for each node of $S_c^{(2)}$. Therefore, the proposed method above to interpolate $\boldsymbol{\lambda}_T$ is difficult to achieve.

• In this work, we will choose a simple alternative process, which will allow us to work with arbitrary meshes. The weak form [7.21] involves λ_N and $\boldsymbol{\lambda}_T$ separately. Here, in a computational scenario, it is advantageous to have only one vector which combines the two together. The idea here is therefore to introduce a new vector $\boldsymbol{\lambda}$, which we do not require necessarily in theory, and which combines the two unknowns λ_N and $\boldsymbol{\lambda}_T$ into one:

Definition.– *The multiplier vector* $\boldsymbol{\lambda}$ is defined by:

$$\boxed{\boldsymbol{\lambda} = \lambda_N\boldsymbol{\nu} - \boldsymbol{\lambda}_T} \quad \text{with } \boldsymbol{\lambda}_T \perp \boldsymbol{\nu} \qquad [8.14]$$

where we recall that $\boldsymbol{\nu}$ is the unit normal to $S_c^{(2)}$ at $\mathbf{y}$.

We have:

$$\boldsymbol{\lambda} = \lambda_N\boldsymbol{\nu} - \boldsymbol{\lambda}_T \text{ with } \boldsymbol{\lambda}_T \perp \boldsymbol{\nu} \quad \Leftrightarrow \quad \boxed{\lambda_N = \boldsymbol{\lambda}.\boldsymbol{\nu}} \text{ and } \boxed{\boldsymbol{\lambda}_T = -(\mathbf{I} - \boldsymbol{\nu}\otimes\boldsymbol{\nu}).\boldsymbol{\lambda}} \qquad [8.15]$$

where $(\mathbf{I} - \boldsymbol{\nu}\otimes\boldsymbol{\nu})$ is the projection operator onto the tangent plane.

According to definition [8.14], the vector $\boldsymbol{\lambda}$ is constructed using three unknowns: the contact stresses λ_N and $\boldsymbol{\lambda}_T$, as well as the normal vector $\boldsymbol{\nu}$, which itself is deduced from the unknown displacement fields $\mathbf{U}^{(1)}$ and $\mathbf{U}^{(2)}$. Each quantity in question is a function of point $\mathbf{X}$ and time t (in quasi-statics, the time t means the computation step n).

By fixing the normal fields $\boldsymbol{\nu}$, equation [8.15] shows that knowing $(\lambda_N, \boldsymbol{\lambda}_T)$ (three scalars) is equivalent to knowing $\boldsymbol{\lambda}$ (also three scalars).

In the weak form [7.21], we make the decision to replace the unknowns λ_N and $\boldsymbol{\lambda}_T$ with the unknown $\boldsymbol{\lambda}$, that is, they are replaced by their expressions on the right-hand side of [8.15]. This method of calculating $\boldsymbol{\lambda}_T = -(\mathbf{I} - \boldsymbol{\nu}\otimes\boldsymbol{\nu}).\boldsymbol{\lambda}$ automatically satisfies the hypothesis [7.20], namely that $\boldsymbol{\lambda}_T \perp \boldsymbol{\nu}$. It is the Cartesian components of the multiplier vector $\boldsymbol{\lambda}$ (the components relative to the fixed orthonormal basis $(\mathbf{e}_1, \mathbf{e}_2, \mathbf{e}_3)$), which will be interpolated.

If only frictionless contact problems were considered, then there would be just one unknown contact factor, namely λ_N. We would then be content with a simpler weak form than the one presented in [7.21], which would only contain the scalar λ_N without the need of the $\boldsymbol{\lambda} = \lambda_N\boldsymbol{\nu}$ vector. It is because we consider contact in general – with or without friction – that we must construct the vector $\boldsymbol{\lambda} = \lambda_N \boldsymbol{\nu} - \boldsymbol{\lambda}_T$, which causes both λ_N and $\boldsymbol{\nu}$ to appear together. Definition [8.14] is not devoid of any physical meaning – it is in fact quite natural, having been inspired by the result of [7.25] proven theoretically: the fields $(\lambda_N, \boldsymbol{\lambda}_T)$, solutions of the weak form and such that $\boldsymbol{\lambda}_T \perp \boldsymbol{\nu}$, satisfies $\lambda_N = T_N$ and $\boldsymbol{\lambda}_T = \mathbf{T}_T$, and thus the physical meaning of the vector $\boldsymbol{\lambda}$ at convergence: $\boldsymbol{\lambda} = \lambda_N \boldsymbol{\nu} - \boldsymbol{\lambda}_T = T_N \boldsymbol{\nu} - \mathbf{T}_T = \mathbf{T}$.

The intervention of the displacement fields $\mathbf{U}^{(1)}$ and $\mathbf{U}^{(2)}$ in the definition of $\boldsymbol{\lambda}$, via $\boldsymbol{\nu}$, is not a nuisance. This is not the first time that we have seen such a dependency: the definition of $\boldsymbol{\lambda}_T$ already contains the condition from [7.20], that is, that $\boldsymbol{\lambda}_T \perp \boldsymbol{\nu}$, so $\mathbf{U}^{(1)}$ and $\mathbf{U}^{(2)}$ both already feature.

8.6.2. *Interpolation of* $\boldsymbol{\lambda}$

The multiplier $\boldsymbol{\lambda}$ on the contactor surface – or, more precisely, its Cartesian components (the components relative to the fixed orthonormal basis $(\mathbf{e}_1, \mathbf{e}_2, \mathbf{e}_3)$) – can be obtained by interpolating the nodal multipliers:

$$\underset{(3,1)}{\{\boldsymbol{\lambda}\}} = \underset{(3,3nne^{(1)})}{[\mathcal{N}^{e^{(1)}}(\boldsymbol{\xi})]} \; \underset{(3nne^{(1)},1)}{\{\boldsymbol{\Lambda}^{e^{(1)}}\}} \qquad [8.16]$$

As previously mentioned, the element nodal multiplier vectors $\{\boldsymbol{\Lambda}^{e^{(1)}}\}$ are unknowns in the same way that the element nodal displacement vectors $\{\mathbf{U}^{(1),e^{(1)}}\}$ and $\{\mathbf{U}^{(2),e^{(2)}}\}$ are.

In theory, we have the separate unknowns λ_N and $\boldsymbol{\lambda}_T \perp \boldsymbol{\nu}$. Numerically, the unknown is the vector $\boldsymbol{\lambda}$. The column-vector $\{\boldsymbol{\Lambda}^{e^{(1)}}\}$ contains the unknowns $\boldsymbol{\lambda}$ at each node.

According to [8.15] and [8.16], the multipliers λ_N and $\boldsymbol{\lambda}_T$ at each point of the element $e^{(1)}$ are deduced from $\{\boldsymbol{\Lambda}^{e^{(1)}}\}$ via the following chain:

$$\{\boldsymbol{\Lambda}^{e^{(1)}}\} \;\Rightarrow\; \{\boldsymbol{\lambda}\} = [\mathcal{N}^{e^{(1)}}(\boldsymbol{\xi})]\{\boldsymbol{\Lambda}^{e^{(1)}}\} \;\Rightarrow\; \begin{cases} \lambda_N = \boldsymbol{\nu}.\boldsymbol{\lambda} \\ \boldsymbol{\lambda}_T = -(\mathbf{I} - \boldsymbol{\nu} \otimes \boldsymbol{\nu}).\boldsymbol{\lambda} \end{cases} \qquad [8.17]$$

8.6.3. *Interpolation of* $\boldsymbol{\lambda}^*$

The virtual multiplier $\boldsymbol{\lambda}^*$ on the contactor surface is interpolated in the same way as the real multiplier by using the virtual multipliers at each node:

$$\underset{(3,1)}{\{\boldsymbol{\lambda}^*\}} = \underset{(3,3nne^{(1)})}{[\mathcal{N}^{e^{(1)}}(\boldsymbol{\xi})]} \; \underset{(3nne^{(1)},1)}{\{\boldsymbol{\Lambda}^{*,e^{(1)}}\}} \qquad [8.18]$$

8.7. Discretization of the element virtual contact power $(\mathcal{P}^*_{\text{contact}})^{e^{(1)}}$

By using the interpolations previously introduced, we can express each quantity appearing in the element virtual contact power $(\mathcal{P}^*_{\text{contact}})^{e^{(1)}}$ relative to the element $e^{(1)} \subset S^{(1)}_{0c}$ – that is to say, equation [8.3] – as a function of the nodal quantities. According to the interpolations [8.9] and [8.13]:

$$\left[\langle\hat{\lambda}_N\rangle\boldsymbol{\nu} - \left(1 - \langle 1 - \frac{\mu\langle\hat{\lambda}_N\rangle}{\|\hat{\boldsymbol{\lambda}}_T\|}\rangle\right)\hat{\boldsymbol{\lambda}}_T\right] . \left(\mathbf{U}^{(1)*} - \mathbf{U}^{(2)*}\right) =$$

$$\left\{ \begin{array}{c} \{\mathbf{U}^{(1)*,e^{(1)}}\} \\ \{\mathbf{U}^{(2)*,e^{(2)}}\} \end{array} \right\}^T \left\{ \begin{array}{c} \underset{(3nne^{(1)},3)}{[\mathcal{N}^{e^{(1)}}(\boldsymbol{\xi})]^T} \left(\langle\hat{\lambda}_N\rangle \underset{(3,1)}{\{\boldsymbol{\nu}\}} - \left(1 - \langle 1 - \frac{\mu\langle\hat{\lambda}_N\rangle}{\|\hat{\boldsymbol{\lambda}}_T\|}\rangle\right) \underset{(3,1)}{\{\hat{\boldsymbol{\lambda}}_T\}} \right) \\ -\underset{(3nne^{(2)},3)}{[\mathcal{N}^{e^{(2)}}(\boldsymbol{\eta})]^T} \left(\langle\hat{\lambda}_N\rangle \underset{(3,1)}{\{\boldsymbol{\nu}\}} - \left(1 - \langle 1 - \frac{\mu\langle\hat{\lambda}_N\rangle}{\|\hat{\boldsymbol{\lambda}}_T\|}\rangle\right) \underset{(3,1)}{\{\hat{\boldsymbol{\lambda}}_T\}} \right) \end{array} \right\}$$

From the interpolation [8.18]:

$$\left\{ \frac{1}{\epsilon_N} \left(\lambda_N - \langle\hat{\lambda}_N\rangle\right) \mathbf{m} - \frac{1}{\epsilon_T} \left[\boldsymbol{\lambda}_T - \left(1 - \langle 1 - \frac{\mu\langle\hat{\lambda}_N\rangle}{\|\hat{\boldsymbol{\lambda}}_T\|}\rangle\right) \hat{\boldsymbol{\lambda}}_T \right] \right\} . \boldsymbol{\Lambda}^* =$$

$$< \boldsymbol{\Lambda}^{*,e^{(1)}} > \underset{(3nne^{(1)},3)}{[\mathcal{N}^{e^{(1)}}(\boldsymbol{\xi})]^T} \left(\frac{1}{\epsilon_N} \left(\lambda_N - \langle\hat{\lambda}_N\rangle\right) \underset{(3,1)}{\{\mathbf{m}\}} - \frac{1}{\epsilon_T} \left\{ \boldsymbol{\lambda}_T - \left(1 - \langle 1 - \frac{\mu\langle\hat{\lambda}_N\rangle}{\|\hat{\boldsymbol{\lambda}}_T\|}\rangle\right) \hat{\boldsymbol{\lambda}}_T \right\} \right)$$

where $\{\mathbf{m}\}$ is the column-vector that contains the three components of the vector $\mathbf{m}$, relative to the fixed Cartesian basis $(\mathbf{e}_1, \mathbf{e}_2, \mathbf{e}_3)$.

Thus, the element virtual contact power $(\mathcal{P}^*_{\text{contact}})^{e^{(1)}}$ from [8.3] becomes:

$$(\mathcal{P}^*_{\text{contact}})^{e^{(1)}} = \int_{e^{(1)}} \left(\left\{ \begin{array}{c} \{\mathbf{U}^{(1)*,e^{(1)}}\} \\ \{\mathbf{U}^{(2)*,e^{(2)}}\} \end{array} \right\}^T \left\{ \begin{array}{c} [\mathcal{N}^{e^{(1)}}(\boldsymbol{\xi})]^T \left(\langle\hat{\lambda}_N\rangle\{\boldsymbol{\nu}\} - \left(1 - \langle 1 - \frac{\mu\langle\hat{\lambda}_N\rangle}{\|\hat{\boldsymbol{\lambda}}_T\|}\rangle\right) \{\hat{\boldsymbol{\lambda}}_T\} \right) \\ -[\mathcal{N}^{e^{(2)}}(\boldsymbol{\eta})]^T \left(\langle\hat{\lambda}_N\rangle\{\boldsymbol{\nu}\} - \left(1 - \langle 1 - \frac{\mu\langle\hat{\lambda}_N\rangle}{\|\hat{\boldsymbol{\lambda}}_T\|}\rangle\right) \{\hat{\boldsymbol{\lambda}}_T\} \right) \end{array} \right\} \right.$$
$$\left. + < \boldsymbol{\Lambda}^{*,e^{(1)}} > [\mathcal{N}^{e^{(1)}}(\boldsymbol{\xi})]^T \left(\frac{1}{\epsilon_N} \left(\lambda_N - \langle\hat{\lambda}_N\rangle\right) \{\mathbf{m}\} - \frac{1}{\epsilon_T} \left\{ \boldsymbol{\lambda}_T - \left(1 - \langle 1 - \frac{\mu\langle\hat{\lambda}_N\rangle}{\|\hat{\boldsymbol{\lambda}}_T\|}\rangle\right) \hat{\boldsymbol{\lambda}}_T \right\} \right) \right) dS_0 \qquad [8.19]$$

Integration must be performed with respect to the variable $\mathbf{X} \in e^{(1)}$, or equivalently, with respect to the variable $\boldsymbol{\xi}$ belonging to the parent-element of $e^{(1)}$. Table 8.1 shows how each quantity in the integrand is a function of the element $e^{(1)}$ and the point $\boldsymbol{\xi}$.

For each row of the table, we start with the quantities in the first column, before then deducing the other quantities in the second column via the equations listed in the last column. The starting point of the table is $(e^{(1)}, \xi)$, and we deduce the different quantities successively by going from one row to another.

Knowing	We obtain	By using
$e^{(1)}, \xi$	(i) $[\mathcal{N}^{e^{(1)}}(\xi)]$ (ii) $\{\mathbf{X}^{e^{(1)}}\}$ (iii) $\mathbf{X}$	The mesh of $S^{(1)}_{0c}$ The interpolation [8.6]
$\xi, \{\mathbf{U}^{(1),e^{(1)}}\}$	$\mathbf{U}^{(1)}, \mathbf{x}$	The interpolations [8.7]–[8.8]
$\mathbf{x}, \psi^{(2)}(.,t)$	(i) $\eta, e^{(2)}$ (ii) $[\mathcal{N}^{e^{(2)}}(\eta)]$ (iii) $\{\mathbf{Y}^{e^{(2)}}\}$ (iv) $\mathbf{Y}$ (v) $\{\mathbf{U}^{(2),e^{(2)}}\}$ (vi) $\mathbf{U}^{(2)}, \mathbf{y}$ (vii) The tangent vectors $\mathbf{a}_1$, $\mathbf{a}_2$ at $\mathbf{y}$ to the surface $S^{(2)}_c$ (viii) The normal vector $\boldsymbol{\nu}$ at $\mathbf{y}$ to the surface $S^{(2)}_c$ (ix) The vector $\mathbf{m}$	The projection process [2.20]. This passage is specific to contact The mesh of $S^{(2)}_{0c}$ The interpolation [8.10] The extraction of $\psi^{(2)}(.,t)$ The interpolations [8.11]–[8.12] [2.9]: $\mathbf{a}_1 = \dfrac{\partial \psi^{(2)}}{\partial \xi^{(2)1}}(\eta,t)$, $\mathbf{a}_2 = \dfrac{\partial \psi^{(2)}}{\partial \xi^{(2)2}}(\eta,t)$ [2.10] : $\boldsymbol{\nu} = \dfrac{\mathbf{a}_1 \times \mathbf{a}_2}{\Vert\mathbf{a}_1 \times \mathbf{a}_2\Vert}$ The choice $\mathbf{m} = \boldsymbol{\nu}$, made later in [9.67]
$\mathbf{x}, \mathbf{y}$	The proximity g	$g = -\boldsymbol{\nu}.(\mathbf{x} - \mathbf{y})$
$\mathbf{y}$, $\boldsymbol{\phi}^{(2)}(\mathbf{Y}_{n-1}, t)$	The slip velocity $\mathbf{V}_T$	The discrete expression [8.56] which will be proven later. $\boldsymbol{\phi}^{(2)}(\mathbf{Y}_{n-1}, t)$ is obtained from ξ as explained in Table 8.2
$\xi, \{\boldsymbol{\Lambda}^{e^{(1)}}\}$	The multiplier vector $\boldsymbol{\lambda}$	The interpolation [8.16] Since $\{\boldsymbol{\Lambda}^{e^{(1)}}\}$ is a function of t, the vector $\boldsymbol{\lambda}$ is a function of (ξ, t)
$\boldsymbol{\lambda}, \boldsymbol{\nu}$	The multipliers $\lambda_N, \boldsymbol{\lambda}_T$	The equation [8.17] Since $\boldsymbol{\lambda}$ depends on (ξ, t), and $\boldsymbol{\nu}$ depends on ξ, the multipliers $\lambda_N, \boldsymbol{\lambda}_T$ are functions of (ξ, t). We obtain what we wrote initially: $\lambda_N = \lambda_N(\mathbf{X}, t)$ and $\boldsymbol{\lambda}_T = \boldsymbol{\lambda}_T(\mathbf{X}, t)$
λ_N, g	$\hat{\lambda}_N \equiv \lambda_N + \epsilon_N g$	
$\boldsymbol{\lambda}_T, \mathbf{V}_T$	$\hat{\boldsymbol{\lambda}}_T \equiv \boldsymbol{\lambda}_T + \epsilon_T \mathbf{V}_T$	

Table 8.1. *How to successively obtain the different quantities in the integrand of* [8.19] *from* $(e^{(1)}, \xi)$

For a given element $e^{(1)}$, we will retain the following dependencies from Table 8.1:

1. The points $\mathbf{X}$, $\mathbf{Y}$, $\mathbf{x}$, $\mathbf{y}$, η are functions of ξ via the following chain (see Figure 8.2):

ξ	$\rightarrow$	$\mathbf{X}$ by the interpolation [8.6];
$\xi, \{\mathbf{U}^{(1),e^{(1)}}\}$	$\rightarrow$	$\mathbf{x}$ by the interpolations [8.7] – –[8.8];
$\mathbf{x}, \psi^{(2)}(., t)$	$\rightarrow$	$\eta, e^{(2)}$ by the projection operator [2.20] (specific passage to contact);
η	$\rightarrow$	$\mathbf{Y}$ by the interpolation [8.10];
$\eta, \{\mathbf{U}^{(2),e^{(2)}}\}$	$\rightarrow$	$\mathbf{y}$ by the interpolation [8.12].

The element $e^{(2)} \subset S_{0c}^{(2)}$ which contains the pre-image $\mathbf{Y}$ of $\mathbf{y} = \text{proj}_{S_c^{(2)}}\mathbf{x}$ depends on the point $\mathbf{x}$, and thus the element $e^{(2)}$ depends on point ξ. We know that one or more such elements $e^{(2)}$ can exist and, in the case where several elements exist, it suffices just to choose one arbitrarily.

Note that the integral over the element $e^{(1)}$ may involve several $e^{(2)} \subset S_{0c}^{(2)}$ elements, as shown in Figure 8.3.

2. The multipliers λ_N, $\boldsymbol{\lambda}_T$ are functions of ξ.

3. On the one hand, the vectors $\hat{\lambda}_N$ and $\hat{\boldsymbol{\lambda}}_T$ depend on the displacements via g and $\mathbf{V}_T$ and, on the other hand, on the multipliers λ_N and $\boldsymbol{\lambda}_T$.

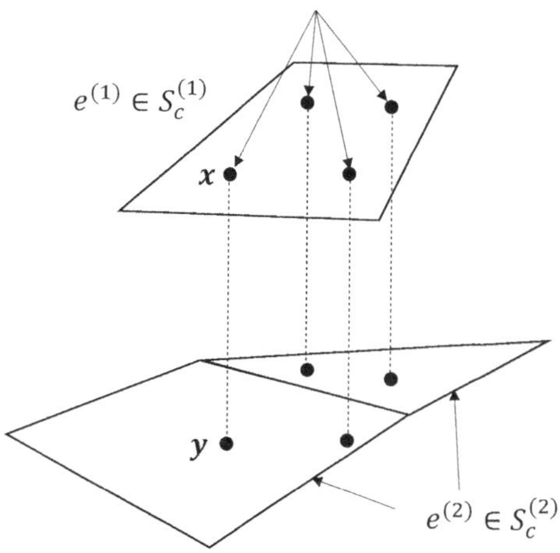

Figure 8.3. *An element $e^{(1)} \subset S_{0c}^{(1)}$ may involve several elements $e^{(2)} \subset S_{0c}^{(2)}$. To simplify the figure, we have written $e^{(1)}$ and $e^{(2)}$ to represent the respective images of $e^{(1)}$ and $e^{(2)}$ after a displacement of the bodies. For a color version of this figure, see www.iste.co.uk/levan/contact.zip*

Furthermore, we may also recall the following two important facts regarding the dependencies of the virtual nodal quantities that appear in the integrand of [8.19]:

1. On the one hand, the column-vectors $\{\mathbf{U}^{(1)*,e^{(1)}}\}$ and $\{\mathbf{\Lambda}^{*,e^{(1)}}\}$, which are virtual nodal quantities relative to the element $e^{(1)}$, do not depend on the points $\boldsymbol{\xi}$, $\boldsymbol{\eta}$, $\mathbf{X}$, $\mathbf{Y}$, $\mathbf{x}$, nor $\mathbf{y}$.

2. On the other hand, the column-vector $\{\mathbf{U}^{(2)*,e^{(2)}}\}$, which is relative to the element $e^{(2)}$, does depend on the point $\boldsymbol{\xi}$ via the element $e^{(2)}$.

Now, let us look at the dependencies vis-à-vis the unknowns of the problem:

- With the help of Table 8.1. we can check that some of the variables in the integrand of [8.19] do depend on the nodal vectors $\{\mathbf{U}^{(1),e^{(1)}}(t)\}$, $\{\mathbf{U}^{(2),e^{(2)}}(t)\}$, and $\{\mathbf{\Lambda}^{e^{(1)}}(t)\}$.

- In anticipation of the discussion that follows after Table 8.2 later on, we see that the slip velocity $\mathbf{V}_T$ depends on $\{\mathbf{U}^{(1),e^{(1)}}(t)\}$, $\{\mathbf{U}^{(2),e^{(2)}}(t)\}$, and also on $\{\mathbf{U}^{(2),e^{(2)}_{n-1}}(t)\}$, where the element $e^{(2)}_{n-1}$ is defined in [8.54] later.

Note in passing that the dependence relative to the time instant t – which is implicit in the aforementioned nodal vectors – is ubiquitous, and therefore does not deserve any special attention.

• Since the column-vectors $\{\mathbf{U}^{(1)*,e^{(1)}}\}$ and $\{\mathbf{\Lambda}^{*,e^{(1)}}\}$ of [8.19] are independent of the integration variable $\boldsymbol{\xi}$, they can be taken outside of the integral. On the other hand, we must leave the column-vector $\{\mathbf{U}^{(2)*,e^{(2)}}\}$ inside the integrand because, as we have just observed, the element $e^{(2)}$ varies relative to point $\boldsymbol{\xi}$. We have then:

$$
\begin{aligned}
(\mathcal{P}^*_{\text{contact}})^{e^{(1)}} = \; & < \mathbf{U}^{(1)*,e^{(1)}} > \underbrace{\int_{e^{(1)}} [\mathcal{N}^{e^{(1)}}(\boldsymbol{\xi})]^T \left(\langle \hat{\lambda}_N \rangle \{\boldsymbol{\nu}\} - \left(1 - \langle 1 - \frac{\mu \langle \hat{\lambda}_N \rangle}{\|\hat{\boldsymbol{\lambda}}_T\|} \rangle \right) \{\hat{\boldsymbol{\lambda}}_T\} \right) dS_0}_{(a)} \\
& - \underbrace{\int_{e^{(1)}} < \mathbf{U}^{(2)*,e^{(2)}} > [\mathcal{N}^{e^{(2)}}(\boldsymbol{\eta})]^T \left(\langle \hat{\lambda}_N \rangle \{\boldsymbol{\nu}\} - \left(1 - \langle 1 - \frac{\mu \langle \hat{\lambda}_N \rangle}{\|\hat{\boldsymbol{\lambda}}_T\|} \rangle \right) \{\hat{\boldsymbol{\lambda}}_T\} \right) dS_0}_{(b)} \\
+ < \mathbf{\Lambda}^{*,e^{(1)}} > & \underbrace{\int_{e^{(1)}} [\mathcal{N}^{e^{(1)}}(\boldsymbol{\xi})]^T \left(\frac{1}{\epsilon_N} \left(\lambda_N - \langle \hat{\lambda}_N \rangle \right) \{\mathbf{m}\} - \frac{1}{\epsilon_T} \left\{ \boldsymbol{\lambda}_T - \left(1 - \langle 1 - \frac{\mu \langle \hat{\lambda}_N \rangle}{\|\hat{\boldsymbol{\lambda}}_T\|} \rangle \right) \hat{\boldsymbol{\lambda}}_T \right\} \right) dS_0}_{(c)}
\end{aligned}
$$

The column-vectors (a) and (c) in the previous expression, which are multiplicands of $< \mathbf{U}^{(1)*,e^{(1)}} >$ and $< \mathbf{\Lambda}^{*,e^{(1)}} >$, respectively, may be defined in a classical way so that they can be thought of as being element nodal vectors. On the other hand, due to its non-classical form, the (b) term requires a slightly more advanced and specialized treatment.

With this in mind, from now on, we will make the decision to apply the same specific treatment to the entire term on the right-hand side of [8.19], which is usually reserved solely for the single (b) term, in order to standardize our presentation. For terms other than (b), the reader can check that our treatment is indeed equivalent to the classical one used in the finite element method.

Therefore, the element virtual contact power $(\mathcal{P}^*_{\text{contact}})^{e^{(1)}}$ in [8.19] relative to the element $e^{(1)}$ can be written in the discrete form:

$$
\boxed{(\mathcal{P}^*_{\text{contact}})^{e^{(1)}} = \int_{e^{(1)}} \left(\begin{Bmatrix} \{\mathbf{U}^{(1)*,e^{(1)}}\} \\ \{\mathbf{U}^{(2)*,e^{(2)}}\} \end{Bmatrix}^T \begin{Bmatrix} \{\mathbf{\Phi}^{e^{(1)}}_{\text{contact}}\} \\ \{\mathbf{\Phi}^{e^{(2)}}_{\text{contact}}\} \end{Bmatrix} - < \mathbf{\Lambda}^{*,e^{(1)}} > \{\mathbf{R}^{e^{(1)}}_{\Lambda}\} \right) dS_0} \qquad [8.20]
$$

where we have adopted the following notation and definitions:

DEFINITION.–

1. *The element contact force vectors* (N/m^2 in SI units) are, by definition:

$$\underset{(3nne^{(1)},1)}{\left\{\boldsymbol{\Phi}^{e^{(1)}}_{\text{contact}}\right\}} \equiv \underset{(3nne^{(1)},3)}{[\mathcal{N}^{e^{(1)}}(\boldsymbol{\xi})]^T} \left(\langle\hat{\lambda}_N\rangle \underset{(3,1)}{\{\boldsymbol{\nu}\}} - \left(1 - \langle 1 - \frac{\mu\langle\hat{\lambda}_N\rangle}{\|\hat{\boldsymbol{\lambda}}_T\|}\rangle\right) \underset{(3,1)}{\{\hat{\boldsymbol{\lambda}}_T\}} \right)$$
$$\underset{(3nne^{(2)},1)}{\left\{\boldsymbol{\Phi}^{e^{(2)}}_{\text{contact}}\right\}} \equiv -\underset{(3nne^{(2)},3)}{[\mathcal{N}^{e^{(2)}}(\boldsymbol{\eta})]^T} \left(\langle\hat{\lambda}_N\rangle \underset{(3,1)}{\{\boldsymbol{\nu}\}} - \left(1 - \langle 1 - \frac{\mu\langle\hat{\lambda}_N\rangle}{\|\hat{\boldsymbol{\lambda}}_T\|}\rangle\right) \underset{(3,1)}{\{\hat{\boldsymbol{\lambda}}_T\}} \right) \quad [8.21]$$

The top term is a column-vector with $3nne^{(1)}$ components, whereas the bottom term is a column-vector with $3nne^{(2)}$ components.

2. *The element residual contact law vector* (with units m!) is, by definition:

$$\underset{(3nne^{(1)},1)}{\left\{\mathbf{R}^{e^{(1)}}_{\Lambda}\right\}} \equiv \underset{(3nne^{(1)},3)}{[\mathcal{N}^{e^{(1)}}(\boldsymbol{\xi})]^T} \left(\frac{1}{\epsilon_N} \left(\langle\hat{\lambda}_N\rangle - \lambda_N\right) \underset{(3,1)}{\{\mathbf{m}\}} + \frac{1}{\epsilon_T} \underset{(3,1)}{\left\{ \boldsymbol{\lambda}_T - \left(1 - \langle 1 - \frac{\mu\langle\hat{\lambda}_N\rangle}{\|\hat{\boldsymbol{\lambda}}_T\|}\rangle\right) \hat{\boldsymbol{\lambda}}_T \right\}} \right) \quad [8.22]$$

This is also a column-vector with $3nne^{(1)}$ components.

3. To explain the recurring coefficient $1 - \langle 1 - \frac{\mu\langle\hat{\lambda}_N\rangle}{\|\hat{\boldsymbol{\lambda}}_T\|}\rangle$, let us introduce the following terminology:

   (a) When $\hat{\lambda}_N < 0$, we say that we are in the *algorithmic gap* case, where the term *algorithmic* is used to distinguish this from the real "gap" case when $g < 0$ (or when $\lambda_N = 0$).

   (b) When $\hat{\lambda}_N \geq 0$ and $\|\hat{\boldsymbol{\lambda}}_T\| \geq \mu\hat{\lambda}_N$, we say that we are in the *algorithmic slip* case, to distinguish this from the real "slip" case when $g = 0$ (or when $\lambda_N \geq 0$) and $\|\boldsymbol{\lambda}_T\| = \mu\lambda_N$.

   (c) When $\hat{\lambda}_N \geq 0$ and $\|\hat{\boldsymbol{\lambda}}_T\| < \mu\hat{\lambda}_N$, we say that we are in the *algorithmic stick* case, to distinguish this from the real "stick" case when $g = 0$ (or when $\lambda_N \geq 0$) and $\|\boldsymbol{\lambda}_T\| < \mu\lambda_N$.

4. We can then express the coefficient $1 - \langle 1 - \frac{\mu\langle\hat{\lambda}_N\rangle}{\|\hat{\boldsymbol{\lambda}}_T\|}\rangle$ according to these three cases:

$$1 - \langle 1 - \frac{\mu\langle\hat{\lambda}_N\rangle}{\|\hat{\boldsymbol{\lambda}}_T\|}\rangle = \begin{cases} 0 & \text{if } \hat{\lambda}_N < 0 \quad \text{(algorithmic gap)} \\ \dfrac{\mu\hat{\lambda}_N}{\|\hat{\boldsymbol{\lambda}}_T\|} & \text{if } \hat{\lambda}_N \geq 0 \text{ and } \|\hat{\boldsymbol{\lambda}}_T\| \geq \mu\hat{\lambda}_N \quad \text{(algorithmic slip contact)} \\ 1 & \text{if } \hat{\lambda}_N \geq 0 \text{ and } \|\hat{\boldsymbol{\lambda}}_T\| < \mu\hat{\lambda}_N \quad \text{(algorithmic stick contact)} \end{cases} \quad [8.23]$$

REMARK.– In the classical finite element method, a column-vector is said to be "element" – for example, the element internal force vector – when it is either relative or corresponds to a given entire element. Here, the term "element", which applies to the vectors $\left\{\mathbf{\Phi}^{e^{(1)}}_{\text{contact}}\right\}$, $\left\{\mathbf{\Phi}^{e^{(2)}}_{\text{contact}}\right\}$ and $\left\{\mathbf{R}^{e^{(1)}}_{\Lambda}\right\}$, does not mean quite the same thing and therefore should be understood as an extension of the classical term. This remark will make more sense when we come to section 8.8.3, where we speak of "the assembly" of these vectors. ■

There is the question of what the element contact force vectors $\left\{\mathbf{\Phi}^{e^{(1)}}_{\text{contact}}\right\}$, $\left\{\mathbf{\Phi}^{e^{(2)}}_{\text{contact}}\right\}$ depend on, and what the element residual contact law vector $\left\{\mathbf{R}^{e^{(1)}}_{\Lambda}\right\}$ depends on.

1. According to the expressions [8.21] and [8.22], these vectors depend on the element $e^{(1)}$ and the integration point ξ, where $e^{(1)}$ is *known* and ξ is *given*, with a *known* value.

2. On the other hand, according to the discussion that follows after Table 8.1, the same vectors also depend on the following nodal vectors, which are *unknowns* of the problem:

   (a) the nodal displacement vectors $\{\mathbf{U}^{(1),e^{(1)}}\}$, $\{\mathbf{U}^{(2),e^{(2)}}\}$ and $\{\mathbf{U}^{(2),e^{(2)}_{n-1}}\}$ of the elements $e^{(1)}$, $e^{(2)}$ and $e^{(2)}_{n-1}$ (the last element is defined in [8.54] in connection to the slip velocity $\mathbf{V}_T$);

   (b) the element nodal multiplier vector $\{\mathbf{\Lambda}^{e^{(1)}}\}$.

We can express these dependencies on the unknowns of the problem by writing:

$$\begin{aligned}
\mathbf{\Phi}^{e^{(1)}}_{\text{contact}} &= \mathbf{\Phi}^{e^{(1)}}_{\text{contact}}(\mathbf{U}^{(1),e^{(1)}}, \mathbf{U}^{(2),e^{(2)}}, \mathbf{U}^{(2),e^{(2)}_{n-1}}, \mathbf{\Lambda}^{e^{(1)}}) \\
\mathbf{\Phi}^{e^{(2)}}_{\text{contact}} &= \mathbf{\Phi}^{e^{(2)}}_{\text{contact}}(\mathbf{U}^{(1),e^{(1)}}, \mathbf{U}^{(2),e^{(2)}}, \mathbf{U}^{(2),e^{(2)}_{n-1}}, \mathbf{\Lambda}^{e^{(1)}}) \\
\mathbf{R}^{e^{(1)}}_{\Lambda} &= \mathbf{R}^{e^{(1)}}_{\Lambda}(\mathbf{U}^{(1),e^{(1)}}, \mathbf{U}^{(2),e^{(2)}}, \mathbf{U}^{(2),e^{(2)}_{n-1}}, \mathbf{\Lambda}^{e^{(1)}})
\end{aligned} \quad [8.24]$$

To abbreviate, let us introduce the following notation:

NOTATION.– We denote by $\mathbf{U}^e$ the column-vector that combines the nodal displacement vectors $\{\mathbf{U}^{(1),e^{(1)}}\}$, $\{\mathbf{U}^{(2),e^{(2)}}\}$ and $\{\mathbf{U}^{(2),e^{(2)}_{n-1}}\}$ of the elements $e^{(1)}$, $e^{(2)}$ and $e^{(2)}_{n-1}$ together:

$$\boxed{\mathbf{U}^e \equiv \left(\mathbf{U}^{(1),e^{(1)}}, \mathbf{U}^{(2),e^{(2)}}, \mathbf{U}^{(2),e^{(2)}_{n-1}}\right)} \quad [8.25]$$

Hence, the superscript e denotes the three elements, $e^{(1)}$, $e^{(2)}$ and $e^{(2)}_{n-1}$.

With this notation, the dependencies [8.24] can be abbreviated as follows:

$$\begin{aligned}
\mathbf{\Phi}^{e^{(1)}}_{\text{contact}} &= \mathbf{\Phi}^{e^{(1)}}_{\text{contact}}(\mathbf{U}^e, \mathbf{\Lambda}^{e^{(1)}}) \\
\mathbf{\Phi}^{e^{(2)}}_{\text{contact}} &= \mathbf{\Phi}^{e^{(2)}}_{\text{contact}}(\mathbf{U}^e, \mathbf{\Lambda}^{e^{(1)}}) \\
\mathbf{R}^{e^{(1)}}_{\Lambda} &= \mathbf{R}^{e^{(1)}}_{\Lambda}(\mathbf{U}^e, \mathbf{\Lambda}^{e^{(1)}})
\end{aligned}$$

• There are two cases in particular which deserve special attention when calculating $(\mathcal{P}^*_{\text{contact}})^{e^{(1)}}$ using the formula [8.20]. Consider one of the summation terms in [8.20], corresponding to a point $\mathbf{x}$ of the contactor surface.

1. If there are several projection points of $\mathbf{x}$ on the target surface (see Figure 8.4a), then it suffices to choose one arbitrarily to calculate the term in question, since we know that if contact is established at the point $\mathbf{x}$, then its projection point will be unique.

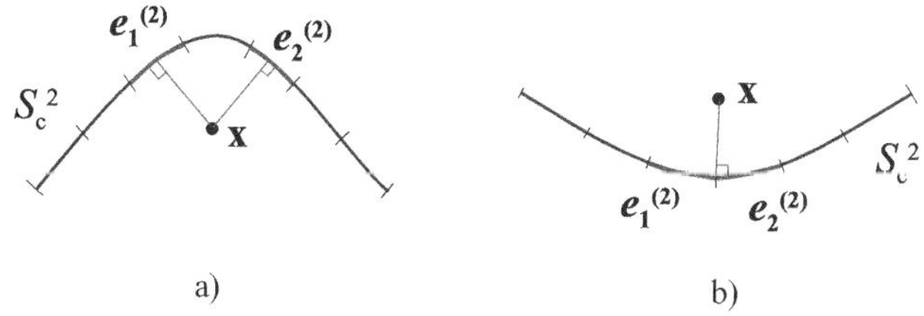

Figure 8.4. *(a) Non-unique projection points; (b) a projection point belonging to the join of several elements of the target surface. For a color version of this figure, see www.iste.co.uk/levan/contact.zip*

2. If the projection point of $\mathbf{x}$ on the target surface belongs to the join of several elements of the target surface (see Figure 8.4b), then we can still choose an arbitrary element from these to calculate the constitutive term in [8.20], but as the projection point can switch incessantly between one element of the target surface to another between each iteration, the calculation risks non-convergence. In this case, the solution is to calculate $(\mathcal{P}^*_{\text{contact}})^{e^{(1)}}$ by taking the average of each contribution from the various elements surrounding the projection point:

$$(\mathcal{P}^*_{\text{contact}})^{e^{(1)}} =$$
$$\sum_{\text{integ pt }\xi} \frac{1}{Ne^{(2)}} \sum_{e^{(2)}} \left[\langle \hat{\lambda}_N \rangle \boldsymbol{\nu} - \left(1 - \langle 1 - \frac{\mu \langle \hat{\lambda}_N \rangle}{\|\hat{\boldsymbol{\lambda}}_T\|} \rangle \right) \hat{\boldsymbol{\lambda}}_T \right] . \left(\mathbf{U}^{(1)*}(\mathbf{X}) - \mathbf{U}^{(2)*}(\mathbf{Y}(\mathbf{X})) \right) j(\xi) +$$
$$\sum_{\text{integ pt }\xi} \frac{1}{Ne^{(2)}} \sum_{e^{(2)}} \left\{ \frac{1}{\epsilon_N} \left(\lambda_N - \langle \hat{\lambda}_N \rangle \right) \mathbf{m} - \frac{1}{\epsilon_T} \left[\boldsymbol{\lambda}_T - \left(1 - \langle 1 - \frac{\mu \langle \hat{\lambda}_N \rangle}{\|\hat{\boldsymbol{\lambda}}_T\|} \rangle \right) \hat{\boldsymbol{\lambda}}_T \right] \right\} . \boldsymbol{\Lambda}^* j(\xi)$$

where $Ne^{(2)}$ is the number of $e^{(2)}$ elements surrounding the projection point, and $j(\xi)$ will be defined later on in [8.44]. This process is different from that found in [106], where [8.20] is still applied but instead $\boldsymbol{\nu}$ is taken to be the average of the normal to each element surrounding the projection point.

8.7.1. *Explicit expressions for* $\{\boldsymbol{\Phi}^{e^{(1)}}_{contact}\}$, $\{\boldsymbol{\Phi}^{e^{(2)}}_{contact}\}$ *and* $\{R^{e^{(1)}}_{\Lambda}\}$ *in the three cases: algorithmic* gap, *algorithmic* slip *and algorithmic* stick

Let us consider a given integration point on a particular element $e^{(1)}$ of $S^{(1)}_{0c}$. We will investigate the vectors $\{\boldsymbol{\Phi}^{e^{(1)}}_{\text{contact}}\}$, $\{\boldsymbol{\Phi}^{e^{(2)}}_{\text{contact}}\}$ and $\{\mathbf{R}^{e^{(1)}}_{\Lambda}\}$ of [8.21] and [8.22] by distinguishing between three different cases at the considered integration point: the algorithmic gap, the algorithmic slip and the algorithmic stick cases.

Algorithmic gap *case at the given integration point:* $\hat{\lambda}_N < 0$

In this case, we deduce from [8.21] and [8.22] that $\{\boldsymbol{\Phi}^{e^{(1)}}_{\text{contact}}\}$, $\{\boldsymbol{\Phi}^{e^{(2)}}_{\text{contact}}\}$ and $\{\mathbf{R}^{e^{(1)}}_{\Lambda}\}$ are independent of the coefficient of friction μ, as expected:

$$\boxed{\begin{aligned} \left\{\boldsymbol{\Phi}^{e^{(1)}}_{\text{contact}}(\mathbf{U}^e, \boldsymbol{\Lambda}^{e^{(1)}})\right\} &= \{\mathbf{0}\} \\ \left\{\boldsymbol{\Phi}^{e^{(2)}}_{\text{contact}}(\mathbf{U}^e, \boldsymbol{\Lambda}^{e^{(1)}})\right\} &= \{\mathbf{0}\} \end{aligned}} \qquad [8.26]$$

$$\left\{\mathbf{R}_{\Lambda}^{e^{(1)}}\right\} = [\mathcal{N}^{e^{(1)}}(\xi)]^T \left(-\frac{\lambda_N}{\epsilon_N}\{\mathbf{m}\} + \frac{1}{\epsilon_T}\{\boldsymbol{\lambda}_T\} \right) \qquad [8.27]$$

Algorithmic slip *case at the given integration point:* $\hat{\lambda}_N \geq 0$ *and* $||\hat{\boldsymbol{\lambda}}_T|| \geq \mu\hat{\lambda}_N$

• If, at the given point $\hat{\boldsymbol{\lambda}}_T \neq \mathbf{0}$ and if there is algorithmic slip, then $1 - \langle 1 - \frac{\mu\langle\hat{\lambda}_N\rangle}{||\hat{\boldsymbol{\lambda}}_T||}\rangle = \frac{\mu\hat{\lambda}_N}{||\hat{\boldsymbol{\lambda}}_T||}$.

From which:

$$\begin{aligned} \left\{\boldsymbol{\Phi}_{\text{contact}}^{e^{(1)}}(\mathbf{U}^e, \boldsymbol{\Lambda}^{e^{(1)}})\right\} &= [\mathcal{N}^{e^{(1)}}(\xi)]^T \left(\hat{\lambda}_N\{\boldsymbol{\nu}\} - \mu\hat{\lambda}_N \frac{\{\hat{\boldsymbol{\lambda}}_T\}}{||\hat{\boldsymbol{\lambda}}_T||} \right) \\ \left\{\boldsymbol{\Phi}_{\text{contact}}^{e^{(2)}}(\mathbf{U}^e, \boldsymbol{\Lambda}^{e^{(1)}})\right\} &= -[\mathcal{N}^{e^{(2)}}(\eta)]^T \left(\hat{\lambda}_N\{\boldsymbol{\nu}\} - \mu\hat{\lambda}_N \frac{\{\hat{\boldsymbol{\lambda}}_T\}}{||\hat{\boldsymbol{\lambda}}_T||} \right) \end{aligned} \qquad [8.28]$$

$$\left\{\mathbf{R}_{\Lambda}^{e^{(1)}}\right\} = [\mathcal{N}^{e^{(1)}}(\xi)]^T \left(g\{\mathbf{m}\} + \frac{1}{\epsilon_T}\left\{\boldsymbol{\lambda}_T - \mu\hat{\lambda}_N \frac{\hat{\boldsymbol{\lambda}}_T}{||\hat{\boldsymbol{\lambda}}_T||}\right\} \right) \qquad [8.29]$$

• Let us examine the case when $\hat{\boldsymbol{\lambda}}_T = \mathbf{0}$ at the given point.

The algorithmic slip case is the only one of the three contact cases where the expressions for vectors $\{\boldsymbol{\Phi}_{\text{contact}}^{e^{(1)}}\}$, $\{\boldsymbol{\Phi}_{\text{contact}}^{e^{(2)}}\}$ and $\{\mathbf{R}_{\Lambda}^{e^{(1)}}\}$ contain $\hat{\boldsymbol{\lambda}}_T$ in the denominators, and hence they are no longer valid when $\hat{\boldsymbol{\lambda}}_T = \mathbf{0}$.

Note first that, according to the hypotheses $\hat{\lambda}_N \geq 0$ and $||\hat{\boldsymbol{\lambda}}_T|| \geq \mu\hat{\lambda}_N$, the equality $\hat{\boldsymbol{\lambda}}_T = \mathbf{0}$ implies that $\hat{\lambda}_N = 0$. Thus, there is only one point in the $(\hat{\lambda}_N, \hat{\boldsymbol{\lambda}}_T)$ space, corresponding to the algorithmic slip case and with $\hat{\boldsymbol{\lambda}}_T = \mathbf{0}$, namely the point $(\hat{\lambda}_N = 0, \hat{\boldsymbol{\lambda}}_T = \mathbf{0})$, that is, at the origin **O**, which is the vertex of the Coulomb cone presented in Figure 8.5.

In comparison, there are infinitely many points in the $(\hat{\lambda}_N, \hat{\boldsymbol{\lambda}}_T)$ space that correspond to the algorithmic stick case and with $\hat{\boldsymbol{\lambda}}_T = \mathbf{0}$, namely the points that lie on the positive semi-axis of $\hat{\lambda}_N$.

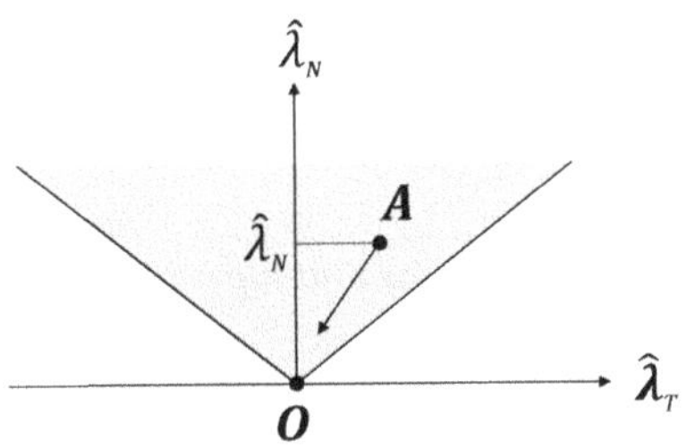

Figure 8.5. *Coulomb cone in the $(\hat{\lambda}_N, \hat{\boldsymbol{\lambda}}_T)$ space. The point **O** corresponds to algorithmic slip case and with $\hat{\boldsymbol{\lambda}}_T = \mathbf{0}$, whereas the point **A** corresponds to algorithmic stick case and with $\hat{\lambda}_N \to 0$*

The case in which slip occurs and with $\hat{\boldsymbol{\lambda}}_T = \mathbf{0}$ is exceptional, but not impossible. The following proposition provides us with a situation when such a case occurs:

PROPOSITION.– [8.30]

Hypotheses:

(i) We are in a static regime.

(ii) Initially, the two bodies are barely touching one another at the point in question, that is to say, $g = 0$ at the given point.

Then, at the considered point and at the initial time, we are in the algorithmic slip case with $\hat{\lambda}_N = 0$ and $\hat{\boldsymbol{\lambda}}_T = \mathbf{0}$.

PROOF.

– At the initial time instance, the contact stresses vanish: $\lambda_N = 0$, $\boldsymbol{\lambda}_T = \mathbf{0}$.

– On the other hand, and in anticipating expression [9.19] later on, we know that during the first iteration of each computation step, the slip velocity $\mathbf{V}_T$ is zero.

Consequently, $\hat{\lambda}_N = \lambda_N + \epsilon_N g = 0$ and $\hat{\boldsymbol{\lambda}}_T = \boldsymbol{\lambda}_T + \epsilon_T \mathbf{V}_T = \mathbf{0}$.

We then fall into the algorithmic slip case at the given integration point, as we satisfy $\hat{\lambda}_N \geq 0$ and $||\hat{\boldsymbol{\lambda}}_T|| \geq \mu\hat{\lambda}_N$. ■

An example that satisfies hypothesis (ii) is one that consists of two bodies whose initial contact surfaces $S_{0c}^{(1)}$ and $S_{0c}^{(2)}$ are plane and are barely touching. Hypothesis (ii) is then satisfied for any point of $S_{0c}^{(1)}$.

REMARK.– Proposition [8.30] is valid in static instances but not necessarily in dynamic ones. Indeed, in a dynamic instance, if hypothesis (ii) is satisfied, then we see that $\hat{\lambda}_N = 0$; but in general, as we do not have $\mathbf{V}_T = \mathbf{0}$ for the first iteration of each computation step, we do not have that $\hat{\boldsymbol{\lambda}}_T = \mathbf{0}$. We remain in the algorithmic slip case, as we have $\hat{\lambda}_N \geq 0$ and $||\hat{\boldsymbol{\lambda}}_T|| \geq \mu\hat{\lambda}_N$, but we do not enter into the frame where $\hat{\boldsymbol{\lambda}}_T = \mathbf{0}$ assumed here. ■

In the algorithmic slip case at the given integration point and with $\hat{\boldsymbol{\lambda}}_T = \mathbf{0}$, it has to be admitted that the expressions for the vectors $\{\boldsymbol{\Phi}_{\text{contact}}^{e(1)}\}$, $\{\boldsymbol{\Phi}_{\text{contact}}^{e(2)}\}$ and $\{\mathbf{R}_{\Lambda}^{e(1)}\}$ can be deduced from [8.28] and [8.29] by formally setting:

$$\frac{\mu\hat{\lambda}_N}{||\hat{\boldsymbol{\lambda}}_T||} = 1 \qquad [8.31]$$

We thus obtain:

$$\left\{\boldsymbol{\Phi}_{\text{contact}}^{e(1)}(\mathbf{U}^e, \boldsymbol{\Lambda}^{e(1)})\right\} = \{\mathbf{0}\}$$
$$\left\{\boldsymbol{\Phi}_{\text{contact}}^{e(2)}(\mathbf{U}^e, \boldsymbol{\Lambda}^{e(1)})\right\} = \{\mathbf{0}\}$$
$$\left\{\mathbf{R}_{\Lambda}^{e(1)}\right\} = [\mathcal{N}^{e(1)}(\xi)]^T \left(g\{\mathbf{m}\} + \frac{1}{\epsilon_T}\{\boldsymbol{\lambda}_T\}\right) = [\mathcal{N}^{e(1)}(\xi)]^T (g\{\mathbf{m}\} - \{\mathbf{V}_T\})$$

The last equality has been obtained by noting that, as $\mathbf{0} = \hat{\boldsymbol{\lambda}}_T = \boldsymbol{\lambda}_T + \epsilon_T \mathbf{V}_T$, we have $\dfrac{\boldsymbol{\lambda}_T}{\epsilon_T} = -\mathbf{V}_T$.

It can be seen that the expressions found for $\{\boldsymbol{\Phi}_{\text{contact}}^{e(1)}\}$ and $\{\boldsymbol{\Phi}_{\text{contact}}^{e(2)}\}$ are identical to those below in [8.32] for the algorithmic stick case, written when $\hat{\lambda}_N \to 0$ (so $\hat{\boldsymbol{\lambda}}_T \to \mathbf{0}$). On the other hand, the expression for $\{\mathbf{R}_{\Lambda}^{e(1)}\}$ is identical to that of [8.33] for the same algorithmic stick case. To summarize, it turns out that the equations are the same for the following two contact regimes:

1. algorithmic slip case at the given integration point and $\hat{\boldsymbol{\lambda}}_T = \mathbf{0}$ (the point **O** in Figure 8.5);
2. algorithmic stick case at the given integration point and $\hat{\lambda}_N \to 0$ (the point **A** in the same figure).

Algorithmic stick *case at the given integration point:* $\hat{\lambda}_N \geq 0$ *and* $\|\hat{\boldsymbol{\lambda}}_T\| < \mu\hat{\lambda}_N$

- If, at the given point $\hat{\boldsymbol{\lambda}}_T \neq \mathbf{0}$ and if there is algorithmic stick, then $1 - \langle 1 - \dfrac{\mu\langle\hat{\lambda}_N\rangle}{\|\hat{\boldsymbol{\lambda}}_T\|}\rangle = 1$.

From which:

$$\boxed{\begin{array}{rcl} \left\{\mathbf{\Phi}^{e^{(1)}}_{\text{contact}}(\mathbf{U}^e, \mathbf{\Lambda}^{e^{(1)}})\right\} & = & [\mathcal{N}^{e^{(1)}}(\boldsymbol{\xi})]^T \left(\hat{\lambda}_N\{\boldsymbol{\nu}\} - \{\hat{\boldsymbol{\lambda}}_T\}\right) \\ \left\{\mathbf{\Phi}^{e^{(2)}}_{\text{contact}}(\mathbf{U}^e, \mathbf{\Lambda}^{e^{(1)}})\right\} & = & -[\mathcal{N}^{e^{(2)}}(\boldsymbol{\eta})]^T \left(\hat{\lambda}_N\{\boldsymbol{\nu}\} - \{\hat{\boldsymbol{\lambda}}_T\}\right) \end{array}} \quad \text{independent of } \mu! \qquad [8.32]$$

$$\boxed{\left\{\mathbf{R}^{e^{(1)}}_{\Lambda}\right\} = [\mathcal{N}^{e^{(1)}}(\boldsymbol{\xi})]^T \left(g\{\mathbf{m}\} - \{\mathbf{V}_T\}\right)} \quad \text{independent of } \mu! \qquad [8.33]$$

- In the algorithmic stick case, it is possible to have $\hat{\boldsymbol{\lambda}}_T = \mathbf{0}$ (we can check that $\|\hat{\boldsymbol{\lambda}}_T = \mathbf{0}\| < \mu\hat{\lambda}_N$ if $\hat{\lambda}_N$ is > 0); however, the case $\hat{\boldsymbol{\lambda}}_T = \mathbf{0}$ at the given point does not pose any difficulty. To obtain the corresponding expressions for $\{\mathbf{\Phi}^{e^{(1)}}_{\text{contact}}\}$, $\{\mathbf{\Phi}^{e^{(2)}}_{\text{contact}}\}$ and $\{\mathbf{R}^{e^{(1)}}_{\Lambda}\}$, it suffices to reuse the previous relations by setting $\hat{\boldsymbol{\lambda}}_T = \mathbf{0}$.

8.8. System of matrix equations for the contact problem

8.8.1. *Global nodal vectors*

To proceed with assembly, we will need to explain how the element column-vectors [8.5] are extracted from other larger column-vectors, which are said to be global vectors, in the sense that they refer to the entire structure or to the entire contactor surface. These vectors are defined as follows.

NOTATION AND DEFINITIONS.– [8.34]

1. NODAL DISPLACEMENTS:

   – $\{\mathbf{U}^{(1)}\}$: the column-vector containing the three components of the displacement vector $\mathbf{U}^{(1)}$ for each node of body 1.

     It is a vector with $n_U^{(1)}$ components, where $\boxed{n_U^{(1)}}$ is equal to three times the total number of nodes of body 1;

   – $\{\mathbf{U}^{(2)}\}$: the column-vector containing the three components of the displacement vector $\mathbf{U}^{(2)}$ for each node of body 2.

     It is a vector with $n_U^{(2)}$ components, where $\boxed{n_U^{(2)}}$ is equal to three times the total number of nodes of body 2;

   – $\{\mathbf{U}\} \equiv \begin{Bmatrix} \{\mathbf{U}^{(1)}\} \\ \{\mathbf{U}^{(2)}\} \end{Bmatrix}$: the column-vector combining the nodal displacements of bodies 1 and 2.

{**U**} is called *the nodal displacement vector*. It is a column-vector with n_U components, where $\boxed{n_U = n_U^{(1)} + n_U^{(2)}}$ is equal to three times the total number of nodes of bodies 1 and 2.

2. NODAL VIRTUAL VELOCITIES:

– The column-vectors relating to the nodal virtual velocities, denoted $\{\mathbf{U}^{(1)*}\}$, $\{\mathbf{U}^{(2)*}\}$ and $\{\mathbf{U}^*\} \equiv \left\{ \begin{matrix} \{\mathbf{U}^{(1)*}\} \\ \{\mathbf{U}^{(2)*}\} \end{matrix} \right\}$, are defined in an analogous manner.

$\{\mathbf{U}^*\}$ is called *the nodal virtual velocity vector*. It has the same number of components n_U as {**U**}.

3. NODAL MULTIPLIERS:

– $\{\mathbf{\Lambda}\}$: the column-vector combining the three components of the multiplier vector λ at the nodes of $S_{0c}^{(1)}$.

$\{\mathbf{\Lambda}\}$ is called *the nodal multiplier vector*. It is an unknown column-vector in the problem, in the same way that the column-vector of the nodal displacements {**U**} is. If we denote the number of nodes on the contactor surface $S_{0c}^{(1)}$ by $\boxed{nnsc1}$, then $\{\mathbf{\Lambda}\}$ has size $\boxed{n_\Lambda = 3\ nnsc1}$.

– An analogous definition exists for *the vector of nodal virtual multipliers*, which we denote by $\{\mathbf{\Lambda}^*\}$.

According to the convention outlined in section 8.3, we recall that each component in question is defined relative to the fixed orthonormal basis $(\mathbf{e}_1, \mathbf{e}_2, \mathbf{e}_3)$.

REMARK.– It can be seen that each notation $\{\mathbf{U}^{(1)}\}$, $\{\mathbf{U}^{(2)}\}$, $\{\mathbf{U}^{(1)*}\}$ and $\{\mathbf{U}^{(2)*}\}$ has two different meanings. For example, for $\{\mathbf{U}^{(1)}\}$:

1. in [8.4], $\{\mathbf{U}^{(1)}\}$ denotes the three-component column-vector combining the three components of the displacement vector $\mathbf{U}^{(1)}$;

2. here, $\{\mathbf{U}^{(1)}\}$ is the $n_U^{(1)}$-component column-vector combining the three components of the displacement vector $\mathbf{U}^{(1)}$ for every node of body 1.

To distinguish between the two quantities properly, strictly speaking we should:

1. denote the displacement vector with an arrow above it, that is, as $\vec{U}^{(1)}$, and also denote the column-vector that represents the displacement vector with an arrow too, that is, as $\{\vec{U}^{(1)}\}$;

2. or, alternatively, denote the column-vector that represents the nodal displacements using the blackboard bold typeface, that is, as $\{\mathbb{U}^{(1)}\}$.

However, we will not make this distinction, since in practice the context is clear enough to know whether it is a column-vector with three or with $n_U^{(1)}$ components, so that there is no source of possible confusion. Additionally, in the finite element method, the three-component column-vector representing the displacement vector only appears at the beginning, and it is the nodal displacement column-vectors which play a recurring role thereafter. ■

DEFINITION.– The element nodal displacement vectors, the element nodal multiplier vectors, as well as their virtual counterparts are extracted from the global nodal vectors via the following relations:

$$\begin{array}{lll|lll} \{\mathbf{U}^{(1),e^{(1)}}\} &=& [\mathbb{L}^{e^{(1)}}]\{\mathbf{U}^{(1)}\} & \{\mathbf{U}^{(1)*,e^{(1)}}\} &=& [\mathbb{L}^{e^{(1)}}]\{\mathbf{U}^{(1)*}\} \\ \{\mathbf{U}^{(2),e^{(2)}}\} &=& [\mathbb{L}^{e^{(2)}}]\{\mathbf{U}^{(2)}\} & \{\mathbf{U}^{(2)*,e^{(2)}}\} &=& [\mathbb{L}^{e^{(2)}}]\{\mathbf{U}^{(2)*}\} \\ \{\mathbf{\Lambda}^{e^{(1)}}\} &=& [\mathbb{L}^{e^{(1)}}]\{\mathbf{\Lambda}\} & \{\mathbf{\Lambda}^{*,e^{(1)}}\} &=& [\mathbb{L}^{e^{(1)}}]\{\mathbf{\Lambda}^*\} \end{array} \quad [8.35]$$

where the symbol $\mathbb{L}$ designates the classical finite element localization matrices (these are binary matrices whose components are either 0 or 1).

8.8.2. *Discretization of the classical terms*

In the weak form [7.21], the terms relating to the virtual powers of the internal forces, the external forces *other than the contact forces*, and the acceleration quantities are discretized in a classical way. We obtain:

1. The virtual power of the internal forces:

$$\mathcal{P}^*_{int} = -<\mathbf{U}^*>\{\mathbf{\Psi}(\mathbf{U})\} \quad [8.36]$$

   where $\{\mathbf{\Psi}(\mathbf{U})\}$ is the internal force vector, and where we have abbreviated the nodal displacement vector as $\mathbf{U}$ instead of $\{\mathbf{U}\}$.

2. The virtual power of the external forces, *not including any contact force*:

$$\mathcal{P}^*_{ext} =<\mathbf{U}^*>\{\mathbf{\Phi}\} \quad [8.37]$$

   where $\{\mathbf{\Phi}\}$ is called the external force vector *not including contact forces*, which is due to the various external forces that are exerted on the bodies *not including any contact force*. These loads are of two types:

   (a) The *dead* loads, which are defined to be the forces that are independent of the displacement (but which can explicitly depend on time). Self-weight is an example of such a dead load.

   (b) The *follower* loads, which are defined to be the forces that depend on the displacement. In practice, examples of such forces are *pressures*, that is, forces applied to the boundary surfaces of the bodies and which remain normal to the deformed surfaces. These pressures fall into different categories themselves: prescribed pressures, ideal gas pressures (encountered in the problems involving hermetic membrane structures and containing ideal gases) and hydrostatic pressures.

   To simplify the presentation, we will assume that none of the displacements are prescribed, so that the external forces do not include any unknown constraint forces (or reaction forces), which would otherwise render some components of $\{\mathbf{\Phi}\}$ incalculable. Now, let us examine the arguments of $\{\mathbf{\Phi}\}$:

   1. In the cases with dead loads only, the external force vector $\{\mathbf{\Phi}\}$ may depend on t but not on $\{\mathbf{U}\}$. In statics, t is a kinematic parameter used to describe the evolution of the given dead loads during each computation step.

2. In the cases with follower loads, the external force vector $\{\mathbf{\Phi}\}$ may also depend on $\{\mathbf{U}\}$, just as the internal force vector $\{\mathbf{\Psi}\}$ does.

To accompany either case, we will write $\{\mathbf{\Phi}(\mathbf{U},t)\}$ in general.

3. The virtual power of the acceleration quantities that appear in the RHS of [7.21]:

$$\mathcal{P}^*_{\text{acceleration}} = <\mathbf{U}^*> [\mathbf{M}]\{\ddot{\mathbf{U}}\} \tag{8.38}$$

where $[\mathbf{M}]$ is the mass matrix of the two bodies and where $\{\ddot{\mathbf{U}}\}$ is the nodal acceleration vector.

8.8.3. *Assembly of element virtual contact powers*

By inserting the element virtual contact power [8.20] into [8.2], we obtain:

$$\begin{aligned}
\mathcal{P}^*_{\text{contact}} &= \sum_{e^{(1)}\subset S^{(1)}_{0c}} \left(\mathcal{P}^*_{\text{contact}}\right)^{e^{(1)}} \\
&= \sum_{e^{(1)}\subset S^{(1)}_{0c}} \int_{e^{(1)}} \left(\left\{ \begin{matrix} \{\mathbf{U}^{(1)*,e^{(1)}}\} \\ \{\mathbf{U}^{(2)*,e^{(2)}}\} \end{matrix} \right\}^T \left\{ \begin{matrix} \{\mathbf{\Phi}^{e^{(1)}}_{\text{contact}}\} \\ \{\mathbf{\Phi}^{e^{(2)}}_{\text{contact}}\} \end{matrix} \right\} - <\mathbf{\Lambda}^{*,e^{(1)}}> \{\mathbf{R}^{e^{(1)}}_{\Lambda}(\mathbf{U}^e,\mathbf{\Lambda}^{e^{(1)}})\} \right) dS_0
\end{aligned}$$

The assembly of the column-vectors is carried out symbolically thanks to the localization matrices defined in [8.35]:

$$\begin{aligned}
\mathcal{P}^*_{\text{contact}} = & \sum_{e^{(1)}\subset S^{(1)}_{0c}} \int_{e^{(1)}} <\mathbf{U}^{(1)*,e^{(1)}}> \{\mathbf{\Phi}^{e^{(1)}}_{\text{contact}}\} dS_0 + \sum_{e^{(1)}\subset S^{(1)}_{0c}} \int_{e^{(1)}} <\mathbf{U}^{(2)*,e^{(2)}}> \{\mathbf{\Phi}^{e^{(2)}}_{\text{contact}}\} dS_0 \\
& - \sum_{e^{(1)}\subset S^{(1)}_{0c}} \int_{e^{(1)}} <\mathbf{\Lambda}^{*,e^{(1)}}> \{\mathbf{R}^{e^{(1)}}_{\Lambda}(\mathbf{U}^e,\mathbf{\Lambda}^{e^{(1)}})\} dS_0 \\
= & <\{\mathbf{U}^{(1)*}\}> \sum_{e^{(1)}\subset S^{(1)}_{0c}} \int_{e^{(1)}} [\mathbb{L}^{e^{(1)}}]^T \{\mathbf{\Phi}^{e^{(1)}}_{\text{contact}}\} dS_0 \\
& + <\{\mathbf{U}^{(2)*}\}> \sum_{e^{(1)}\subset S^{(1)}_{0c}} \int_{e^{(1)}} [\mathbb{L}^{e^{(2)}}]^T \{\mathbf{\Phi}^{e^{(2)}}_{\text{contact}}\} dS_0 \\
& - <\mathbf{\Lambda}^*> \sum_{e^{(1)}\subset S^{(1)}_{0c}} \int_{e^{(1)}} [\mathbb{L}^{e^{(1)}}]^T \{\mathbf{R}^{e^{(1)}}_{\Lambda}(\mathbf{U}^e,\mathbf{\Lambda}^{e^{(1)}})\} dS_0
\end{aligned}$$

Thus, we arrive at the discretized form of the virtual contact power [8.1]:

$$\boxed{\mathcal{P}^*_{\text{contact}} = \left\{ \begin{matrix} \{\mathbf{U}^{(1)*}\} \\ \{\mathbf{U}^{(2)*}\} \end{matrix} \right\}^T \{\mathbf{\Phi}_{\text{contact}}(\mathbf{U},\mathbf{\Lambda})\} - <\mathbf{\Lambda}^*> \{\mathbf{R}_{\Lambda}(\mathbf{U},\mathbf{\Lambda})\}} \tag{8.39}$$

where we have adopted the following definitions:

DEFINITIONS.–

1. *The contact force vector*, denoted by $\{\boldsymbol{\Phi}_{\text{contact}}(\mathbf{U},\boldsymbol{\Lambda})\}$ (with units $N/m^2 \times m^2 = N$), which is defined using the element contact force vectors [8.21] to be:

$$\{\boldsymbol{\Phi}_{\text{contact}}(\mathbf{U},\boldsymbol{\Lambda})\} = \left\{ \begin{array}{l} \boldsymbol{\Phi}^{(1)}_{\text{contact}} \equiv \displaystyle\sum_{e^{(1)} \subset S^{(1)}_{0c}} \int_{e^{(1)}} [\mathbb{L}^{e^{(1)}}]^T \left\{ \boldsymbol{\Phi}^{e^{(1)}}_{\text{contact}}(\mathbf{U}^e, \boldsymbol{\Lambda}^{e^{(1)}}) \right\} dS_0 \\ \boldsymbol{\Phi}^{(2)}_{\text{contact}} \equiv \displaystyle\sum_{e^{(1)} \subset S^{(1)}_{0c}} \int_{e^{(1)}} [\mathbb{L}^{e^{(2)}}]^T \left\{ \boldsymbol{\Phi}^{e^{(2)}}_{\text{contact}}(\mathbf{U}^e, \boldsymbol{\Lambda}^{e^{(1)}}) \right\} dS_0 \end{array} \right\} \begin{array}{l} n^{(1)}_U \\ \\ n^{(2)}_U \end{array} \qquad [8.40]$$

$\{\boldsymbol{\Phi}^{(1)}_{\text{contact}}\}$, $\{\boldsymbol{\Phi}^{(2)}_{\text{contact}}\}$ and $\{\boldsymbol{\Phi}_{\text{contact}}\}$, are column-vectors with $n^{(1)}_U$, $n^{(2)}_U$, and n_U components, respectively. The number to the right of each vector in equation [8.40] is the number of components in each vector, respectively.

2. *The residual contact law vector*, denoted by $\{\mathbf{R}_\Lambda(\mathbf{U},\boldsymbol{\Lambda})\}$ (with units $m \times m^2 = m^3$!), is defined using the element residual contact law vector from [8.22] to be:

$$\{\mathbf{R}_\Lambda(\mathbf{U},\boldsymbol{\Lambda})\} = \sum_{e^{(1)} \subset S^{(1)}_{0c}} \int_{e^{(1)}} [\mathbb{L}^{e^{(1)}}]^T \{\mathbf{R}^{e^{(1)}}_\Lambda(\mathbf{U}^e, \boldsymbol{\Lambda}^{e^{(1)}})\} dS_0 \qquad [8.41]$$

It is a column-vector with n_Λ components.

The vectors $\{\boldsymbol{\Phi}_{\text{contact}}(\mathbf{U},\boldsymbol{\Lambda})\}$ and $\{\mathbf{R}_\Lambda(\mathbf{U},\boldsymbol{\Lambda})\}$ depend on the column-vectors $\{\mathbf{U}\}$ and $\{\boldsymbol{\Lambda}\}$. Let us recall that the nodal multiplier vector $\{\boldsymbol{\Lambda}\}$ is an unknown in the problem, in a similar way that the nodal displacement vector $\{\mathbf{U}\}$ is.

1. In definitions $[8.40]_1$ and [8.41], we can bring the matrix $[\mathbb{L}^{e^{(1)}}]^T$ outside of the integral over $e^{(1)}$, so that then $\boldsymbol{\Phi}^{(1)}_{\text{contact}}$ and $\mathbf{R}_\Lambda(\mathbf{U},\boldsymbol{\Lambda})$ become:

$$\begin{aligned} \boldsymbol{\Phi}^{(1)}_{\text{contact}} &= \sum_{e^{(1)} \subset S^{(1)}_{0c}} [\mathbb{L}^{e^{(1)}}]^T \int_{e^{(1)}} \left\{ \boldsymbol{\Phi}^{e^{(1)}}_{\text{contact}}(\mathbf{U}^e, \boldsymbol{\Lambda}^{e^{(1)}}) \right\} dS_0 \\ \mathbf{R}_\Lambda(\mathbf{U},\boldsymbol{\Lambda}) &= \sum_{e^{(1)} \subset S^{(1)}_{0c}} [\mathbb{L}^{e^{(1)}}]^T \int_{e^{(1)}} \{\mathbf{R}^{e^{(1)}}_\Lambda(\mathbf{U}^e, \boldsymbol{\Lambda}^{e^{(1)}})\} dS_0 \end{aligned} \qquad [8.42]$$

Multiplying the integral in the RHS from the left with the matrix $[\mathbb{L}^{e^{(1)}}]^T$ has the effect of extending this integral to a vector of the same size as the global nodal vector, which is in the LHS. Thus, the column-vectors $\boldsymbol{\Phi}^{(1)}_{\text{contact}}$ and $\mathbf{R}_\Lambda(\mathbf{U},\boldsymbol{\Lambda})$ are *the assemblies* of the element vectors $\int_{e^{(1)}} \left\{ \boldsymbol{\Phi}^{e^{(1)}}_{\text{contact}}(\mathbf{U}^e, \boldsymbol{\Lambda}^{e^{(1)}}) \right\} dS_0$ and $\int_{e^{(1)}} \{\mathbf{R}^{e^{(1)}}_\Lambda(\mathbf{U}^e, \boldsymbol{\Lambda}^{e^{(1)}})\} dS_0$, respectively. That is to say, assemblies in the usual finite element sense.

2. The situation becomes a bit more complicated for $\boldsymbol{\Phi}^{(2)}_{\text{contact}}$, which is defined by $[8.40]_2$. As seen in the dependency chain from Table 8.1, the element $e^{(2)}$ varies depending on the point $\boldsymbol{\xi}$ and, consequently, we cannot bring the matrix $[\mathbb{L}^{e^{(2)}}]^T$ outside of the integral

over $e^{(1)}$. The column-vector $\mathbf{\Phi}^{(2)}_{\text{contact}}$ is not therefore the assembly of the element vector $\int_{e^{(1)}} \left\{\mathbf{\Phi}^{e^{(2)}}_{\text{contact}}(\mathbf{U}^e, \mathbf{\Lambda}^{e^{(1)}})\right\} dS_0$.

This subtlety was mentioned by Laursen [77], who uses a different contact formulation than ours. It is of a geometrical origin and we must have it in mind when dealing with any contact problem, whenever the meshes of the contactor and target surfaces are not concordant.

To see the differences between $\mathbf{\Phi}^{(1)}_{\text{contact}}$ and $\mathbf{\Phi}^{(2)}_{\text{contact}}$ more clearly, we will discuss their numerical calculation. By using a numerical integration method (for example, the Gaussian integration scheme), we can decompose the integral over an element $e^{(1)} \subset S^{(1)}_{0c}$ into a linear combination of the integrand evaluated at various integration points $\boldsymbol{\xi}$ of $e^{(1)}$:

$$\int_{e^{(1)}} [\cdots] dS_0 = \sum_{\text{integ pt } \boldsymbol{\xi}} [\cdots] j(\boldsymbol{\xi}) \qquad [8.43]$$

where $j(\boldsymbol{\xi})$ is the product of the area Jacobian at the point $\boldsymbol{\xi}$ with the weight factor $w(\boldsymbol{\xi})$ corresponding to the same point:

$$\boxed{j(\boldsymbol{\xi}) = \|\mathbf{X}_{,\xi^1}(\boldsymbol{\xi}) \times \mathbf{X}_{,\xi^2}(\boldsymbol{\xi})\| \, w(\boldsymbol{\xi})} \qquad [8.44]$$

The column-vector $\mathbf{\Phi}^{(1)}_{\text{contact}}$ is then obtained by the following standard assembly scheme:

LOOP OVER THE ELEMENTS $e^{(1)}$ OF THE SURFACE $S^{(1)}_{0c}$

LOOP OVER THE INTEGRATION POINTS $\boldsymbol{\xi}$ OF EACH ELEMENT $e^{(1)}$

Calculate the vector $\left\{\mathbf{\Phi}^{e^{(1)}}_{\text{contact}}(\mathbf{U}^e, \mathbf{\Lambda}^{e^{(1)}})\right\} j(\boldsymbol{\xi})$ at the point $\boldsymbol{\xi}$.

Combine the result into a vector.

END OF LOOP

This loop produces the integral $\int_{e^{(1)}} \left\{\mathbf{\Phi}^{e^{(1)}}_{\text{contact}}(\mathbf{U}^e, \mathbf{\Lambda}^{e^{(1)}})\right\} dS_0$.

Assemble the vector obtained in $\mathbf{\Phi}^{(1)}_{\text{contact}}$ at the location corresponding to the element $e^{(1)}$. This assembly is carried out symbolically by the addition of the term:

$$[\mathbb{L}^{e(1)}]^T \int_{e^{(1)}} \left\{\mathbf{\Phi}^{e^{(1)}}_{\text{contact}}(\mathbf{U}^e, \mathbf{\Lambda}^{e^{(1)}})\right\} dS_0$$

END OF LOOP

while the column-vector $\mathbf{\Phi}^{(2)}_{\text{contact}}$ is obtained by the following *non-classical assembly scheme*:

LOOP OVER THE ELEMENTS $e^{(1)}$ OF THE SURFACE $S^{(1)}_{0c}$

LOOP OVER THE INTEGRATION POINTS $\boldsymbol{\xi}$ OF EACH ELEMENT $e^{(1)}$

Calculate the vector $\left\{\mathbf{\Phi}^{e^{(2)}}_{\text{contact}}(\mathbf{U}^e, \mathbf{\Lambda}^{e^{(1)}})\right\} j(\boldsymbol{\xi})$ at the point $\boldsymbol{\xi}$.

Assemble this vector into $\mathbf{\Phi}^{(2)}_{\text{contact}}$ at the location corresponding to the element $e^{(1)}$. This assembly is carried out symbolically by the addition of the term:

$$[\mathbb{L}^{e(2)}]^T \left\{\mathbf{\Phi}^{e^{(2)}}_{\text{contact}}(\mathbf{U}^e, \mathbf{\Lambda}^{e^{(1)}})\right\} j(\boldsymbol{\xi})$$

END OF LOOP

END OF LOOP

For $\mathbf{\Phi}^{(1)}_{\text{contact}}$, we finish calculating the integral over $e^{(1)}$ before then assembling it. For $\mathbf{\Phi}^{(2)}_{\text{contact}}$, we calculate the integral at each integration point $\boldsymbol{\xi}$ before then assembling it progressively.

In order to standardize both the presentation and the programming, in the following, we decide to treat $\mathbf{\Phi}^{(1)}_{\text{contact}}$ and $\mathbf{R}_{\Lambda}(\mathbf{U}, \mathbf{\Lambda})$ similarly to $\mathbf{\Phi}^{(2)}_{\text{contact}}$, that is, we will work with the relations $[8.40]_1$ and [8.41] of $\mathbf{\Phi}^{(1)}_{\text{contact}}$ and $\mathbf{R}_{\Lambda}(\mathbf{U}, \mathbf{\Lambda})$, rather than with [8.42].

8.8.4. *System of matrix equations*

In using the discrete relations [8.36], [8.37], [8.38] and [8.39], the discretized expression of the weak form [7.21] can be written as:

$$< \mathbf{U}^* > \Big([\mathbf{M}]\{\ddot{\mathbf{U}}\} + \{\mathbf{\Psi}(\mathbf{U})\} - \{\mathbf{\Phi}((\mathbf{U}))\} - \{\mathbf{\Phi}_{\text{contact}}(\mathbf{U},\mathbf{\Lambda})\}\Big) + < \mathbf{\Lambda}^* > \{\mathbf{R}_\Lambda(\mathbf{U},\mathbf{\Lambda})\} = 0 \quad [8.45]$$

Theorem.–
The weak form [7.21] is valid $\forall \mathbf{U}^{(1)*}, \mathbf{U}^{(2)*}, \boldsymbol{\lambda}^*$ $\Rightarrow$ the discretized form [8.45] is valid $\forall\{\mathbf{U}^*\},\{\mathbf{\Lambda}^*\}$. [8.46]

Proof. The discretized form [8.45] is a transformation of the weak form [7.21] obtained by replacing the virtual fields $\mathbf{U}^{(1)*}$, $\mathbf{U}^{(2)*}$ and $\boldsymbol{\lambda}^*$ with the *particular* fields that are defined piecewise, namely, [8.9], [8.13] and [8.18]: $\{\mathbf{U}^{(1)*}\} = [\mathcal{N}^{e^{(1)}}(\xi)]\{\mathbf{U}^{(1)*,e^{(1)}}\}$, $\{\mathbf{U}^{(2)*}\} = [\mathcal{N}^{e^{(2)}}(\eta)]\{\mathbf{U}^{(2)*,e^{(2)}}\}$ and $\{\boldsymbol{\lambda}^*\} = [\mathcal{N}^{e^{(1)}}(\xi)]\{\mathbf{\Lambda}^{*,e^{(1)}}\}$.

As the weak form is valid $\forall \mathbf{U}^{(1)*}$, $\forall \mathbf{U}^{(2)*}$ and $\forall \boldsymbol{\lambda}^*$, it remains so when restricted to the *particular* fields mentioned. If we focus on just displacement fields, then it is a classical result in finite elements. ■

Since [8.45] is valid when $\{\mathbf{U}^*\}$ and $\{\mathbf{\Lambda}^*\}$ are arbitrary, we arrive at the following system of equations, valid at any instant within the time interval $[0, t_{max}]$ considered:

$$\forall t \in [0, t_{max}], \quad \boxed{\begin{array}{lll} [\mathbf{M}]\{\ddot{\mathbf{U}}\} + \underset{(n_U,1)}{\{\mathbf{\Psi}(\mathbf{U})\}} & = & \{\mathbf{\Phi}(\mathbf{U},t)\} + \{\mathbf{\Phi}_{\text{contact}}(\mathbf{U},\mathbf{\Lambda})\} \\ \underset{(n_\Lambda,1)}{\{\mathbf{R}_\Lambda(\mathbf{U},\mathbf{\Lambda})\}} & = & \{\mathbf{0}\} \end{array}} \quad [8.47]$$

The unknowns are the nodal displacement vector $\{\mathbf{U}\} = \left\{ \begin{array}{c} \{\mathbf{U}^{(1)}\} \\ \{\mathbf{U}^{(2)}\} \end{array} \right\}$ of bodies 1 and 2, as well as the nodal multiplier vector $\{\mathbf{\Lambda}\}$ on $S_{0c}^{(1)}$.

The internal force vector $\{\mathbf{\Psi}(\mathbf{U})\}$ and the external force vector (*excluding any contact force*) $\{\mathbf{\Phi}(\mathbf{U},t)\}$ are classical. The vectors $\{\mathbf{\Phi}_{\text{contact}}(\mathbf{U},\mathbf{\Lambda})\}$ and $\{\mathbf{R}_\Lambda(\mathbf{U},\mathbf{\Lambda})\}$ are new and are specific to contact problems.

Definitions.– [8.48]

1. The time-dependent differential equation $[8.47]_1$ is called *the (semi-discrete) equations of motion of the bodies.*
2. The algebraic relation $[8.47]_2$ is called the *discrete contact laws.*
3. The system [8.47] is called *the system of equations for the dynamic contact problem.*

The system [8.47] is made up of (strongly) nonlinear and coupled equations. It will be solved in Chapter 9 in the static case, and then in Chapter 11 in the dynamic case.

In the weak form [7.21], the unknowns are λ_N and λ_T, but not the vector $\boldsymbol{\lambda}$. In the relation [8.47], the unknown is $\{\mathbf{\Lambda}\}$ which contains the components of the vector $\boldsymbol{\lambda}$ at each node.

We should not forget that relation $[8.47]_2$ – where $\{\mathbf{R}_\Lambda\}$ is the assembly of $\{\mathbf{R}_\Lambda^{e^{(1)}}\}$ from [8.22] – is just the discrete version of the contact laws [7.3] and [7.9], that is to say, $\lambda_N = \langle\hat{\lambda}_N\rangle$ and $\lambda_T = \left(1 - \langle 1 - \frac{\mu\langle\hat{\lambda}_N\rangle}{\|\hat{\lambda}_T\|}\rangle\right)\hat{\lambda}_T$ on $S_{0c}^{(1)}$. This justifies the name given to $\{\mathbf{R}_\Lambda\}$.

• Note that the converse to [8.46] does *not* hold, that is, that the discretized form of [8.45] is valid $\forall\{\mathbf{U}^*\}, \{\mathbf{\Lambda}^*\} \quad \not\Rightarrow \quad$ the weak form [7.21] is valid $\forall\mathbf{U}^{(1)*}$, $\mathbf{U}^{(2)*}$ and λ^*.

Indeed, the particular fields [8.9], [8.13] and [8.18], mentioned in the proof of [8.46], make up only a strict subset of the set of all possible fields: there exist virtual fields $\mathbf{U}^{(1)*}$, $\mathbf{U}^{(2)*}$ and λ^*, which are not of the form [8.9], [8.13] and [8.18].

The discretized form [8.45] is a consequence of the weak form [7.21]; it is not equivalent to it. Thus, in general, the solution [8.47] of the system is only an approximation of the exact solution arising from the continuous weak form. This is a well-known property of the finite element method. To improve the approximate solution, it is necessary to enrich the set of functions [8.9], [8.13] and [8.18] by refining the mesh and/or by increasing the degree of the interpolation functions. We will return to this subject in more detail in section 8.9.

• The column-vectors $\{\mathbf{\Psi}(\mathbf{U})\}$, $\{\mathbf{\Phi}(\mathbf{U},t)\}$, $\{\mathbf{\Phi}_{\text{contact}}(\mathbf{U},\mathbf{\Lambda})\}$ and $\{\mathbf{R}_\Lambda(\mathbf{U},\mathbf{\Lambda})\}$ in [8.47] are functions of $\{\mathbf{U}\}$ and $\{\mathbf{\Lambda}\}$. Although not essential, we shall specify the dependencies of the contact force vector $\{\mathbf{\Phi}_{\text{contact}}\}$ and of the residual vector $\{\mathbf{R}_\Lambda\}$.

According to [8.24] and [8.40], the vectors $\{\mathbf{\Phi}^{(1)}_{\text{contact}}\}$ and $\{\mathbf{\Phi}^{(2)}_{\text{contact}}\}$ in fact depend only on the *restrictions* $\mathbf{U}^{(1)}{}_{|S^{(1)}_{0c}}$ and $\mathbf{U}^{(2)}{}_{|S^{(2)}_{0c}}$ of $\mathbf{U}^{(1)}$ and $\mathbf{U}^{(2)}$ to the contact surfaces $S^{(1)}_{0c}$ and $S^{(2)}_{0c}$, respectively. We have therefore:

$$\{\mathbf{\Phi}_{\text{contact}}(\mathbf{U},\mathbf{\Lambda})\} = \left\{\begin{array}{l} \mathbf{\Phi}^{(1)}_{\text{contact}}(\mathbf{U}^{(1)}{}_{|S^{(1)}_{0c}}, \mathbf{U}^{(2)}{}_{|S^{(2)}_{0c}}, \mathbf{\Lambda}) \\ \\ \mathbf{\Phi}^{(2)}_{\text{contact}}(\mathbf{U}^{(1)}{}_{|S^{(1)}_{0c}}, \mathbf{U}^{(2)}{}_{|S^{(2)}_{0c}}, \mathbf{\Lambda}) \end{array}\right\} \begin{array}{l} n_U^{(1)} \\ \\ n_U^{(2)} \end{array}$$

In the same way, the residual vector $\{\mathbf{R}_\Lambda\}$ depends only on the *restrictions* $\mathbf{U}^{(1)}{}_{|S^{(1)}_{0c}}$ and $\mathbf{U}^{(2)}{}_{|S^{(2)}_{0c}}$:

$$\{\mathbf{R}_\Lambda(\mathbf{U},\mathbf{\Lambda})\} = \{\mathbf{R}_\Lambda(\mathbf{U}^{(1)}{}_{|S^{(1)}_{0c}}, \mathbf{U}^{(2)}{}_{|S^{(2)}_{0c}}, \mathbf{\Lambda})\}$$

Therefore, we can specify the dependence of each variable in the system of equations [8.47] by splitting the internal force vector $\{\mathbf{\Psi}\}$ and the external force (excluding any contact forces) vector $\{\mathbf{\Phi}\}$ into two parts indexed by the exponents (1) and (2), as was done for $\{\mathbf{\Phi}_{\text{contact}}\}$ as follows:

$$\boxed{\forall t, \quad \begin{array}{lcl} [\mathbf{M}^{(1)}]\{\ddot{\mathbf{U}}^{(1)}\} + \underset{(n_U^{(1)},1)}{\{\mathbf{\Psi}^{(1)}(\mathbf{U}^{(1)})\}} & = & \{\mathbf{\Phi}^{(1)}(\mathbf{U}^{(1)},t)\} + \{\mathbf{\Phi}^{(1)}_{\text{contact}}(\mathbf{U}^{(1)}{}_{|S^{(1)}_{0c}}, \mathbf{U}^{(2)}{}_{|S^{(2)}_{0c}}, \mathbf{\Lambda})\} \\ [\mathbf{M}^{(2)}]\{\ddot{\mathbf{U}}^{(2)}\} + \underset{(n_U^{(2)},1)}{\{\mathbf{\Psi}^{(2)}(\mathbf{U}^{(2)})\}} & = & \{\mathbf{\Phi}^{(2)}(\mathbf{U}^{(2)},t)\} + \{\mathbf{\Phi}^{(2)}_{\text{contact}}(\mathbf{U}^{(1)}{}_{|S^{(1)}_{0c}}, \mathbf{U}^{(2)}{}_{|S^{(2)}_{0c}}, \mathbf{\Lambda})\} \\ \underset{(n_\Lambda,1)}{\{\mathbf{R}_\Lambda(\mathbf{U}^{(1)}{}_{|S^{(1)}_{0c}}, \mathbf{U}^{(2)}{}_{|S^{(2)}_{0c}}, \mathbf{\Lambda})\}} & = & \{\mathbf{0}\} \end{array}}$$

These relations clearly show that the coupling in each equation takes place via the nodal displacements $\mathbf{U}^{(1)}{}_{|S^{(1)}_{0c}}$ and $\mathbf{U}^{(2)}{}_{|S^{(2)}_{0c}}$ of the contact surfaces, $S^{(1)}_{0c}$ and $S^{(2)}_{0c}$, as well as via the multipliers $\mathbf{\Lambda}$ for the nodes of $S^{(1)}_{0c}$.

8.9. Abnormal contact stresses

The contact laws seen in Chapter 5 imply the following properties:

- At the non-contact points, the contact stress vector vanishes: $\lambda_N = 0$ and $\boldsymbol{\lambda}_T = \mathbf{0}$.
- At the contact points, the normal component λ_N of the contact stress vector is either positive or zero. Without friction, we must have that $\boldsymbol{\lambda}_T = \mathbf{0}$, whereas, with friction, we must satisfy that $||\boldsymbol{\lambda}_T|| \leq \mu\lambda_N$.

However, when solving computationally, it should be noted that we can encounter contact stresses which do not satisfy these properties, and thus are said to be abnormal, at certain computation steps and at certain nodes. Namely:

- contact stresses which are *non-zero* outside of the contact area;
- at certain contact points, the normal component λ_N of the contact stress vector is negative. When there is friction, the tangential stress may sometimes be such that $||\boldsymbol{\lambda}_T|| > \mu\lambda_N$.

When we draw the contact stress vectors in post-processing, it is easy to recognize the few vectors whose normal component λ_N is negative, as can be clearly seen in a figure. On the other hand, the tangential stresses which do not satisfy $||\boldsymbol{\lambda}_T|| \leq \mu\lambda_N$ are not often easily detectable to the naked eye, and we have to perform a post-processing calculation to spot them. In the sequel, we will explain the origins of abnormal stresses.

8.9.1. *First cause of abnormal contact stresses*

The first cause of abnormal contact stresses comes from *the approximate nature of the discretized weak form*, which has been noticed already after obtaining the system of equations [8.47]. In classical structural mechanics (contactless mechanics), it is known that the finite element method provides numerical results (displacements, stresses in the bodies) which are not exact, since the discretized principle of virtual powers (PVP) does not exactly satisfy the governing local equations. The weak form [7.21], which is a generalization of the PVP to the contact case, cannot escape from this fact.

Let us first recall the known facts in classical structural mechanics.

1. The motion equation is equivalent to the PVP, insofar as the latter is written for any virtual velocity field.
2. The discretized PVP – which is obtained using the finite element method – is not equivalent to the PVP, since it is written for only a finite number of nodal virtual velocities.
3. The matrix equation of motion, which is equivalent to the discretized PVP, is therefore not equivalent to the PVP.

Thus, in general, the numerical results that are obtained from solving the matrix equation of motion do not satisfy the motion equations exactly. In statics, for example, the calculated stresses do not satisfy the local equilibrium nor the stress boundary conditions exactly in general. However, each finite element is in equilibrium under the system of element nodal forces.

In structural contact mechanics, the same reasoning can be applied to the contact laws.

1. In Chapter 7, we transformed the contact laws [7.3] and [7.9] – that is, $\lambda_N = \langle \hat{\lambda}_N \rangle$ and $\lambda_T = \left(1 - \langle 1 - \frac{\mu \langle \hat{\lambda}_N \rangle}{\|\hat{\lambda}_T\|} \rangle \right) \hat{\lambda}_T$ on the contactor surface $S_{0c}^{(1)}$ – into the equivalent weak form:

$$\forall \lambda^*, \int_{S_{0c}^{(1)}} \left\{ \frac{1}{\epsilon_N} (\lambda_N - \langle \hat{\lambda}_N \rangle) \, \mathbf{m} - \frac{1}{\epsilon_T} \left[\lambda_T - \left(1 - \langle 1 - \frac{\mu \langle \hat{\lambda}_N \rangle}{\|\hat{\lambda}_T\|} \rangle \right) \hat{\lambda}_T \right] \right\} . \lambda^* dS_0 = 0 \qquad [8.49]$$

Equivalence holds here, since the weak form is written for *any arbitrary* virtual multiplier field λ^*.

2. In this chapter, we have discretized the previous relation using the finite element method to yield the discretized weak form (see [8.45]):

$$\forall \{\mathbf{\Lambda}^*\}, \quad < \mathbf{\Lambda}^* > \{\mathbf{R}_\Lambda(\mathbf{U}, \mathbf{\Lambda})\} = 0 \qquad [8.50]$$

The vector $\{\mathbf{R}_\Lambda(\mathbf{U}, \mathbf{\Lambda})\}$, called the residual contact law vector, is defined by [8.41]:

$$\{\mathbf{R}_\Lambda(\mathbf{U}, \mathbf{\Lambda})\} = \sum_{e^{(1)} \subset S_{0c}^{(1)}} \int_{e^{(1)}} [\mathbb{L}^e]^T \{\mathbf{R}_\Lambda^{e^{(1)}}(\mathbf{U}^e, \mathbf{\Lambda}^{e^{(1)}})\} dS_0$$

where $\left\{\mathbf{R}_\Lambda^{e^{(1)}}(\mathbf{U}^e, \mathbf{\Lambda}^{e^{(1)}})\right\}$ is the element residual contact law vector, and was defined in [8.22]:

$$\left\{\mathbf{R}_\Lambda^{e^{(1)}}(\mathbf{U}^e, \mathbf{\Lambda}^{e^{(1)}})\right\} \equiv$$

$$[\mathcal{N}^{e^{(1)}}(\xi)]^T \left(\frac{1}{\epsilon_N} \left(\langle \hat{\lambda}_N \rangle - \lambda_N \right) \{\mathbf{m}\} + \frac{1}{\epsilon_T} \left\{ \lambda_T - \left(1 - \langle 1 - \frac{\mu \langle \hat{\lambda}_N \rangle}{\|\hat{\lambda}_T\|} \rangle \right) \hat{\lambda}_T \right\} \right)$$

The discretized weak form [8.50] is *not* equivalent to the weak continuous form [8.49] exactly, since it is valid for any column-vector $\{\mathbf{\Lambda}^*\}$ that contains *finitely* many nodal virtual multipliers. Instead of working with arbitrary λ^* fields, we restricted ourselves in [8.50] to fields λ^* of the form [8.18] $\{\lambda^*\} = [\mathcal{N}^{e^{(1)}}(\xi)]\{\mathbf{\Lambda}^{*,e^{(1)}}\}$, that is, to interpolations of nodal virtual multipliers.

3. The discretized weak form [8.50] is equivalent to the discretized contact laws $[8.47]_2$:

$$\{\mathbf{R}_\Lambda(\mathbf{U}, \mathbf{\Lambda})\} = \{\mathbf{0}\} \qquad [8.47_2\text{bis}]$$

which is the discrete version of the contact laws [7.3] and [7.9].

The chain of implications can therefore be summarized as follows:

$$\begin{array}{rcl} \text{The contact laws [7.3] and [7.9]} & \Leftrightarrow & \text{the weak form [8.49]} \\ & \left\{ \begin{array}{c} \Rightarrow \\ \not\Leftarrow \end{array} \right\} & \text{the discretized weak form [8.50]} \\ & \Leftrightarrow & \text{the discretized contact laws } [8.47]_2 \end{array}$$

From this analysis, again we see that the discrete contact laws $[8.47]_2$ are *not* equivalent to the contact laws [7.3] and [7.9] exactly, but rather that they are only the discrete and approximate versions of them. Consequently, when solving computationally, it is possible that at some certain computation steps and at certain nodes with abnormal contact, stresses may appear.

To verify the contact laws in a better way, one solution consists of *refining the contactor surface* $S_{0c}^{(1)}$ (which entails a refinement of the contactor body mesh), or to *increase the degree of the interpolation functions* on the finite elements of $S_{0c}^{(1)}$. For example, by increasing the number of nodes for $S_{0c}^{(1)}$, the number of nodal virtual multipliers in $\{\mathbf{\Lambda}^*\}$ will also increase, thus enriching the space of virtual multipliers $\boldsymbol{\lambda}^*$ and allowing for the weak discretized form [8.50], which is valid $\forall\{\mathbf{\Lambda}^*\}$, to converge closer toward the weak form [8.49], which is valid $\forall\boldsymbol{\lambda}^*$.

8.9.2. *Second cause of abnormal contact stresses*

We have just seen that the contact laws are not well-defined if the interpolation of the multiplier field is not sufficiently refined. We are now going to witness a scenario in which no interpolation is able to represent the multipliers correctly, which is the case where there are elements *partially in contact*. A finite element of the contactor surface $S_{0c}^{(1)}$ is partially in contact when some of the element is in contact with the target surface, whereas the remainder of the element is not in contact (we may also say that the element is *in partial contact*).

In general, the portion of the surface $S_{0c}^{(1)}$ which is actually in contact varies throughout the calculation. At a given computation step, there may be elements of $S_{0c}^{(1)}$ that are completely in contact, others that are completely out of contact, and yet there may be others in partial contact.

For an element in partial contact, there is a section on which the contact stress vector $\boldsymbol{\lambda}$ is non-zero, but also another section on which $\boldsymbol{\lambda}$ vanishes. The contact stress vector $\boldsymbol{\lambda}$ is described by *two different expressions* on the element (it does not matter if $\boldsymbol{\lambda}$ is continuous or not). However, as seen in [8.16], the contact stress vector $\boldsymbol{\lambda}$ at a current point of the element is obtained by interpolating the multipliers at each node. Being of polynomial type, such an interpolation cannot represent a multi-valued expression of the function $\boldsymbol{\lambda}$ on the element.

Therefore, when there are elements partially in contact, the contact stresses found on these elements must *not* be correct. These are the stresses obtained by the computational scheme, which tries to represent a multi-valued function using a classical single-valued function to the best of its abilities. For this type of element, we very often encounter *abnormal* contact stresses in the sense defined above, in particular for stresses $\lambda_N < 0$.

In a problem where the effective contact surface evolves over each step, there is no way to avoid elements in partial contact. Fortunately, by refining the mesh, the sizes of these elements decreases, and consequently the negative influence of their abnormal stresses will be diminished.

In classical structural mechanics, we can observe that there are situations analogous to the one encountered here, where the fields have a multi-valued expression when restricted to certain finite elements. In plasticity, for example, the displacement field has two different expressions within a partially yielded element (an element which straddles between the elastic and the plastic regions). In such an element, the polynomial-type interpolation of the displacement field is incapable of approximating the real field precisely. To reduce calculation errors, it is necessary to use sufficiently fine meshes.

8.9.3. *Third cause of abnormal contact stresses*

Lastly, another cause of abnormal contact stresses is due to *the difficulty of accurately calculating the contact terms* using the numerical Gaussian integration formula [8.43]. The contact terms in question are as follows:

– the element contact force vectors $\{\mathbf{\Phi}^{e(1)}_{\text{contact}}\}$ and $\{\mathbf{\Phi}^{e(2)}_{\text{contact}}\}$ in [8.21], the element residual contact law vector $\{\mathbf{R}_{\Lambda}^{e^{(1)}}\}$ in [8.22]. These vectors contain the "positive part" function $\langle\hat{\lambda}_N\rangle$, which is a continuous function but with a discontinuous gradient;

- the element contact tangent matrix $[\mathbf{K}^e_{\text{contact}}]$ that we will encounter in Chapter 9. In addition to the function $\langle\hat{\lambda}_N\rangle$, this matrix contains the Heaviside functions $H(\hat{\lambda}_N)$ and $H(1 - \dfrac{\mu\langle\hat{\lambda}_N\rangle}{||\hat{\lambda}_T||})$ which are discontinuous (see section 9.6.6).

When the points of discontinuity lie inside of the element $e^{(1)} \subset S^{(1)}_{0c}$, the numerical Gaussian integration scheme does not work well with the "positive part" function $\langle\hat{\lambda}_N\rangle$, nor with the Heaviside functions. The fact that the contact terms cannot be precisely integrated gives rise to the *abnormal* contact stresses – notably the stresses $\lambda_N < 0$ – at certain numerical steps and at certain nodes of $S^{(1)}_{0c}$. Computational experience reveals that:

- If there are too few Gaussian points on the elements of the contactor surface $S^{(1)}_{0c}$, then we may obtain stresses $\lambda_N < 0$ which are too large at some of the nodes of $S^{(1)}_{0c}$. For example, 4×4 Gaussian points on the eight-node quadrilaterals (Q8) for the contactor surface is generally insufficient.

- To reduce the stresses $\lambda_N < 0$, we should *take the greatest possible number of Gaussian points* allowed by the computation code. For example, 6×6 or 7×7 Gaussian points on the Q8's.

We have just exhibited two remedies to alleviate the difficulties that arise when $\lambda_N < 0$: (i) either refine the contactor surface $S^{(1)}_{0c}$ (or increase the degree of the interpolation functions on the elements of $S^{(1)}_{0c}$) (ii) or consider using a greater number of Gaussian points for the elements of $S^{(1)}_{0c}$ that is closer to the permitted maximum. Of course, the best remedy is to combine the two individual ones; however, numerical testing has revealed that it is the second remedy which has a greater positive effect:

- With a coarse mesh, we can obtain good results, provided that a large number of Gaussian points on $S^{(1)}_{0c}$ are considered.

- Conversely, if there are not enough Gaussian points on $S^{(1)}_{0c}$, then we obtain large stresses $\lambda_N < 0$ and consequently refining the mesh is not sufficient to remove these abnormal stresses.

In plasticity, the internal force vector and the tangent stiffness matrix contain the "positive part" and the Heaviside function, whose argument is the yield function. In this case too, it does not suffice to use the Gaussian integration scheme to correctly evaluate both the vector and the matrix within the partially yielded elements.

8.10. Projection calculation: contact detection

A basic and recurring task in contact calculus consists of determining the projection point $\mathbf{y}$ on the target surface $S^{(2)}_c$ that comes from a given integration point $\mathbf{x}$ on the contactor surface $S^{(1)}_c$. This purely geometric operation is important since, once the projection point $\mathbf{y}$ is known, we can then calculate the other kinematic quantities.

There are several reasons why the projection calculation may constitute a large amount of the overall computation time. First of all, it must be carried out for each integration point on the contactor surface $S_c^{(1)}$. Then, for large deformations, it must be carried out for each iteration of each calculation step, since portions of the surface in contact can change significantly over time and are not known a priori. Finally, the projection calculation is itself made up of an iterative loop, except if the elements of the target surface $S_c^{(2)}$ are linear.

Different and more-or-less sophisticated techniques have been developed to reduce the contact search time, especially in the cases involving multi-body systems. A review of these methods can be found in [138, 75]. We will present here a calculation strategy which is intuitive, simple to implement, but not optimal. In order to avoid overburdening the notation, the current position of an element $e^{(1)}$ or $e^{(2)}$ belonging to the reference contact surfaces will be denoted by the same symbol $e^{(1)}$ or $e^{(2)}$, respectively.

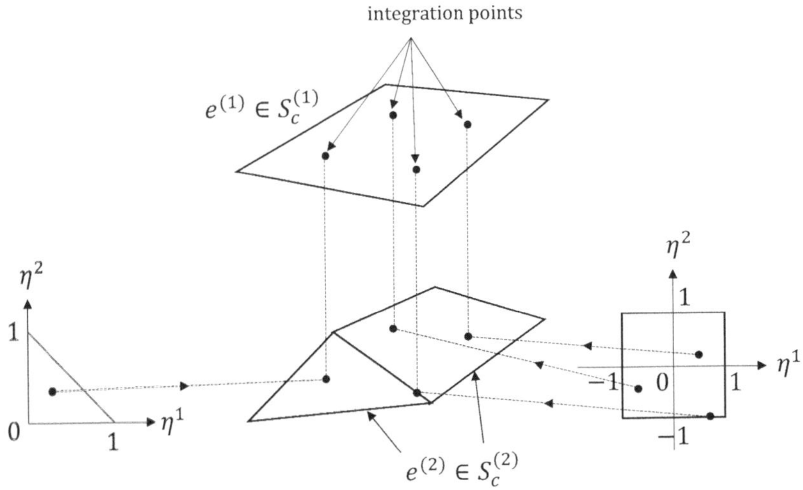

Figure 8.6. *Projection calculation. For a color version of this figure, see www.iste.co.uk/levan/contact.zip*

Figure 8.6 represents an example involving an element $e^{(1)} \subset S_c^{(1)}$ that contains a number of integration points. Let us consider an integration point $\mathbf{x}$ on $e^{(1)}$ and look for the closest point $\mathbf{y}$ on the target surface $S_c^{(2)}$. The calculation process consists of three steps:

(i) Among the nodes of $S_c^{(2)}$, find the one closest to the considered integration point $\mathbf{x}$. If $nnsc1$ and $nnsc2$ are the number of nodes of $S_c^{(1)}$ and $S_c^{(2)}$ respectively, then the number of floating-point operations will be of the order of $O(nnsc1.nnsc2)$.

(ii) Among the elements of $S_c^{(2)}$, identify the one that contains the closest node found in the previous step. Then the sought after projection point $\mathbf{y}$, should it exist, must be contained within one of these elements.

(iii) In each element $e^{(2)} \subset S_c^{(2)}$ found, look for the point $\mathbf{y}$ that realizes the minimum of $\|\mathbf{x} - \mathbf{x}^{(2)}\|, \mathbf{x}^{(2)} \in e^{(2)}$. We are reduced to calculating the pre-image $\boldsymbol{\eta}$ of $\mathbf{y}$, which is the root of the algebraic equation [2.20]:

$$(\mathbf{x} - \boldsymbol{\psi}^{(2)}(\boldsymbol{\eta})).\frac{\partial \boldsymbol{\psi}^{(2)}}{\partial \xi^{(2)\alpha}}(\boldsymbol{\eta}) = 0 \qquad [8.51]$$

The hypothesis of [2.20], namely that the parameterization $\boldsymbol{\xi}^{(2)} \mapsto \mathbf{x}^{(2)} = \boldsymbol{\psi}^{(2)}(\boldsymbol{\xi}^{(2)})$ of the surface $S_c^{(2)}$ is known, is automatically satisfied here since we know that, within a finite element, this parameterization is replaced by the shape interpolation [8.12], $\boldsymbol{\eta} \mapsto \{\mathbf{y}\} = [\mathcal{N}^{e^{(2)}}(\boldsymbol{\eta})]\{\mathbf{y}^{e^{(2)}}\}$, defined element-by-element.

Now we look for a point $\boldsymbol{\eta} = (\eta^1, \eta^2)$ which satisfies [8.51], in addition to the following inequalities:

$$\begin{array}{ll} \text{If } e^{(2)} \text{ is a triangle:} & \eta^1 \geq 0, \eta^2 \geq 0 \text{ and } \eta^1 + \eta^2 \leq 1 \\ \text{If } e^{(2)} \text{ is a quadrangle:} & |\eta^1| \leq 1 \text{ and } |\eta^2| \leq 1 \end{array} \qquad [8.52]$$

If the considered element $e^{(2)}$ is nonlinear, then equation [8.51] is also nonlinear and hence must be solved iteratively up to a specified precision.

If a root $\boldsymbol{\eta}$ is found, then the desired projection point $\mathbf{y}$ is given by $\mathbf{y} = \boldsymbol{\psi}^{(2)}(\boldsymbol{\eta})$. We can then calculate the proximity g, and then store the other data that are used for the rest of the calculation, such as $\boldsymbol{\eta}$, $\mathbf{y}$, g and the number of the element $e^{(2)}$ containing $\mathbf{y}$.

In some cases, if we cannot find a root $\boldsymbol{\eta}$ that satisfies [8.51]–[8.52], then either the projection point does not exist or it lies outside of the target surface.

8.11. Discrete expression of the slip $\mathbf{V}_T\Delta t$

The slip velocity $\mathbf{V}_T = \dot{\eta}^\alpha \mathbf{a}_\alpha$, which was defined in [2.39]$_2$ and features in [8.21]–[8.22], must be calculated numerically using a time-discrete expression. Before we begin to look for such an expression, notice that the weak form [7.21] remains valid if we replace $\mathbf{V}_T$ by $\mathbf{V}_T\Delta t$, where Δt is a finite time interval chosen arbitrarily.

DEFINITION.– The quantity $\mathbf{V}_T\Delta t$, which has the dimension of a length, is called *the slip*. [8.53]

In computational terms, Δt is the difference between two well-defined time instants as we will see, and the slip $\mathbf{V}_T\Delta t$ is preferable to the slip velocity $\mathbf{V}_T$, since it simplifies the question regarding the physical dimensions: the penalty parameters ϵ_N and ϵ_T then have the same dimension, namely that of a body force (N/m^3 in SI units).

Now, let us place ourselves at a current iteration of the current computation step n, and let us introduce some new notation that is illustrated in Figure 8.7. In order to lighten the writing load, any quantity calculated at the current iteration of the current step n will *not* have an index. On the other hand, any magnitude calculated at the previous step will be indicated using the index $(n - 1)$.

Let us therefore consider two time instants, t and t_{n-1}, that denote the current time instant corresponding to the current iteration of the current step n, and the time instant corresponding to the previous step $(n - 1)$, respectively. Consider a given particle on the contactor surface $S_c^{(1)}$ with reference position $\mathbf{X}$; its position at t_{n-1} is $\mathbf{x}_{n-1} = \boldsymbol{\phi}^{(1)}(\mathbf{X}, t_{n-1})$, its current position at t is

$\mathbf{x} = \boldsymbol{\phi}^{(1)}(\mathbf{X}, t)$ as in Figure 8.7. Let us recall the standard notation defined on the target surface $S_c^{(2)}(t)$ at the time instant t:

1. $\mathbf{y}$ = the projection point of $\mathbf{x}$ on $S_c^{(2)}(t)$;
2. $\mathbf{Y}$ and $\boldsymbol{\eta}$ = the pre-images of $\mathbf{y}$ via the chain of parameterizations [2.19].

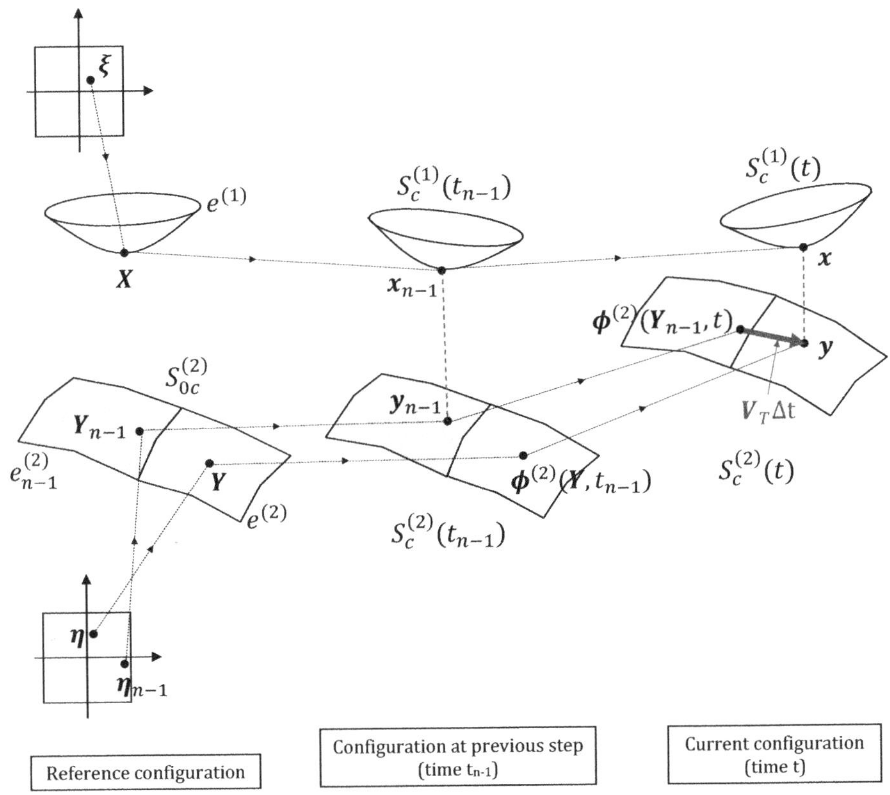

Figure 8.7. *Notation for the slip $\mathbf{V}_T\Delta t$. For a color version of this figure, see www.iste.co.uk/levan/contact.zip*

NOTATION.– We will need the following quantities, considered relative to the time instant t_{n-1}:

$\cdot\ \mathbf{y}_{n-1}$	=	the projection of $\mathbf{x}_{n-1}$ onto $S_c^{(2)}(t_{n-1})$;
$.\ \mathbf{Y}_{n-1}$ and $\boldsymbol{\eta}_{n-1}$	=	the pre-images of $\mathbf{y}_{n-1}$ via the chain of parameterizations [2.19];
$\cdot\ e_{n-1}^{(2)}$	=	the element of $S_{0c}^{(2)}$ containing the point $\mathbf{Y}_{n-1}$.

[8.54]

We will also denote $\boxed{\Delta t \equiv t - t_{n-1}}$.

Observe the following facts:

(i) The points $\mathbf{y}_{n-1}$ and $\mathbf{y}$ are *not* the positions at the respective time instants t_{n-1} and t of the same particle on the target surface.

(ii) The current position of the particle $\mathbf{Y}_{n-1}$ is $\boldsymbol{\phi}^{(2)}(\mathbf{Y}_{n-1}, t)$ and, in general, it *differs* from $\mathbf{y} = \boldsymbol{\phi}^{(2)}(\mathbf{Y}, t)$.

(iii) The elements $e^{(2)}_{n-1}$ and $e^{(2)}$ are *different*, in general. If the relative displacement of the two bodies between the time instants t_{n-1} and t is small, then these two elements can coincide.

With the notation introduced above, we can apply the backward Euler integration scheme to [2.39]$_2$ to obtain the following discretized expression of the slip:

$$\mathbf{V}_T \Delta t = (\eta^\alpha - \eta^\alpha_{n-1})\mathbf{a}_\alpha \qquad [8.55]$$

However, such an expression is valid only in theory, where it is assumed that the parameterization [2.19] of $\boldsymbol{\eta} \mapsto \mathbf{y} = \boldsymbol{\psi}^{(2)}(\boldsymbol{\eta}, t)$ is defined on the entirety of the target surface $S^{(2)}_c$. It cannot be used computationally when the surface $S^{(2)}_c$ is discretized into finite elements and when the parameterization [2.19] of $S^{(2)}_c$ is performed piecewise. Indeed:

- It may be that the points $\boldsymbol{\phi}^{(2)}(\mathbf{Y}_{n-1}, t)$ and $\mathbf{y}$, while being close to one another, do not belong to the same element and that their pre-images $\boldsymbol{\eta}_{n-1}$ and $\boldsymbol{\eta}$ may be very far apart. This situation is illustrated in Figure 8.8(a) for a plane problem, where we can see that equation [8.55] does not provide a good approximation of the slip.

- The converse scenario situation may also exist: the points $\boldsymbol{\phi}^{(2)}(\mathbf{Y}_{n-1}, t)$ and $\mathbf{y}$ do not belong to the same element and are quite distant, and yet their pre-images $\boldsymbol{\eta}_{n-1}$ and $\boldsymbol{\eta}$ coincide. This is illustrated in Figure 8.8(b), where the actual slip is non-zero, whereas applying [8.55] yields $\mathbf{V}_T \Delta t = \mathbf{0}$, since the points $\boldsymbol{\eta}_{n-1}$ and $\boldsymbol{\eta}$ are the same.

A remedy to this would be to extend the target element at the time instant t, so that it contains the point $\boldsymbol{\phi}^{(2)}(\mathbf{y}_{n-1}, t)$ and then recalculate the new value of $\boldsymbol{\eta}_{n-1}$ (see Figure 8.9). Unfortunately, this solution is only applicable if the point $\boldsymbol{\phi}^{(2)}(\mathbf{Y}_{n-1}, t)$ is close enough to the edge of the current target element. In practice, when the slip between the two bodies is large enough, or when the mesh of the target surface is sufficiently fine, then in general, the point $\mathbf{y} \subset S^{(2)}_c$ crosses between several elements of $S^{(2)}_c$ between the times instants t and t_{n-1}, so this solution cannot be implemented.

The moral of the story: we must search for another expression for the slip $\mathbf{V}_T \Delta t$, which does not involve the pre-images, $\boldsymbol{\eta}_{n-1}$ and $\boldsymbol{\eta}$.

THEOREM.–

$$\boxed{\mathbf{V}_T \Delta t = \mathbf{y} - \boldsymbol{\phi}^{(2)}(\mathbf{Y}_{n-1}, t)} \qquad [8.56]$$

PROOF. Let us consider equation [8.55] while taking into account the definition [2.9] of $\mathbf{a}_\alpha$:

$$\mathbf{V}_T \Delta t = (\eta^\alpha - \eta^\alpha_{n-1}) \frac{\partial \boldsymbol{\psi}^{(2)}}{\partial \xi^{(2)\alpha}}(\boldsymbol{\eta}, t)$$

By applying the backward Euler scheme to the variable $\boldsymbol{\eta}$ *at a fixed time instant* t, we obtain:

$$\mathbf{V}_T \Delta t = \boldsymbol{\psi}^{(2)}(\boldsymbol{\eta}, t) - \boldsymbol{\psi}^{(2)}(\boldsymbol{\eta}_{n-1}, t) \qquad [8.57]$$

and thus, thanks to [2.19], QED. ■

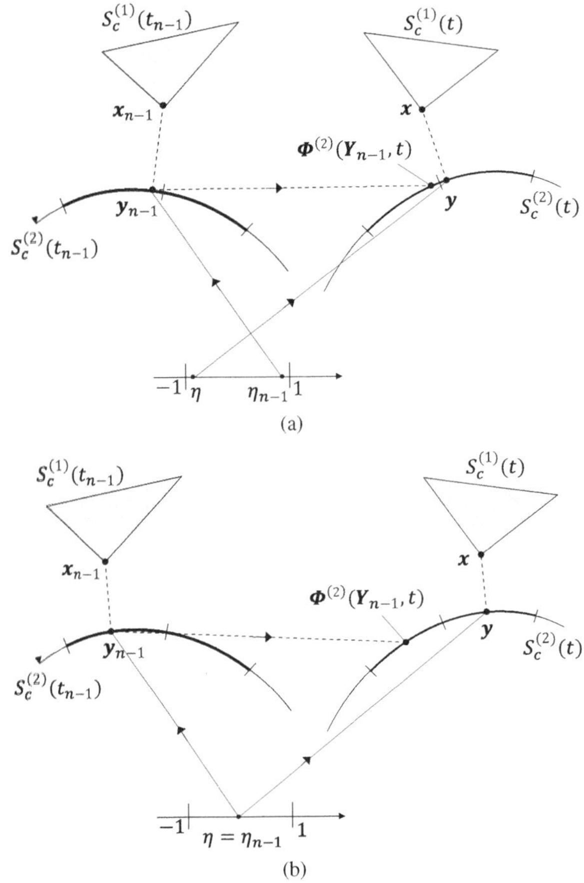

Figure 8.8. *Examples of cases for which [8.55] does not apply. For a color version of this figure, see www.iste.co.uk/levan/contact.zip*

The slip vector $\mathbf{V}_T \Delta t$ given by [8.56] is represented in Figure 8.7. The geometric interpretation of equation [8.56] is clear:

- At the time instant t_{n-1}, we determine the projection point $\mathbf{y}_{n-1}$. The pre-image of this point is the point $\mathbf{Y}_{n-1}$ on the reference target surface.

- At the time instant t, the particle with reference position $\mathbf{Y}_{n-1}$ comes in to occupy the position $\boldsymbol{\phi}^{(2)}(\mathbf{Y}_{n-1}, t)$.

- Physically, the slip $\mathbf{V}_T \Delta t$ is *the oriented arc* drawn on the target surface, which starts at the point $\boldsymbol{\phi}^{(2)}(\mathbf{Y}_{n-1}, t)$ and ends at the point $\mathbf{y}$. Equation [8.56] states that this arc is approximated by *the chord*, $\mathbf{y} - \boldsymbol{\phi}^{(2)}(\mathbf{Y}_{n-1}, t)$.

The expression [8.56] does not face the problem of discontinuity that is inherent in [8.55]. We will use it in order to compute the slip later on.

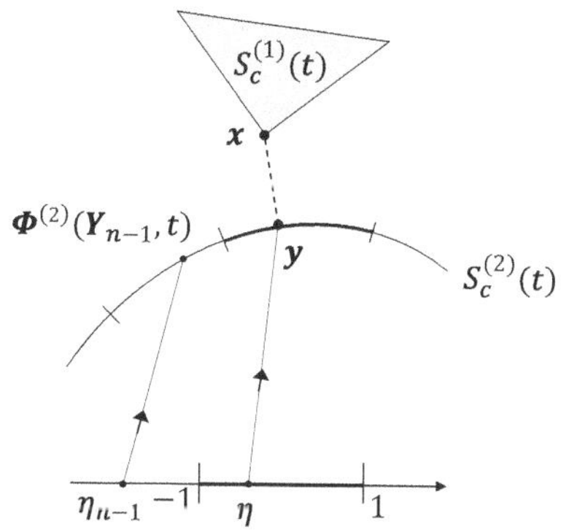

Figure 8.9. *A weak remedy: extending the target element. For a color version of this figure, see www.iste.co.uk/levan/contact.zip*

Just like we did for the point $\mathbf{y}$, the point $\boldsymbol{\phi}^{(2)}(\mathbf{Y}_{n-1},t)$ in [8.56] is obtained via interpolation. If we know which element $e_{n-1}^{(2)}$ contains $\mathbf{Y}_{n-1}$ in addition to the nodal displacement vector $\{\mathbf{U}^{(2),e_{n-1}^{(2)}}(t)\}$ of this element, then we can calculate $\boldsymbol{\phi}^{(2)}(\mathbf{Y}_{n-1},t)$ via interpolation:

$$\boxed{\{\boldsymbol{\phi}^{(2)}(\mathbf{Y}_{n-1},t)\} = \{\mathbf{Y}_{n-1}\} + \{\mathbf{U}^{(2)}(\mathbf{Y}_{n-1},t)\} = [\mathcal{N}^{e_{n-1}^{(2)}}(\boldsymbol{\eta}_{n-1})]\{\mathbf{Y}^{e_{n-1}^{(2)}}\} + [\mathcal{N}^{e_{n-1}^{(2)}}(\boldsymbol{\eta}_{n-1})]\{\mathbf{U}^{(2),e_{n-1}^{(2)}}\}} \quad [8.58]$$

Observe that the quantities at the step $n-1$ depend on the element $e^{(1)}$ and the point $\boldsymbol{\xi}$ too, in a similar fashion to what occurs with quantities defined in the current configuration. Table 8.2 describes this dependence and shows how to obtain the components which determine the slip velocity [8.56] successively from $(e^{(1)},\xi)$. It is an excerpt from Table 8.1 and completes this excerpt.

In other words, if we consider a given element $e^{(1)}$, then each point $\mathbf{Y}_{n-1}$, $\mathbf{y}_{n-1}$ and $\boldsymbol{\eta}_{n-1}$, is a function of $\boldsymbol{\xi}$ via the following chain. See Figure 8.7:

$\boldsymbol{\xi}$ $\rightarrow$ $\mathbf{X}$ via the interpolation [8.6];
$\boldsymbol{\xi}$, $\{\mathbf{U}^{(1),e^{(1)}}(t_{n-1})\}$ $\rightarrow$ $\mathbf{x}_{n-1}$ via the interpolations [8.7]–[8.8];
$\mathbf{x}_{n-1}$, $\boldsymbol{\psi}^{(2)}(.,t_{n-1})$ $\rightarrow$ $\boldsymbol{\eta}_{n-1}$, $e_{n-1}^{(2)}$ via the projection operation [2.20] (*passage specific to contact*);
$\boldsymbol{\eta}_{n-1}$ $\rightarrow$ $\mathbf{Y}_{n-1}$ by the interpolation [8.10] considered on $e_{n-1}^{(2)}$ instead of $e^{(2)}$;
$\boldsymbol{\eta}_{n-1}$, $\{\mathbf{U}^{(2),e_{n-1}^{(2)}}\}$ $\rightarrow$ $\mathbf{y}_{n-1}$ via the interpolation [8.12].

The element $e_{n-1}^{(2)} \subset S_{0c}^{(2)}$, which contains the pre-image $\mathbf{Y}_{n-1}$ of $\mathbf{y}_{n-1} = \text{proj}_{S_c^{(2)}(t_{n-1})}\mathbf{x}_{n-1}$, depends on the point $\mathbf{x}_{n-1}$ and hence on the point $\boldsymbol{\xi}$. In the same way as presented in Figure 8.3, observe that a single element $e^{(1)}$ can give rise to *several* elements $e_{n-1}^{(2)} \subset S_{0c}^{(2)}$.

Now, let us look at the dependence of the slip $\mathbf{V}_T\Delta t = \mathbf{y} - \boldsymbol{\phi}^{(2)}(\mathbf{Y}_{n-1},t)$ [8.56] vis-á-vis the *unknowns of the problem*. By combining Tables 8.1 and 8.2, we observe that:

- the slip depends on $\{\mathbf{U}^{(1),e^{(1)}}(t)\}$, $\{\mathbf{U}^{(2),e^{(2)}}(t)\}$ via $\mathbf{y}$ (see Table 8.1).

- the slip also depends on $\{\mathbf{U}^{(1),e^{(1)}}(t_{n-1})\}$, $\{\mathbf{U}^{(2),e_{n-1}^{(2)}}(t_{n-1})\}$ and $\{\mathbf{U}^{(2),e_{n-1}^{(2)}}(t)\}$ via $\boldsymbol{\phi}^{(2)}(\mathbf{Y}_{n-1},t)$ (see Table 8.2).

Knowing	We obtain	By using
$e^{(1)}, \xi$	(i) $[\mathcal{N}^{e^{(1)}}(\xi)]$ (ii) $\mathbf{X}$	 The interpolation [8.6]
$\mathbf{X}, \boldsymbol{\phi}^{(1)}(., t_{n-1})$	$\mathbf{x}_{n-1}$	$\mathbf{x}_{n-1} = \boldsymbol{\phi}^{(1)}(\mathbf{X}, t_{n-1})$
$\mathbf{x}_{n-1}, \boldsymbol{\psi}^{(2)}(., t_{n-1})$	(i) $\boldsymbol{\eta}_{n-1}, e^{(2)}_{n-1}$ (ii) $[\mathcal{N}^{e^{(2)}_{n-1}}(\boldsymbol{\eta}_{n-1})]$ (iii) $\{\mathbf{Y}^{e^{(2)}_{n-1}}\}$ (iv) $\mathbf{Y}_{n-1}$	The projection process [2.20]. *This passage is specific to contact* The mesh of $S^{(2)}_{0c}$ The interpolation [8.10] written on $e^{(2)}_{n-1}$ in the place of $e^{(2)}$: $\{\mathbf{Y}_{n-1}\} = [\mathcal{N}^{e^{(2)}_{n-1}}(\boldsymbol{\eta}_{n-1})]\{\mathbf{Y}^{e^{(2)}_{n-1}}\}$
$\boldsymbol{\psi}^{(2)}(., t), e^{(2)}_{n-1}$	(i) $\{\mathbf{U}^{(2),e^{(2)}_{n-1}}(t)\}$ (ii) $\mathbf{U}^{(2)}(\mathbf{Y}_{n-1}, t)$ (iii) $\boldsymbol{\phi}^{(2)}(\mathbf{Y}_{n-1}, t)$	The extraction of $\boldsymbol{\psi}^{(2)}(., t)$ The interpolation [8.11]: $\{\mathbf{U}^{(2)}(\mathbf{Y}_{n-1}, t)\} = [\mathcal{N}^{e^{(2)}_{n-1}}(\boldsymbol{\eta}_{n-1})]\{\mathbf{U}^{(2),e^{(2)}_{n-1}}\}$ Relation [8.58]: $\boldsymbol{\phi}^{(2)}(\mathbf{Y}_{n-1}, t) = \mathbf{Y}_{n-1} + \mathbf{U}^{(2)}(\mathbf{Y}_{n-1}, t)$
$\mathbf{y}, \boldsymbol{\phi}^{(2)}(\mathbf{Y}_{n-1}, t)$	The slip $\mathbf{V}_T$	The discrete equation [8.56] The point $\mathbf{y}$ is obtained from $(e^{(1)}, \xi)$ according to Table 8.1

Table 8.2. *How we obtain the ingredients for the slip velocity* [8.56] *from* $(e^{(1)}, \xi)$

Among these vectors, $\{\mathbf{U}^{(1),e^{(1)}}(t_{n-1})\}$ and $\{\mathbf{U}^{(2),e^{(2)}_{n-1}}(t_{n-1})\}$ are calculated at the time instant t_{n-1} and are known to us. If we are only interested in the *unknown* vectors, then we can say that the slip $\mathbf{V}_T \Delta t$ depends on the vectors $\mathbf{U}^e \equiv \left(\mathbf{U}^{(1),e^{(1)}}, \mathbf{U}^{(2),e^{(2)}}, \mathbf{U}^{(2),e^{(2)}_{n-1}}\right)$ (the notation $\mathbf{U}^e$ was defined in [8.25]).

8.12. Physical units

Apart from the coefficient of friction μ which is dimensionless, the other contact quantities have their own physical units and this fact should be kept in mind when performing any numerical computation. Let us recall the SI units for some contact quantities that have been introduced, before completing them with the units of other quantities.

- The SI unit of the penalty parameters ϵ_N and ϵ_T is $\mathrm{N/m^3}$.

- The SI unit of the multipliers λ_N and $\boldsymbol{\lambda}_T$ is $\mathrm{N/m^2}$.

- The SI unit of the element contact force vectors $\left\{\boldsymbol{\Phi}^{e^{(1)}}_{\text{contact}}\right\}$ and $\left\{\boldsymbol{\Phi}^{e^{(2)}}_{\text{contact}}\right\}$ [8.21] is $\mathrm{N/m^2}$.

Therefore, since we know that the unit of the surface element dS_0 is m^2, the unit of the contact force vector $\{\boldsymbol{\Phi}_{\text{contact}}\}$ in [8.40] is N.

- The SI unit of the element residual contact law vector $\left\{\mathbf{R}^{e^{(1)}}_{\Lambda}\right\}$ in [8.22] is m!

Thus, the unit of the residual contact law vector $\{\mathbf{R}_{\Lambda}\}$ in [8.41] is m^3!

• Using the previous results, assuming that the virtual velocities and the virtual multipliers have the same units as the actual displacements and multipliers, respectively, the unit of the element virtual contact power $(\mathcal{P}^*_{\text{contact}})^{e^{(1)}}$ in [8.20] can be determined:

$$(\mathcal{P}^*_{\text{contact}})^{e^{(1)}} = \int_{e^{(1)}} \Bigg(\underbrace{\begin{Bmatrix} \{\mathbf{U}^{(1)*,e^{(1)}}\} \\ \{\mathbf{U}^{(2)*,e^{(2)}}\} \end{Bmatrix}^T}_{m} \underbrace{\begin{Bmatrix} \{\mathbf{\Phi}^{e^{(1)}}_{\text{contact}}\} \\ \{\mathbf{\Phi}^{e^{(2)}}_{\text{contact}}\} \end{Bmatrix}}_{N/m^2} - \underbrace{< \mathbf{\Lambda}^{*,e^{(1)}} >}_{N/m^2} \underbrace{\{\mathbf{R}^{e^{(1)}}_{\Lambda}\}}_{m} \Bigg) \underbrace{dS_0}_{m^2}$$

Hence, we see that the unit of $(\mathcal{P}^*_{\text{contact}})^{e^{(1)}}$ is Nm.

• The units of each term in the system of equations [8.47] for the contact problem are:

$$\underbrace{\mathbf{M}.}_{kg}\ \underbrace{\ddot{\mathbf{U}}}_{m/s^2} + \underbrace{\mathbf{\Psi}}_{N} = \mathbf{\Phi} + \underbrace{\mathbf{\Phi}_{\text{contact}}}_{N} \qquad \underbrace{\mathbf{R}_{\Lambda}}_{m^3} = \mathbf{0}$$

8.13. Chapter summary

In this chapter, we obtained the following two key results:

1. We have discretized the weak form [7.21] using the finite element method to obtain the system of equations [8.47] of the contact problem, which is a coupled nonlinear system with the unknowns $\mathbf{U}$ (the nodal displacement vector) and $\mathbf{\Lambda}$ (the nodal multiplier vector):

$$\begin{cases} \mathbf{M}.\ddot{\mathbf{U}} + \mathbf{\Psi}(\mathbf{U}) = \mathbf{\Phi}(\mathbf{U}, t) + \mathbf{\Phi}_{\text{contact}}(\mathbf{U}, \mathbf{\Lambda}) \\ \mathbf{R}_{\Lambda}(\mathbf{U}, \mathbf{\Lambda}) = \mathbf{0} \end{cases}$$

   In this system, the vectors $\mathbf{\Phi}_{\text{contact}}(\mathbf{U}, \mathbf{\Lambda})$ and $\mathbf{R}_{\Lambda}(\mathbf{U}, \mathbf{\Lambda})$ represent the contribution that arises from contact, whereas the other vectors and matrices are classical.

2. Furthermore, we obtained the discrete slip expression [8.56]: $\mathbf{V}_T \Delta t = \mathbf{y} - \boldsymbol{\phi}^{(2)}(\mathbf{Y}_{n-1}, t)$, which enters into the construction of the contact force vector and the residual contact law vector (as well as that of the contact tangent matrix, which we will encounter again in Chapter 9). This expression does not depend on the mesh of the target surface, and therefore it does not experience any problems regarding discontinuity when the projection point $\mathbf{y}$ crosses between the elements of the mesh.

The method used to solve the system of equations [8.47] will be discussed in the following chapters, first in the static case and then in the dynamical case.

9

Solution of the Quasi-static Contact Problem

In the previous chapter, we obtained the system [8.47] of equations governing the contact problem. It is a coupled nonlinear system with the unknowns $\mathbf{U}$ (the nodal displacement vector) and $\boldsymbol{\Lambda}$ (the nodal multiplier vector), which presents two difficulties: the presence of derivatives with respect to time, in this case the acceleration $\ddot{\mathbf{U}}$, and the strongly nonlinear nature of the equations.

In this chapter, we will solve the system [8.47] of equations by restricting it to the *static* (or *quasi-static*) frame. The inertial term subsequently disappears and the system reduces to one consisting of nonlinear algebraic equations, allowing us to concentrate solely on the nonlinear aspect of the problem. We will explain how to solve this system using the iterative Newton–Raphson method, and we will establish the expression for the contact tangent matrix, which is the essential ingredient for each numerical iteration.

The dynamical contact problem will be solved in Chapter 11.

9.1. System of equations for the static contact problem

In quasi-statics, the inertial term $\mathbf{M}.\ddot{\mathbf{U}}$ vanishes and the differential system [8.47] reduces to the coupled nonlinear *algebraic* equations with the unknowns $(\mathbf{U}, \boldsymbol{\Lambda})$:

$$\boxed{\begin{array}{l}\boldsymbol{\Psi}(\mathbf{U}) = \boldsymbol{\Phi}(\mathbf{U}, t) \; + \boldsymbol{\Phi}_{\text{contact}}(\mathbf{U}, \boldsymbol{\Lambda}) \\ \mathbf{R}_{\Lambda}(\mathbf{U}, \boldsymbol{\Lambda}) = \mathbf{0}\end{array}} \qquad [9.1]$$

DEFINITIONS.– [9.2]

1. The equation $[9.1]_1$ is called the *(discrete) equilibrium equation.*

   Recall that equation $[9.1]_2$, which is none other than $[8.47]_2$, was defined in [8.48] as *the discretized contact laws.*

2. The system [9.1] is called *the system of equations for the static contact problem.*

Recall that $\boldsymbol{\Phi}$ denotes the external force vector (*excluding any contact forces*). It comes from the dead loads applied to the mechanical system and the follower forces (such as pressures), which a priori explains its dependence on both t and $\mathbf{U}$.

1. The meaning of the variable t in $\boldsymbol{\Phi}$: The given dead loads or pressure values can either be constant or vary. When they do vary, their variations can be described using one or more

parameters called the external load parameters, and which are generically denoted by η. If there are prescribed displacements which vary in the problem, then the set of parameters η also contains the prescribed displacement parameters.

In statics, we formally describe the variation of the parameters η using a function $t \mapsto \eta(t)$, where t is *not* the physical time but rather *a kinematic time*. We can simply take t to be equal to the computation step number.

2. In the presence of the follower forces applied to the bodies, the force vector $\mathbf{\Phi}$ will additionally depend on the displacement $\mathbf{U}$.

Equivalently, we could write $\mathbf{\Phi}(\mathbf{U}, \eta)$, but we prefer the notation $\mathbf{\Phi}(\mathbf{U}, t)$ since the symbol t is more familiar.

In general, equations [9.1] are strongly nonlinear, and the nature of these nonlinearities are diverse: geometric nonlinearities (due to large displacements, large rotations and follower forces), material nonlinearities (e.g. from large elastic strains or plasticity) and especially the nonlinearities of boundary conditions arising from contact.

• We will need a residual vector $\mathbf{R}$, which is the equivalent of the residual contact law vector $\mathbf{R}_\Lambda$, and which is defined as follows:

DEFINITION.– *The residual*, denoted by $\mathbf{R}$, is the function of $(\mathbf{U}, \mathbf{\Lambda}, t)$, which is defined to be the difference between the LHS and the RHS of the equilibrium equation $[9.1]_1$:

$$\boxed{\mathbf{R}(\mathbf{U}, \mathbf{\Lambda}, t) \equiv \mathbf{\Psi}(\mathbf{U}) - \mathbf{\Phi}(\mathbf{U}, t) - \mathbf{\Phi}_{\text{contact}}(\mathbf{U}, \mathbf{\Lambda})} \qquad [9.3]$$

With this definition, we can rewrite the system [9.1] of the static contact problem in a residual form:

$$\boxed{\begin{array}{l} \mathbf{R}(\mathbf{U}, \mathbf{\Lambda}, t) \equiv \mathbf{\Psi}(\mathbf{U}) - \mathbf{\Phi}(\mathbf{U}, t) - \mathbf{\Phi}_{\text{contact}}(\mathbf{U}, \mathbf{\Lambda}) = \mathbf{0} \\ \mathbf{R}_\Lambda(\mathbf{U}, \mathbf{\Lambda}) = \mathbf{0} \end{array}} \qquad [9.4]$$

Hence, solving the system of equations [9.1] is equivalent to the residuals $\mathbf{R}$ and $\mathbf{R}_\Lambda$ vanishing.

• If we were to solve the problem analytically, we would search for a function $\eta \mapsto (\mathbf{U}(\eta), \mathbf{\Lambda}(\eta))$, which satisfies [9.1] or [9.4] on a continuous set of η. Here, since we aim to solve the problem numerically, such a function is *not* realizable. The numerical solution of the system [9.1] or [9.4] should be done in an *incremental* way (that is to say, *step-by-step*).

We will try to determine *the values* of $(\mathbf{U}, \mathbf{\Lambda})$ for several successive values of η, values of $(\mathbf{U}, \mathbf{\Lambda})$, that we denote by

$$(\mathbf{U}_1, \mathbf{\Lambda}_1), \ldots, (\mathbf{U}_n, \mathbf{\Lambda}_n), \ldots$$

The numerical solution is therefore obtained by using several *steps*. The solution value at the $(n-1)$th step is used to calculate that for the nth step. The whole computational process is carried out as described in the following loop:

Incremental loop $n \geq 1$

- Knowing:

  (i) the solution for the previous $(n-1)$th step: displacement $\mathbf{U}_{n-1}$, multiplier $\mathbf{\Lambda}_{n-1}$ (and possibly many other plastic quantities);

  (ii) the external load applied on the mechanical system at the nth step. More precisely, the dead load values and those of the follower pressures.

  Calculate the solution at the nth step: displacement $\mathbf{U}_n$, multiplier $\mathbf{\Lambda}_n$ (and possibly other plastic quantities).

  This calculation requires a nonlinear system of equations to be solved, for which iterative Newton–Raphson scheme will be used to do this.

End of the incremental loop

Therefore, using the known initial values $\mathbf{U}_0, \mathbf{\Lambda}_0$, we will numerically construct a sequence of values converging toward the solution $(\mathbf{U}, \mathbf{\Lambda})$ for the various successive steps:

Step	0	1	2	$\cdots$	$n-1$	n	$\cdots$
Value of $\mathbf{U}$	$\mathbf{U}_0$	$\mathbf{U}_1$	$\mathbf{U}_2$	$\cdots$	$\mathbf{U}_{n-1}$	$\mathbf{U}_n$	$\cdots$
Value of $\mathbf{\Lambda}$	$\mathbf{\Lambda}_0$	$\mathbf{\Lambda}_1$	$\mathbf{\Lambda}_2$	$\cdots$	$\mathbf{\Lambda}_{n-1}$	$\mathbf{\Lambda}_n$	$\cdots$

9.2. Incremental loop initialization: the vectors $\mathbf{U}_0, \mathbf{\Lambda}_0$

For each step $n \geq 1$, we have assumed that $\mathbf{U}_{n-1}, \mathbf{\Lambda}_{n-1}$ are known.

1. If $n > 1$, that is, $n - 1 > 0$, then $(\mathbf{U}_{n-1}, \mathbf{\Lambda}_{n-1})$ is just simply the value of the pair $(\mathbf{U}, \mathbf{\Lambda})$ obtained from the preceding step $n - 1 > 0$.
2. If $n = 1$, then $(\mathbf{U}_{n-1}, \mathbf{\Lambda}_{n-1}) = (\mathbf{U}_0, \mathbf{\Lambda}_0)$. The value $(\mathbf{U}_0, \mathbf{\Lambda}_0)$, which is used to initialize the incremental loop, is not yet known to us and must be specified. Let us assume that:
   - the initial displacement $\mathbf{U}_0$ is zero: $\mathbf{U}_0 = \mathbf{0}$;
   - at the beginning of the calculations, the bodies being studied are not in contact, so that the initial values of the multipliers $\mathbf{\Lambda}$ are zero: $\mathbf{\Lambda}_0 = \mathbf{0}$.

9.3. Calculation of step $n \geq 1$: calculating $\mathbf{U}_n, \mathbf{\Lambda}_n$

9.3.1. *Principle of the iterative Newton–Raphson scheme*

Let us consider a given step $n \geq 1$ and determine the value $(\mathbf{U}_n, \mathbf{\Lambda}_n)$ of the solution $(\mathbf{U}, \mathbf{\Lambda})$ for this step. If we write either system [9.1] or [9.4] so that they are satisfied for the nth step, then we obtain two equations for the two unknowns $(\mathbf{U}_n, \mathbf{\Lambda}_n)$:

$$\begin{aligned} &\mathbf{R}(\mathbf{U}_n, \mathbf{\Lambda}_n, t_n) = \mathbf{0} \\ &\mathbf{R}_\Lambda(\mathbf{U}_n, \mathbf{\Lambda}_n) = \mathbf{0} \end{aligned} \qquad [9.5]$$

We will solve this nonlinear system with the help of the iterative Newton–Raphson scheme.

To start, suppose we are at an iteration $i \geq 1$ of the nth step and that, from the previous $(i-1)$th iteration, we obtained the approximation $(\mathbf{U}_n^{i-1}, \mathbf{\Lambda}_n^{i-1})$ of the desired solution, an approximation which does not satisfy the system [9.5] (if not, then $(\mathbf{U}_n^{i-1}, \mathbf{\Lambda}_n^{i-1})$ would be a solution and hence we would continue onto the next step).

Now, we will look for a new approximation $(\mathbf{U}_n^i, \mathbf{\Lambda}_n^i)$ of the solution that satisfies [9.5]:

$$\begin{aligned} &\mathbf{R}(\mathbf{U}_n^i, \mathbf{\Lambda}_n^i, t_n) = \mathbf{0} \\ &\mathbf{R}_\Lambda(\mathbf{U}_n^i, \mathbf{\Lambda}_n^i) = \mathbf{0} \end{aligned} \qquad [9.6]$$

Let us recall here the idea behind the Newton–Raphson scheme when applied to an example involving a scalar function $f(x)$. To determine x^i satisfying $f(x^i) = 0$, let us introduce the term $\delta x \equiv x^i - x^{i-1}$, where x^{i-1} is the solution known from the previous iteration. Since the knowledge of x^i is equivalent to that of δx, we are reduced to solving $f(x^{i-1} + \delta x) = 0$ for the unknown δx. Let us expand f as a Taylor series up to its first-order term in a neighborhood of x^{i-1}:

$$0 = f(x^i) = f(x^{i-1} + \delta x) = f(x^{i-1}) + \frac{\partial f}{\partial x}(x^{i-1})\, \delta x + o(|\delta x|)$$

If we ignore the second-order and greater terms, then the previous equation becomes $f(x^{i-1}) + \frac{\partial f}{\partial x}(x^{i-1})\delta x = 0$, which is an linear equation depending on the unknown δx and is therefore easy to solve. Once δx is known, we deduce x^i from $x^i = x^{i-1} + \delta x$.

Let us now introduce the quantities $\delta\mathbf{U}$ and $\delta\mathbf{\Lambda}$, which are defined as follows:

$$\forall i \geq 1, \boxed{\begin{aligned} \delta\mathbf{U} &= \mathbf{U}_n^i - \mathbf{U}_n^{i-1} \\ \delta\mathbf{\Lambda} &= \mathbf{\Lambda}_n^i - \mathbf{\Lambda}_n^{i-1} \end{aligned}} \qquad [9.7]$$

Knowing $(\mathbf{U}_n^i, \mathbf{\Lambda}_n^i)$ is equivalent to knowing $(\delta\mathbf{U}, \delta\mathbf{\Lambda})$. The notation [9.7] enables us to transform [9.6] into another equivalent system with the unknowns $\delta\mathbf{U}$ and $\delta\mathbf{\Lambda}$:

$$\begin{aligned} &\mathbf{R}(\mathbf{U}_n^{i-1} + \delta\mathbf{U}, \mathbf{\Lambda}_n^{i-1} + \delta\mathbf{\Lambda}, t_n) = \mathbf{0} \\ &\mathbf{R}_\Lambda(\mathbf{U}_n^{i-1} + \delta\mathbf{U}, \mathbf{\Lambda}_n^{i-1} + \delta\mathbf{\Lambda}) = \mathbf{0} \end{aligned}$$

Following the Newton–Raphson scheme, let us expand these equations into their Taylor series up to the first-order terms in a neighborhood of the solution $(\mathbf{U}_n^{i-1}, \mathbf{\Lambda}_n^{i-1})$, which is known from the previous iteration:

$$\begin{aligned} \mathbf{0} &= \mathbf{R}(\mathbf{U}_n^{i-1} + \delta\mathbf{U}, \mathbf{\Lambda}_n^{i-1} + \delta\mathbf{\Lambda}, t_n) \\ &= \mathbf{R}(\mathbf{U}_n^{i-1}, \mathbf{\Lambda}_n^{i-1}, t_n) + \left.\frac{\partial\mathbf{R}}{\partial\mathbf{U}}\right|_{(\mathbf{U}_n^{i-1}, \mathbf{\Lambda}_n^{i-1}, t_n)} .\delta\mathbf{U} + \left.\frac{\partial\mathbf{R}}{\partial\mathbf{\Lambda}}\right|_{(\mathbf{U}_n^{i-1}, \mathbf{\Lambda}_n^{i-1}, t_n)} .\delta\mathbf{\Lambda} + o(|\delta\mathbf{U}|, |\delta\mathbf{\Lambda}|) \\ \mathbf{0} &= \mathbf{R}_\Lambda(\mathbf{U}_n^{i-1} + \delta\mathbf{U}, \mathbf{\Lambda}_n^{i-1} + \delta\mathbf{\Lambda}) \\ &= \mathbf{R}_\Lambda(\mathbf{U}_n^{i-1}, \mathbf{\Lambda}_n^{i-1}) + \left.\frac{\partial\mathbf{R}_\Lambda}{\partial\mathbf{U}}\right|_{(\mathbf{U}_n^{i-1}, \mathbf{\Lambda}_n^{i-1})} .\delta\mathbf{U} + \left.\frac{\partial\mathbf{R}_\Lambda}{\partial\mathbf{\Lambda}}\right|_{(\mathbf{U}_n^{i-1}, \mathbf{\Lambda}_n^{i-1})} .\delta\mathbf{\Lambda} + o(|\delta\mathbf{U}|, |\delta\mathbf{\Lambda}|) \end{aligned}$$

By ignoring the second-order and higher terms, we come to a linear system to be solved for the two unknowns $\delta\mathbf{U}, \delta\mathbf{\Lambda}$ for the ith iteration:

$$\begin{bmatrix} \dfrac{\partial\mathbf{R}}{\partial\mathbf{U}} & \dfrac{\partial\mathbf{R}}{\partial\mathbf{\Lambda}} \\ \dfrac{\partial\mathbf{R}_\Lambda}{\partial\mathbf{U}} & \dfrac{\partial\mathbf{R}_\Lambda}{\partial\mathbf{\Lambda}} \end{bmatrix}_{(\mathbf{U}_n^{i-1}, \mathbf{\Lambda}_n^{i-1}, t_n)} \begin{Bmatrix} \delta\mathbf{U} \\ \delta\mathbf{\Lambda} \end{Bmatrix} = - \begin{Bmatrix} \mathbf{R}_n^{i-1} \\ \mathbf{R}_{\Lambda,n}^{i-1} \end{Bmatrix} \qquad [9.8]$$

where *the residuals* $\mathbf{R}_n^{i-1}$ and $\mathbf{R}_{\Lambda,n}^{i-1}$ are defined by

NOTATION.–

$$
\boxed{\begin{array}{lll} \mathbf{R}_n^{i-1} & \equiv \mathbf{R}(\mathbf{U}_n^{i-1}, \mathbf{\Lambda}_n^{i-1}, t_n) \underset{[9.3]}{=} & \mathbf{\Psi}(\mathbf{U}_n^{i-1}) - \mathbf{\Phi}(\mathbf{U}_n^{i-1}, t_n) - \mathbf{\Phi}_{\text{contact}}(\mathbf{U}_n^{i-1}, \mathbf{\Lambda}_n^{i-1}) \\ \mathbf{R}_{\Lambda,n}^{i-1} & \equiv \mathbf{R}_\Lambda(\mathbf{U}_n^{i-1}, \mathbf{\Lambda}_n^{i-1}) & \end{array}} \qquad [9.9]
$$

9.3.2. *Tangent matrix*

The iterative Newton–Raphson scheme shows that the matrix of system [9.8] is an essential ingredient in the solution, and thus deserves its own name.

DEFINITION.– *The global tangent matrix* of the coupled system [9.4] of the unknowns $(\mathbf{U}, \mathbf{\Lambda})$ is defined by

$$
[\mathbf{K}_T] \equiv \begin{bmatrix} \dfrac{\partial \mathbf{R}}{\partial \mathbf{U}} & \dfrac{\partial \mathbf{R}}{\partial \mathbf{\Lambda}} \\ \dfrac{\partial \mathbf{R}_\Lambda}{\partial \mathbf{U}} & \dfrac{\partial \mathbf{R}_\Lambda}{\partial \mathbf{\Lambda}} \end{bmatrix} = \begin{array}{c} \\ \{\mathbf{U}\} \\ \\ \{\mathbf{\Lambda}\} \end{array} \begin{array}{c} \begin{array}{cc} \qquad\qquad < \mathbf{U} > & \qquad\qquad < \mathbf{\Lambda} > \end{array} \\ \left[\begin{array}{c|c} \mathbf{K} \equiv \underbrace{\dfrac{\partial \mathbf{\Psi}}{\partial \mathbf{U}} - \dfrac{\partial \mathbf{\Phi}}{\partial \mathbf{U}}}_{\mathbf{K}_{classical}} - \dfrac{\partial \mathbf{\Phi}_{\text{contact}}}{\partial \mathbf{U}} & -\dfrac{\partial \mathbf{\Phi}_{\text{contact}}}{\partial \mathbf{\Lambda}} \\ \hline \dfrac{\partial \mathbf{R}_\Lambda}{\partial \mathbf{U}} & \dfrac{\partial \mathbf{R}_\Lambda}{\partial \mathbf{\Lambda}} \end{array}\right] \end{array} \qquad [9.10]
$$

It is a matrix composed of four blocks with the dimensions $\begin{array}{c} \\ n_U \\ n_\Lambda \end{array} \begin{array}{c} n_U \;\; n_\Lambda \\ \left[\begin{array}{c|c} \cdots & \cdots \\ \hline \cdots & \cdots \end{array}\right] \end{array}$.

The symbols $\mathbf{U}$, $\mathbf{\Lambda}$ *that border the matrix refer to the variables that each of the different submatrices corresponds to. Hence, for example, the rows of the submatrix* $-\dfrac{\partial \mathbf{\Phi}_{contact}}{\partial \mathbf{\Lambda}}$ *correspond to the displacement variables* $\mathbf{U}$*, whereas its columns correspond to the multiplier variables* $\mathbf{\Lambda}$*. The numbers* n_U, n_Λ *(defined in* [8.34]*) that border the matrix refer to size of each submatrix.*

The internal force vector $\mathbf{\Psi}(\mathbf{U})$ and the external force vector (*excluding any contact force*) $\mathbf{\Phi}(\mathbf{U})$ are classical in the sense that they are quantities that already existed in contactless mechanics. The matrix $\mathbf{K}_{classical} \equiv \dfrac{\partial \mathbf{\Psi}}{\partial \mathbf{U}} - \dfrac{\partial \mathbf{\Phi}}{\partial \mathbf{U}}$ that arises from this is the classical stiffness matrix, which appears in any nonlinear contactless mechanical problem, and therefore we will not go into any further detail regarding its calculation here.

It suffices to consider the next part of $\mathbf{K}_T$, denoted by $[\mathbf{K}_{\text{contact}}]$, which is new and arises from the terms $\mathbf{\Phi}_{\text{contact}}$ and $\mathbf{R}_\Lambda$ of [9.4]:

DEFINITION.– *The contact tangent matrix* is defined by

$$
[\mathbf{K}_{\text{contact}}] \equiv \begin{array}{c} \\ \{\mathbf{U}\} \\ \\ \{\mathbf{\Lambda}\} \end{array} \begin{array}{c} \begin{array}{cc} \qquad < \mathbf{U} > & \qquad\qquad < \mathbf{\Lambda} > \end{array} \\ \left[\begin{array}{c|c} \mathbf{K}_{UU} \equiv -\dfrac{\partial \mathbf{\Phi}_{\text{contact}}}{\partial \mathbf{U}} & \mathbf{K}_{U\Lambda} \equiv -\dfrac{\partial \mathbf{\Phi}_{\text{contact}}}{\partial \mathbf{\Lambda}} \\ \hline \mathbf{K}_{\Lambda U} \equiv \dfrac{\partial \mathbf{R}_\Lambda}{\partial \mathbf{U}} & \mathbf{K}_{\Lambda\Lambda} \equiv \dfrac{\partial \mathbf{R}_\Lambda}{\partial \mathbf{\Lambda}} \end{array}\right] \end{array} = \begin{array}{c} \\ n_U \\ n_\Lambda \end{array} \begin{array}{c} n_U \;\; n_\Lambda \\ \left[\begin{array}{c|c} \cdots & \cdots \\ \hline \cdots & \cdots \end{array}\right] \end{array} \qquad [9.11]
$$

$[\mathbf{K}_{\text{contact}}]$ is a square matrix of size $n_U + n_\Lambda$ $(n_\Lambda = 3nnsc1)$.

$[\mathbf{K}_{UU}]$ is added to the classical (contactless) tangent matrix $[\mathbf{K}_{classical}]$ to form the matrix that is denoted by $[\mathbf{K}]$ in [9.10].

We will sometimes decompose $[\mathbf{K}_{\text{contact}}]$ into finer sub-blocks as follows:

$$[\mathbf{K}_{\text{contact}}] = \begin{matrix} \\ \{\mathbf{U}^{(1)}\} \\ \{\mathbf{U}^{(2)}\} \\ \{\boldsymbol{\Lambda}^{e^{(1)}}\} \end{matrix} \begin{matrix} <\mathbf{U}^{(1)}> \quad <\mathbf{U}^{(2)}> \quad <\boldsymbol{\Lambda}> \\ \left[\begin{array}{cc|c} \mathbf{K}_{U^{(1)}U^{(1)}} & \mathbf{K}_{U^{(1)}U^{(2)}} & \mathbf{K}_{U^{(1)}\Lambda} \\ \mathbf{K}_{U^{(2)}U^{(1)}} & \mathbf{K}_{U^{(2)}U^{(2)}} & \mathbf{K}_{U^{(2)}\Lambda} \\ \mathbf{K}_{\Lambda U^{(1)}} & \mathbf{K}_{\Lambda U^{(2)}} & \mathbf{K}_{\Lambda\Lambda} \end{array}\right] \end{matrix} = \begin{matrix} \\ n_{U^{(1)}} \\ n_{U^{(2)}} \\ n_{\Lambda} \end{matrix} \begin{matrix} n_{U^{(1)}} \quad n_{U^{(2)}} \quad n_{\Lambda} \\ \left[\begin{array}{cc|c} \cdots & \cdots & \cdots \\ \hline \cdots & \cdots & \cdots \\ \hline \cdots & \cdots & \cdots \end{array}\right] \end{matrix} \qquad [9.12]$$

The matrices $\mathbf{K}_{UU}$, $\mathbf{K}_{U\Lambda}$, $\mathbf{K}_{\Lambda U}$ and $\mathbf{K}_{\Lambda\Lambda}$ are specific to the contact problem, and their expressions will be determined later.

With the previous definitions, we can rewrite the global tangent matrix as follows:

$$[\mathbf{K}_T] = \left[\begin{array}{c|c} \mathbf{K}_{classical} & \\ \hline & \mathbf{0} \end{array}\right] + [\mathbf{K}_{\text{contact}}] = \begin{matrix} \\ \{\mathbf{U}\} \\ \{\boldsymbol{\Lambda}\} \end{matrix} \begin{matrix} <\mathbf{U}> \qquad\qquad <\boldsymbol{\Lambda}> \\ \left[\begin{array}{c|c} \mathbf{K} = \mathbf{K}_{classical} + \mathbf{K}_{UU} & \mathbf{K}_{U\Lambda} \\ \hline \mathbf{K}_{\Lambda U} & \mathbf{K}_{\Lambda\Lambda} \end{array}\right] \end{matrix} \qquad [9.13]$$

and also rewrite the system [9.8] as follows:

$$\left[\begin{array}{c|c} \mathbf{K} = \mathbf{K}_{classical} + \mathbf{K}_{UU} & \mathbf{K}_{U\Lambda} \\ \hline \mathbf{K}_{\Lambda U} & \mathbf{K}_{\Lambda\Lambda} \end{array}\right] \begin{Bmatrix} \delta\mathbf{U} \\ \delta\boldsymbol{\Lambda} \end{Bmatrix} = -\begin{Bmatrix} \mathbf{R}_n^{i-1} \\ \mathbf{R}_{\Lambda,n}^{i-1} \end{Bmatrix} \qquad [9.14]$$

9.3.3. *Block matrix inversion*

There exist cases in which the system [9.14] is not invertible in general:

– In the case of frictional contact in which stick exists at every contact point, the block $\mathbf{K}_{\Lambda\Lambda}$, which will be made explicit in [9.84], vanishes. The matrix of the system [9.14] is therefore positive semi-definite.

– In the frictionless case and with the algorithmic slip contact at every contact point, we will see in [9.94] that the matrix $[\mathbf{K}^e_{\Lambda\Lambda}]_{slip}$ is not invertible.

In each of these cases, direct solving methods such as Gauss's triangulation will fail, and it is necessary to solve the system blockwise. Let us eliminate $\delta\mathbf{U}$ in $[9.14]_1$ by assuming that the matrix $\mathbf{K}$ is invertible:

$$\delta\mathbf{U} = -\mathbf{K}^{-1}.\mathbf{R}_n^{i-1} - \mathbf{K}^{-1}.\mathbf{K}_{U\Lambda}.\delta\boldsymbol{\Lambda} \qquad [9.15]$$

Substituting this equation into $[9.14]_2$, we obtain an equation with just a single unknown $\delta\boldsymbol{\Lambda}$:

$$\left(\mathbf{K}_{\Lambda\Lambda} - \mathbf{K}_{\Lambda U}.\mathbf{K}^{-1}.\mathbf{K}_{U\Lambda}\right).\delta\boldsymbol{\Lambda} = -\mathbf{R}_{\Lambda,n}^{i-1} - \mathbf{K}_{\Lambda U}.\mathbf{K}^{-1}.\mathbf{R}_n^{i-1} \qquad [9.16]$$

The matrix $(\mathbf{K}_{\Lambda\Lambda} - \mathbf{K}_{\Lambda U}.\mathbf{K}^{-1}.\mathbf{K}_{U\Lambda})$, the Schur complement to the matrix $\mathbf{K}$, is invertible in practice. Inverting the linear system [9.16] gives $\delta\boldsymbol{\Lambda}$ and then $\delta\mathbf{U}$ due to [9.15]. Once $\delta\mathbf{U}$ and $\delta\boldsymbol{\Lambda}$

have been obtained, we correct the displacements and the contact multipliers with the help of [9.7]:

$$\boxed{\begin{aligned} \mathbf{U}_n^i &= \mathbf{U}_n^{i-1} + \delta\mathbf{U} \\ \mathbf{\Lambda}_n^i &= \mathbf{\Lambda}_n^{i-1} + \delta\mathbf{\Lambda} \end{aligned}} \qquad [9.17]$$

Then we continue onto the next iteration with the system [9.14] where, this time, the matrices and residuals are calculated using the updated approximation $(\mathbf{U}_n^i, \mathbf{\Lambda}_n^i)$ in the place of $(\mathbf{U}_n^{i-1}, \mathbf{\Lambda}_n^{i-1})$. We continue the iterations until the residuals $\mathbf{R}_n^{i-1}$ and $\mathbf{R}_{\Lambda,n}^{i-1}$ are less than the specific predetermined tolerances, before finally terminating the current step n.

9.3.4. *Iterative loop initialization: the vectors U_n^0 and Λ_n^0*

At each iteration $i \geq 1$ of step n, we need to know what the pair $(\mathbf{U}_n^{i-1}, \mathbf{\Lambda}_n^{i-1})$ is.

- If $i > 1$, that is, $i - 1 > 0$, then $(\mathbf{U}_n^{i-1}, \mathbf{\Lambda}_n^{i-1})$ is simply the value of the pair $(\mathbf{U}, \mathbf{\Lambda})$ obtained from the previous iteration $i - 1 > 0$.
- If $i = 1$, then $(\mathbf{U}_n^{i-1}, \mathbf{\Lambda}_n^{i-1}) = (\mathbf{U}_n^0, \mathbf{\Lambda}_n^0)$. The value $(\mathbf{U}_n^0, \mathbf{\Lambda}_n^0)$, which serves as an estimate used to initialize the iterative loop, is not yet known to us and thus must be specified. We simply choose:

$$\boxed{\begin{aligned} \mathbf{U}_n^0 &= \mathbf{U}_{n-1} \\ \mathbf{\Lambda}_n^0 &= \mathbf{\Lambda}_{n-1} \end{aligned}} \qquad [9.18]$$

- The choice $\mathbf{U}_n^0 = \mathbf{U}_{n-1}$ implies the following interesting consequence regarding the slip:

THEOREM.– At the beginning of each step in statics, the slip is *zero*: $\mathbf{V}_T \Delta t = \mathbf{0}$. [9.19]

PROOF. Borrowing the notation from section 8.11 and referring to Figure 8.7, we see that at the start of step n, as $\mathbf{U}_n^0 = \mathbf{U}_{n-1}$, the point $\mathbf{x}$ is in $\mathbf{x}_{n-1}$, whence $\mathbf{y} = \mathbf{y}_{n-1} = \boldsymbol{\phi}^{(2)}(\mathbf{Y}_{n-1}, t)$. Then by applying [8.56], we get QED. ■

The reader should be aware that the previous theorem is *not* valid in general for dynamics, since in [11.18] we will have to consider $\mathbf{U}_n^0 = \mathbf{U}_{n-1} + h\mathbf{V}_{n-1} + \frac{h^2}{2}\mathbf{A}_{n-1}$ rather than $\mathbf{U}_n^0 = \mathbf{U}_{n-1}$.

9.4. Solution algorithm

The solution of the static contact problem is described using Algorithm 9.1.

Initialize the incremental loop:

$\mathbf{U}_0 = \mathbf{0}$

For contact, set $\mathbf{\Lambda}_0 = \mathbf{0}$.

Knowing $\mathbf{U}_0$, hence the initial positions of $S_c^{(1)}$ and $S_c^{(2)}$, perform a projection calculation to determine the elements $e_0^{(2)}$ and the pre-images $\boldsymbol{\eta}_0$ corresponding to the integration points $\mathbf{X} \in S_{0c}^{(1)}$ [1].

If there is plasticity, initialize the plastic quantities.

Incremental loop $n = 1, 2, \ldots$

$i = 0$

Initialize the iterative loop: predict $\mathbf{U}_n^0 = \mathbf{U}_{n-1}$, $\mathbf{\Lambda}_n^0 = \mathbf{\Lambda}_{n-1}$.

Calculate the external force vector $\mathbf{\Phi}$ due to the dead loads at the step n.

Iterative loop

$i = i + 1$

Knowing $\mathbf{U}_n^{i-1}$, calculate the internal force vector $\mathbf{\Psi}(\mathbf{U}_n^{i-1})$.

If there exist external pressures, then add their contributions in $\mathbf{\Phi}$. We then obtain $\mathbf{\Phi}(\mathbf{U}_n^{i-1}, t_n)$.

Knowing $\mathbf{U}_n^{i-1}$ and $\mathbf{\Lambda}_n^{i-1}$, calculate the external contact force $\mathbf{\Phi}_{\text{contact}}(\mathbf{U}_n^{i-1}, \mathbf{\Lambda}_n^{i-1})$ and the residual contact law vector $\mathbf{R}_{\Lambda,n}^{i-1}$.

This calculation also uses two results from the previous step $n-1$: the $e_{n-1}^{(2)}$ and the $\boldsymbol{\eta}_{n-1}$ corresponding to the integration points $\mathbf{X} \in S_{0c}^{(1)}$ [2].

During this calculation, we perform a projection operation and, among other things, we generate the $e^{(2)}$ and the $\boldsymbol{\eta}$. This information will then be stored at the end of the step.

Calculate the residual $\mathbf{R}_n^{i-1} = \mathbf{\Psi}(\mathbf{U}_n^{i-1}) - \mathbf{\Phi}(\mathbf{U}_n^{i-1}, t_n) - \mathbf{\Phi}_{\text{contact}}(\mathbf{U}_n^{i-1}, \mathbf{\Lambda}_n^{i-1})$.

Check the convergence: if $i > 1$ and if $||\mathbf{R}_n^{i-1}||$ and $||\mathbf{R}_{\Lambda,n}^{i-1}|| \leq$ the given tolerances, then exit the iterative loop.

Calculate the tangent matrices:

$$\text{Calculate } \mathbf{K} = \underbrace{\frac{\partial \mathbf{\Psi}}{\partial \mathbf{U}}(\mathbf{U}_n^{i-1})}_{\text{due to internal forces}} - \underbrace{\frac{\partial \mathbf{\Phi}}{\partial \mathbf{U}}(\mathbf{U}_n^{i-1}, t_n)}_{\text{due to possible external pressures}} + \underbrace{\mathbf{K}_{UU}(\mathbf{U}_n^{i-1}, \mathbf{\Lambda}_n^{i-1})}_{\text{due to contact}}.$$

In addition to $\mathbf{K}_{UU}$, calculate the other contact tangent matrices $\mathbf{K}_{U\Lambda}$, $\mathbf{K}_{\Lambda U}$, $\mathbf{K}_{\Lambda\Lambda}$ at $(\mathbf{U}_n^{i-1}, \mathbf{\Lambda}_n^{i-1})$.

Obtain $\delta\mathbf{\Lambda}$ by solving $\left(\mathbf{K}_{\Lambda\Lambda} - \mathbf{K}_{\Lambda U}.\mathbf{K}^{-1}.\mathbf{K}_{U\Lambda}\right).\delta\mathbf{\Lambda} = -\mathbf{R}_{\Lambda,n}^{i-1} + \mathbf{K}_{\Lambda U}.\mathbf{K}^{-1}.\mathbf{R}_n^{i-1}$.

Deduce $\delta\mathbf{U} = -\mathbf{K}^{-1}.\mathbf{R}_n^{i-1} - \mathbf{K}^{-1}.\mathbf{K}_{U\Lambda}.\delta\mathbf{\Lambda}$.

$$\text{Correct} \begin{cases} \mathbf{U}_n^i = \mathbf{U}_n^{i-1} + \delta\mathbf{U} \\ \mathbf{\Lambda}_n^i = \mathbf{\Lambda}_n^{i-1} + \delta\mathbf{\Lambda} \end{cases}.$$

End of the iterative loop

Update: $\mathbf{U}_n = \mathbf{U}_n^{i-1}$, $\mathbf{\Lambda}_n = \mathbf{\Lambda}_n^{i-1}$ (also update the plastic quantities).

Store some specific results provided by the last projection calculation: the elements $e^{(2)}$ and the pre-images $\boldsymbol{\eta}$ corresponding to the integration points $\mathbf{X} \in S_{0c}^{(1)}$. These quantities will play the role of the $e_{n-1}^{(2)}$ and $\boldsymbol{\eta}_{n-1}$ in the next step.

End of incremental loop

Algorithm 9.1. *Solution algorithm of the static contact problem*

NOTES ON ALGORITHM 9.1.–

[1] During the step $n = 0$, the displacement is zero and $S_c^{(1)} = S_{0c}^{(1)}$, $S_c^{(2)} = S_{0c}^{(2)}$, $\mathbf{x} = \mathbf{X}$, $\mathbf{y} = \mathbf{Y}$.

The element $e_0^{(2)}$ and the pre-image $\boldsymbol{\eta}_0$ obtained at $n = 0$ will play the roles of $e_{n-1}^{(2)}$ and $\boldsymbol{\eta}_{n-1}$ in the next step $n = 1$.

[2] Let us move onto a step $n \geq 1$, and consider a given element $e^{(1)} \subset S_{0c}^{(1)}$ and borrow the notation from section 8.11 as well as from Figure 8.7. Recall that knowledge of the displacement $\mathbf{U}_{n-1}$ at the previous step $n-1$ permits us to calculate the element $e_{n-1}^{(2)}$, the point $\mathbf{Y}_{n-1}$ and the pre-image $\boldsymbol{\eta}_{n-1}$ corresponding to an integration point $\boldsymbol{\xi}$ (or $\mathbf{X}$) of $e^{(1)}$ via the following chain:

$\boldsymbol{\xi} \rightarrow \mathbf{X}$ by the interpolation [8.6];

$\boldsymbol{\xi}, \{\mathbf{U}^{(1),e^{(1)}}(t_{n-1})\} \rightarrow \mathbf{x}_{n-1}$ by the interpolations [8.7] – [8.8];

$\mathbf{x}_{n-1}, \boldsymbol{\psi}^{(2)}(., t_{n-1}) \rightarrow \boldsymbol{\eta}_{n-1}, e_{n-1}^{(2)}$ by the projection operation [2.20] (*passage specific to contact*);

$\boldsymbol{\eta}_{n-1} \rightarrow \mathbf{Y}_{n-1}$ by the interpolation [8.10] set upon $e_{n-1}^{(2)}$ in the place of $e^{(2)}$.

Knowing the element $e_{n-1}^{(2)}$ and the pre-image $\boldsymbol{\eta}_{n-1}$, we calculate the slip by [8.56]: $\mathbf{V}_T \Delta t = \mathbf{y} - \boldsymbol{\phi}^{(2)}(\mathbf{Y}_{n-1}, t)$.

9.5. Calculation method for the tangent matrix

In the RHS of the system [9.14] to be solved at each iteration i of the Newton–Raphson loop, the internal force vector $\{\boldsymbol{\Psi}(\mathbf{U})\}$ and the external force vector (not including the contact force) $\{\boldsymbol{\Phi}(\mathbf{U}, t)\}$ are calculated in a standard manner as done in classical mechanics without contact. Moreover, the contact force vector $\{\boldsymbol{\Phi}_{\text{contact}}(\mathbf{U}, \boldsymbol{\Lambda})\}$ and the residual contact law vector $\{\mathbf{R}_{\Lambda}(\mathbf{U}, \boldsymbol{\Lambda})\}$ can be easily calculated from the element expressions [8.21] and [8.22].

On the other hand, we no longer have an explicit expression for the tangent matrix $[\mathbf{K}_T]$ defined by [9.10] and which appears in the LHS of the system [9.14]. Although the submatrix $\mathbf{K}_{classical} \equiv \dfrac{\partial \boldsymbol{\Psi}}{\partial \mathbf{U}} - \dfrac{\partial \boldsymbol{\Phi}}{\partial \mathbf{U}}$ is classical, the other submatrices that involve the derivatives of the column-vectors $\{\boldsymbol{\Phi}_{\text{contact}}(\mathbf{U}, \boldsymbol{\Lambda})\}$ and $\{\mathbf{R}_{\Lambda}(\mathbf{U}, \boldsymbol{\Lambda})\}$ with respect to the column-vectors $\mathbf{U}, \boldsymbol{\Lambda}$ are not yet known.

In this section, we are going to show the principle behind calculating the tangent matrix $[\mathbf{K}_T]$, or more precisely that behind the contact tangent matrix $[\mathbf{K}_{\text{contact}}]$ defined by [9.11]. The actual calculation of $[\mathbf{K}_{\text{contact}}]$ will be done in the next section.

9.5.1. *Direct method*

The simplest idea is to obtain $[\mathbf{K}_T]$ directly from definition [9.10], that is to say by differentiating the column-vectors $\mathbf{R}$ and $\mathbf{R}_{\Lambda}$ with respect to the column-vectors $\mathbf{U}$ and $\boldsymbol{\Lambda}$. This method works well in mechanical problems without contact, with large deformations with the hyperelastic material laws. Unfortunately, it leads to inextricable calculations for contact problems. To see this, we examine the partial derivative $\dfrac{\partial \mathbf{R}_{\Lambda}}{\partial \mathbf{U}}$, which leads to calculating the partial derivatives of the element residual $\mathbf{R}_{\Lambda}^{e^{(1)}}$ with respect to the element displacements

$\mathbf{U}^{(1),e^{(1)}}$, $\mathbf{U}^{(2),e^{(2)}}$. We have

$$\frac{\partial \mathbf{R}_\Lambda^{e^{(1)}}}{\partial \mathbf{U}^{(1),e^{(1)}}} = [\mathcal{N}^{e^{(1)}}(\xi)]^T$$

$$\left(\frac{1}{\epsilon_N}\frac{\partial}{\partial \mathbf{U}^{(1),e^{(1)}}}\left\{\left(\langle\hat{\lambda}_N\rangle - \lambda_N\right)\{\mathbf{m}\}\right\} + \frac{1}{\epsilon_T}\frac{\partial}{\partial \mathbf{U}^{(1),e^{(1)}}}\left\{\boldsymbol{\lambda}_T - \left(1 - \langle 1 - \frac{\mu\langle\hat{\lambda}_N\rangle}{\|\hat{\boldsymbol{\lambda}}_T\|}\rangle\right)\hat{\boldsymbol{\lambda}}_T\right\}\right)$$

The partial derivative with respect to $\{\mathbf{U}\}$ must be performed at a *fixed* $\{\mathbf{\Lambda}\}$. The condition "$\{\mathbf{\Lambda}\}$ is fixed" implies that the node vectors $\boldsymbol{\lambda}$ are fixed and that the field $\boldsymbol{\lambda}$ is also fixed. However, since $\boldsymbol{\lambda} = \lambda_N \boldsymbol{\nu} - \boldsymbol{\lambda}_T$ and that $\boldsymbol{\nu}$ varies with $\{\mathbf{U}\}$, if we want $\boldsymbol{\lambda}$ to be fixed, then this forces λ_N and $\boldsymbol{\lambda}_T$ to be complicated functions of $\{\mathbf{U}\}$. The derivatives are therefore also complicated.

The examination of the derivative $\dfrac{\partial \mathbf{\Phi}_{\text{contact}}}{\partial \mathbf{U}}$ reveals another difficulty. In the expression [8.40], the point $\boldsymbol{\eta}$ involved in the integrand of $\mathbf{\Phi}^{(2)}_{\text{contact}}$ varies with the point $\mathbf{x}$, and hence with the column-vector of the nodal displacements $\mathbf{U}$. Since we do not know how to express this dependence using an explicit relation, it is difficult to take this dependence into account when calculating the derivative.

Thus, it will be necessary to determine the tangent matrix in an indirect way.

9.5.2. *Indirect method*

The indirect method consists of identifying the tangent matrix $[\mathbf{K}_T]$ starting from the variation of the discretized weak form of the problem. Let us recall the steps which had led us to the discretized expression of the weak form.

1. We start with the continuous expression [7.21] of the weak form, which is written in statics and in abbreviated form as follows:

$$0 = \mathcal{P}^* = \text{functional}\,(\underbrace{\mathbf{U}^{(1)}, \mathbf{U}^{(2)}}_{\text{containing } g,\boldsymbol{\nu},\mathbf{V}_T}, \lambda_N, \boldsymbol{\lambda}_T; \mathbf{U}^{(1)*}, \mathbf{U}^{(2)*}, \boldsymbol{\lambda}^*)$$
$$\equiv \underbrace{\mathcal{P}^*_{\text{acceleration}}}_{=0} - \mathcal{P}^*_{int} - \mathcal{P}^*_{ext} - \mathcal{P}^*_{\text{contact}} \qquad [9.20]$$

2. By interpolation, we express the continuous quantities in terms of the nodal quantities:

$$\begin{cases} \text{vector } \mathbf{U}^{(i)} & \text{in terms of } \{\mathbf{U}\} : \text{use the interpolations [8.7], [8.11] of} \\ & \qquad \mathbf{U}^{(1)}, \mathbf{U}^{(2)} \text{ (standard calculation);} \\ \lambda_N, \boldsymbol{\lambda}_T & \text{in terms of } \{\mathbf{\Lambda}\} : \text{use the interpolation [8.16] of } \boldsymbol{\lambda} \text{ and [8.17];} \\ \text{vector } \mathbf{U}^{(i)*} & \text{in terms of } \{\mathbf{U}^*\} : \text{use the interpolations [8.9], [8.13];} \\ \text{vector } \boldsymbol{\lambda}^* & \text{in terms of } \{\mathbf{\Lambda}^*\} : \text{use the interpolation [8.18] of } \boldsymbol{\lambda}^*. \end{cases} \qquad [9.21]$$

3. The discretization of the functional $\mathcal{P}^*$ has been done in section 8.8, and we found

$$\begin{aligned} \mathcal{P}^*_{int} &= -<\mathbf{U}^*>\{\mathbf{\Psi}(\mathbf{U})\} \\ \mathcal{P}^*_{ext} &= <\mathbf{U}^*>\{\mathbf{\Phi}(\mathbf{U})\} \\ \mathcal{P}^*_{\text{contact}} &= <\mathbf{U}^*>\{\mathbf{\Phi}_{\text{contact}}(\mathbf{U},\mathbf{\Lambda})\} - <\mathbf{\Lambda}^*>\{\mathbf{R}_\Lambda(\mathbf{U},\mathbf{\Lambda})\} \quad \text{(that is [8.39])} \end{aligned}$$

Therefore, we have arrived at the discretized expression for the functional $\mathcal{P}^*$ (i.e. the LHS of [8.45] without the inertial term):

$$\mathcal{P}^* = <\mathbf{U}^*> \underbrace{\left(\{\mathbf{\Psi}(\mathbf{U})\} - \{\mathbf{\Phi}(\mathbf{U})\} - \{\mathbf{\Phi}_{\text{contact}}(\mathbf{U},\mathbf{\Lambda})\}\right)}_{\equiv\{\mathbf{R}(\mathbf{U},\mathbf{\Lambda})\}} + <\mathbf{\Lambda}^*>\{\mathbf{R}_\Lambda(\mathbf{U},\mathbf{\Lambda})\}$$

Or shortly:

$$\mathcal{P}^*_{\text{discretized}} = <\mathbf{U}^*> \{\mathbf{R}(\mathbf{U},\boldsymbol{\Lambda})\} + <\boldsymbol{\Lambda}^*> \{\mathbf{R}_\Lambda(\mathbf{U},\boldsymbol{\Lambda})\} \qquad [9.22]$$

We have written $\mathcal{P}^*_{\text{discretized}}$ in the place of $\mathcal{P}^*$ to recall that we have used the nodal interpolations to express the initial *continuous* weak form as a function of the *nodal* quantities (column-vector of nodal displacements, column-vector of the nodal multipliers).

The indirect calculation of the tangent matrix calls upon the calculus of variations. We begin by introducing the column-vectors $\{\delta\mathbf{U}\}$ and $\{\delta\boldsymbol{\Lambda}\}$, which group together the variations at all the nodes of the structure or the contactor surface:

NOTATION.–

1. VARIATIONS OF THE NODAL DISPLACEMENTS:

   - $\{\delta\mathbf{U}^{(1)}\}$: the column-vector grouping together the triplets of the components of the vector $\delta\mathbf{U}^{(1)}$ (variation of the displacement vector $\mathbf{U}^{(1)}$) at the nodes of body 1;
   - $\{\delta\mathbf{U}^{(2)}\}$: the column-vector grouping together the triplets of the components of the vector $\delta\mathbf{U}^{(2)}$ (variation of the displacement vector $\mathbf{U}^{(2)}$) at the nodes of body 2;
   - $\{\delta\mathbf{U}\} = \left\{ \begin{matrix} \{\delta\mathbf{U}^{(1)}\} \\ \{\delta\mathbf{U}^{(2)}\} \end{matrix} \right\}$: the column-vector grouping together the triplets of the components of the vector $\delta\mathbf{U}^{(1)}$ or $\delta\mathbf{U}^{(2)}$ at the nodes of the entire structure.

2. VARIATIONS OF THE NODAL MULTIPLIERS:

   - $\{\delta\boldsymbol{\Lambda}\}$: the column-vector grouping together the triplets of the components of the vector $\delta\boldsymbol{\lambda}$ (variation of the multiplier vector $\boldsymbol{\lambda}$) at the nodes of $S^{(1)}_{0c}$.

Recall the convention from section 8.3 according to which each component in question is relative to the fixed orthonormal basis $(\mathbf{e}_1, \mathbf{e}_2, \mathbf{e}_3)$.

Let us now consider the *arbitrary and independent* variations $\{\delta\mathbf{U}\}$ and $\{\delta\boldsymbol{\Lambda}\}$ of the column-vectors $\{\mathbf{U}\}$ and $\{\boldsymbol{\Lambda}\}$. If we manage to calculate the variation $\delta\left(\mathcal{P}^*_{\text{discretized}}\right)$ corresponding to $\mathcal{P}^*_{\text{discretized}}$ and put it into the form:

$$\forall\{\delta\mathbf{U}\},\{\delta\boldsymbol{\Lambda}\}, \quad \boxed{\delta\left(\mathcal{P}^*_{\text{discretized}}\right) = \left\{ \begin{matrix} \{\mathbf{U}^*\} \\ \{\boldsymbol{\Lambda}^*\} \end{matrix} \right\}^T [\mathbf{A}] \left\{ \begin{matrix} \{\delta\mathbf{U}\} \\ \{\delta\boldsymbol{\Lambda}\} \end{matrix} \right\}} \qquad [9.23]$$

where $[\mathbf{A}]$ is a square matrix, then $[\mathbf{A}]$ is the tangent matrix $[\mathbf{K}_T]$ that we have been looking for. Indeed, the variation $\delta\left(\mathcal{P}^*_{\text{discretized}}\right)$ is written according to [9.22]:

$$\delta\left(\mathcal{P}^*_{\text{discretized}}\right) = <\mathbf{U}^*> \{\delta\mathbf{R}\} + <\boldsymbol{\Lambda}^*> \{\delta\mathbf{R}_\Lambda\} = \left\{ \begin{matrix} \{\mathbf{U}^*\} \\ \{\boldsymbol{\Lambda}^*\} \end{matrix} \right\}^T \left\{ \begin{matrix} \{\delta\mathbf{R}\} \\ \{\delta\mathbf{R}_\Lambda\} \end{matrix} \right\} \qquad [9.24]$$

where $\{\delta\mathbf{R}\}$ and $\{\delta\mathbf{R}_\Lambda\}$ are functions of $(\mathbf{U},\boldsymbol{\Lambda};\delta\mathbf{U},\delta\boldsymbol{\Lambda})$. The column-vectors $\{\mathbf{U}^*\},\{\boldsymbol{\Lambda}^*\}$ which contain the virtual *nodal* velocities and the virtual *nodal* multipliers do *not* vary with $\{\mathbf{U}\},\{\boldsymbol{\Lambda}\}$!

The relation [9.24] therefore gives by using definition [9.10] of $[\mathbf{K}_T]$:

$$\delta\left(\mathcal{P}^*_{\text{discretized}}\right) = \left\{\begin{array}{c}\{\mathbf{U}^*\}\\ \{\mathbf{\Lambda}^*\}\end{array}\right\}^T [\mathbf{K}_T] \left\{\begin{array}{c}\{\delta\mathbf{U}\}\\ \{\delta\mathbf{\Lambda}\}\end{array}\right\}$$

By comparing this relation to [9.23], we do indeed have the equality $[\mathbf{A}] = [\mathbf{K}_T]$.

With this method, the matrix $[\mathbf{K}_T]$ is obtained in an indirect manner, with the help of a variational calculation followed by an identification, and not by the derivatives as in definition [9.10].

REMARKS.–

(i) The variations $\{\delta\mathbf{U}\}, \{\delta\mathbf{\Lambda}\}$ considered here serve to calculate the tangent matrix $[\mathbf{K}_T]$, and they are considered to be infinitesimals of the first order and have nothing to do with the $\{\delta\mathbf{U}\}, \{\delta\mathbf{\Lambda}\}$ of the iterative Newton–Raphson scheme.

(ii) In this method where we have obtained the matrix $[\mathbf{K}_T]$ by identification, it is important to work with the *arbitrary and independent* variations $\{\delta\mathbf{U}\}, \{\delta\mathbf{\Lambda}\}$ in order to obtain the *complete* expression for $[\mathbf{K}_T]$. ■

• It is necessary therefore to calculate the variation $\delta\left(\mathcal{P}^*_{\text{discretized}}\right)$ corresponding to the variations $\{\delta\mathbf{U}\}$ and $\{\delta\mathbf{\Lambda}\}$. Let us begin with the continuous expression [9.20] for the functional $\mathcal{P}^*$.

– The variation $\{\delta\mathbf{U}\}$ leads to the variations $\delta\mathbf{U}^{(1)}, \delta\mathbf{U}^{(2)}$ of the displacement vectors at each point of bodies 1 and 2 via the following chain, which uses some relations which will be established later:

$\{\delta\mathbf{U}\}$ $\Rightarrow$ the column-vectors $\{\delta\mathbf{U}^{(1)}\}, \{\delta\mathbf{U}^{(2)}\}$,
variations of the displacement column-vectors
$\underset{[9.33]}{\Rightarrow}$ $\forall$ element $e^{(i)}$ of body $i = 1, 2,$ the column-vector
$\{\delta\mathbf{U}^{(i),e^{(i)}}\} = [\mathbb{L}^{e^{(i)}}]\{\delta\mathbf{U}^{(i)}\}$, [9.25]
variation of the element nodal displacements
$\underset{\text{interpolations [9.34] – [9.35]}}{\Rightarrow}$ the vectors $\delta\mathbf{U}^{(i)} = [\mathcal{N}^{e^{(i)}}]\{\delta\mathbf{U}^{(i),e^{(i)}}\}$, variations of the displacement vectors.

– The variation $\{\delta\mathbf{\Lambda}\}$ leads to the variations $\delta\lambda_N, \delta\boldsymbol{\lambda}_T$ of the multipliers at each point of $S^{(1)}_{0c}$ via the following chain, which use some relations which will be established later on:

$\{\delta\mathbf{\Lambda}\}$ $\underset{[9.33]}{\Rightarrow}$ $\forall e^{(1)} \subset S^{(1)}_{0c}$, the column-vector $\{\delta\mathbf{\Lambda}^{e^{(1)}}\} = [\mathbb{L}^{e^{(1)}}]\{\delta\mathbf{\Lambda}\}$,
variation of the element nodal multipliers
$\underset{\text{interpolation [9.37]}}{\Rightarrow}$ the vector $\delta\boldsymbol{\lambda} = [\mathcal{N}^{e^{(1)}}]\{\delta\mathbf{\Lambda}^{e^{(1)}}\}$, variation of the multiplier vector
$\underset{[9.39]}{\Rightarrow}$ $\delta\lambda_N, \delta\boldsymbol{\lambda}_T$

[9.26]

Under the variations $\delta\mathbf{U}^{(1)}$, $\delta\mathbf{U}^{(2)}$, $\delta\lambda_N$, $\delta\lambda_T$ of the fields $\mathbf{U}^{(1)}$, $\mathbf{U}^{(2)}$, λ_N, λ_T, the functional $\mathcal{P}^*$ undergoes the variation:

$$\delta\mathcal{P}^* = \text{functional}\ (\mathbf{U}^{(1)}, \mathbf{U}^{(2)}, \lambda_N, \lambda_T; \quad \delta\mathbf{U}^{(1)}, \delta\mathbf{U}^{(2)}, \delta\lambda_N, \delta\lambda_T; \quad \mathbf{U}^{(1)*}, \mathbf{U}^{(2)*}, \lambda^*; \quad \delta\mathbf{U}^{(2)*}) \tag{9.27}$$

(The variation $\delta\mathbf{U}^{(2)*}$ appears in the functional since, as will be seen in [9.45], $\mathbf{U}^{(2)*}$ varies with $\delta\mathbf{U}^{(1)}$, $\delta\mathbf{U}^{(2)}$!)

Using interpolation, we express the continuous quantities as a function of the nodal quantities:

$$\left[\begin{array}{lll} \text{vector } \delta\mathbf{U}^{(i)} & \text{in terms of } \{\delta\mathbf{U}\} & \multirow{2}{*}{$\Big\}$ as has just been seen in [9.25] and [9.26]} \\ \delta\lambda_N, \delta\lambda_T & \text{in terms of } \{\delta\boldsymbol{\Lambda}\} & \\ \text{vector } \mathbf{U}^{(i)*} & \text{in terms of } \{\mathbf{U}^*\} & \multirow{2}{*}{$\Big\}$ as in [9.21]} \\ \text{vector } \lambda^* & \text{in terms of } \{\boldsymbol{\Lambda}^*\} & \end{array}\right.$$

This allows us to discretize $\delta\mathcal{P}^*$ in [9.27] and to put it into the form:

$$\forall\{\delta\mathbf{U}\}, \{\delta\boldsymbol{\Lambda}\}, \quad \boxed{(\delta\mathcal{P}^*)_{\text{discretized}} = \left\{\begin{array}{c} \{\mathbf{U}^*\} \\ \{\boldsymbol{\Lambda}^*\} \end{array}\right\}^T [\mathbf{B}] \left\{\begin{array}{c} \{\delta\mathbf{U}\} \\ \{\delta\boldsymbol{\Lambda}\} \end{array}\right\}} \tag{9.28}$$

where $[\mathbf{B}]$ is a square matrix to identify.

• Observe that, in starting from the same continuous expression [9.20] for $\mathcal{P}^*$, we have obtained two quantities that are a priori distinct, by following two different paths:

1. Discretize $\mathcal{P}^*$, then calculate the variation $\delta\mathcal{P}^*$ and put it into the form [9.23]. Obtain $\delta\left(\mathcal{P}^*_{\text{discretized}}\right)$.

2. Calculate the variation $\delta\mathcal{P}^*$, then discretize $\delta\mathcal{P}^*$ and put it into the form [9.28]. Obtain $(\delta\mathcal{P}^*)_{\text{discretized}}$.

In fact, the following assumption is implied when calculating the tangent matrix using the indirect method:

HYPOTHESIS.– We can *permute* the spatial discretization operation and the variation operation. In other words, we have the equality:

$$\delta\left(\mathcal{P}^*_{\text{discretized}}\right) = (\delta\mathcal{P}^*)_{\text{discretized}}$$

Therefore, $[\mathbf{B}]$ is equal to the matrix $[\mathbf{A}]$, thus to the tangent matrix $[\mathbf{K}_T]$ we have been looking for.

9.5.3. *Restriction to the contact tangent matrix*

In the tangent matrix [9.10], the submatrix $[\mathbf{K}_{classical}]$ is obtained in a standard manner, exactly as in the problem without contact. Therefore, in the sequel, it suffices to apply the calculation principle described previously while only considering the contact matrix $[\mathbf{K}_{\text{contact}}]$ defined by [9.11]. Recall the steps which have led to the discretized expression of the virtual contact power $\mathcal{P}^*_{\text{contact}}$.

1. Let us start with the continuous expression [8.1] of $\mathcal{P}^*_{\text{contact}}$, which is of the form:

$$\mathcal{P}^*_{\text{contact}} = \int_{S^{(1)}_{0c}} f(\underbrace{\mathbf{U}^{(1)}, \mathbf{U}^{(2)}}_{\text{containing } g, \nu, \mathbf{V}_T}, \lambda_N, \lambda_T; \mathbf{U}^{(1)*}, \mathbf{U}^{(2)*}, \lambda^*) dS_0$$

2. We interpolate the continuous quantities by [9.21], similar to $\mathcal{P}^*$ above.
3. We arrive at the discretized expression [8.39] of the virtual contact power:

$$\mathcal{P}^*_{\text{contact discretized}} = < \mathbf{U}^* > \{\mathbf{\Phi}_{\text{contact}}(\mathbf{U}, \mathbf{\Lambda})\} - < \mathbf{\Lambda}^* > \{\mathbf{R}_\Lambda(\mathbf{U}, \mathbf{\Lambda})\} \quad [9.29]$$

Let us consider now the *arbitrary and independent* variations $\{\delta\mathbf{U}\}$ and $\{\delta\mathbf{\Lambda}\}$ of the column-vectors $\{\mathbf{U}\}$ and $\{\mathbf{\Lambda}\}$. If we manage to calculate the variation $\delta\left(\mathcal{P}^*_{\text{contact discretized}}\right)$ corresponding to $\mathcal{P}^*_{\text{contact discretized}}$ and put it in the form:

$$\forall\{\delta\mathbf{U}\}, \{\delta\mathbf{\Lambda}\}, \quad \boxed{\delta\left(\mathcal{P}^*_{\text{contact discretized}}\right) = -\begin{Bmatrix} \{\mathbf{U}^*\} \\ \{\mathbf{\Lambda}^*\} \end{Bmatrix}^T [\mathbf{A}_{\text{contact}}] \begin{Bmatrix} \{\delta\mathbf{U}\} \\ \{\delta\mathbf{\Lambda}\} \end{Bmatrix}} \quad [9.30]$$

where $[\mathbf{A}_{\text{contact}}]$ is a square matrix, then $[\mathbf{A}_{\text{contact}}]$ is the contact tangent matrix $[\mathbf{K}_{\text{contact}}]$ that we have been looking for. Indeed, the variation $\delta\left(\mathcal{P}^*_{\text{contact discretized}}\right)$ is written according to [9.29]:

$$\delta\left(\mathcal{P}^*_{\text{contact discretized}}\right) = < \mathbf{U}^* > \{\delta\mathbf{\Phi}_{\text{contact}}\} - < \mathbf{\Lambda}^* > \{\delta\mathbf{R}_\Lambda\} = -\begin{Bmatrix} \{\mathbf{U}^*\} \\ \{\mathbf{\Lambda}^*\} \end{Bmatrix}^T \begin{Bmatrix} -\{\delta\mathbf{\Phi}_{\text{contact}}\} \\ \{\delta\mathbf{R}_\Lambda\} \end{Bmatrix}$$

This relation therefore gives, in using definition [9.11] of $[\mathbf{K}_{\text{contact}}]$:

$$\delta\left(\mathcal{P}^*_{\text{contact discretized}}\right) = -\begin{Bmatrix} \{\mathbf{U}^*\} \\ \{\mathbf{\Lambda}^*\} \end{Bmatrix}^T [\mathbf{K}_{\text{contact}}] \begin{Bmatrix} \{\delta\mathbf{U}\} \\ \{\delta\mathbf{\Lambda}\} \end{Bmatrix}$$

In comparing this relation to [9.30], we have indeed the equality $[\mathbf{A}_{\text{contact}}] = [\mathbf{K}_{\text{contact}}]$.

Therefore, the procedure to follow to calculate $[\mathbf{K}_{\text{contact}}]$ is the following:

Calculation steps for the contact matrix. [9.31]

1. Let us consider *arbitrary and independent* variations $\{\delta\mathbf{U}\}$ and $\{\delta\mathbf{\Lambda}\}$ of the column-vectors $\{\mathbf{U}\}$ and $\{\mathbf{\Lambda}\}$.

   These variations imply the variations $\delta\mathbf{U}^{(1)}$, $\delta\mathbf{U}^{(2)}$ of the displacement vectors and the variations $\delta\lambda_N$ and $\delta\lambda_T$ of the multipliers via the chains [9.25] and [9.26].

2. Now let us consider the continuous expression [8.1] of the functional $\mathcal{P}^*_{\text{contact}}$ and calculate the variation $\delta\mathcal{P}^*_{\text{contact}}$ corresponding to the variations $\delta\mathbf{U}^{(1)}$, $\delta\mathbf{U}^{(2)}$, $\delta\lambda_N$, $\delta\lambda_T$.

3. By interpolation, we express the continuous quantities that appear in $\delta\mathcal{P}^*_{\text{contact}}$ as a function of the nodal quantities.

   This allows us to discretize the variation $\delta\mathcal{P}^*_{\text{contact}}$ and to put it into the form (homologous to [9.28]):

$$\forall\{\delta\mathbf{U}\}, \{\delta\mathbf{\Lambda}\}, \quad \boxed{\left(\delta\mathcal{P}^*_{\text{contact}}\right)_{\text{discretized}} = -\begin{Bmatrix} \{\mathbf{U}^*\} \\ \{\mathbf{\Lambda}^*\} \end{Bmatrix}^T [\mathbf{B}_{\text{contact}}] \begin{Bmatrix} \{\delta\mathbf{U}\} \\ \{\delta\mathbf{\Lambda}\} \end{Bmatrix}} \quad [9.32]$$

   where $[\mathbf{B}_{\text{contact}}]$ is a square matrix to identify. The matrix $[\mathbf{B}_{\text{contact}}]$ is equal to $[\mathbf{A}_{\text{contact}}]$, hence to the contact tangent matrix $[\mathbf{K}_{\text{contact}}]$ we have been looking for.

9.6. Calculation of the contact tangent matrix

In this section, we are going to apply the calculation scheme [9.31] to obtain the explicit expression for the contact tangent matrix defined by [9.11]:

$$\boxed{[\mathbf{K}_{\text{contact}}] = \left[\begin{array}{c|c} \mathbf{K}_{UU} & \mathbf{K}_{U\Lambda} \\ \hline \mathbf{K}_{\Lambda U} & \mathbf{K}_{\Lambda\Lambda} \end{array}\right]}$$

9.6.1. *Variations of the arguments of the functional $\mathcal{P}^*_{\text{contact}}$*

Let us begin with the first step described in [9.31]. We consider *arbitrary and independent* variations $\{\delta\mathbf{U}\}$ and $\{\delta\mathbf{\Lambda}\}$ of the column-vectors $\{\mathbf{U}\}$ and $\{\mathbf{\Lambda}\}$. These variations lead to the variations $\delta\mathbf{U}^{(1)}$, $\delta\mathbf{U}^{(2)}$ of the displacement vectors and the variations $\delta\lambda_N$ and $\delta\lambda_T$ of the multipliers via the chains [9.25] and [9.26]. We will explain these chains in detail, even if the one relative to displacement variations is classical.

Variations of the element nodal quantities

The column-vectors which group together the variations at the nodes of an element are denoted as follows.

NOTATION.–

1. VARIATIONS OF THE ELEMENT NODE DISPLACEMENTS:

   – $\{\delta\mathbf{U}^{(1),e^{(1)}}\}$: the column-vector grouping together the triplets of components of the vector $\delta\mathbf{U}^{(1)}$ at the nodes of element $e^{(1)}$;

   – in an analogous way, $\{\delta\mathbf{U}^{(2),e^{(2)}}\}$: the column-vector grouping together the triplets of the components of the vector $\delta\mathbf{U}^{(2)}$ at the nodes of element $e^{(2)}$;

   – we shall see that we will also need $\{\delta\mathbf{U}^{(2),e^{(2)}_{n-1}}\}$: the column-vector grouping together the triplets of components of the vector $\delta\mathbf{U}^{(2)}$ at the nodes of element $e^{(2)}_{n-1}$, where $e^{(2)}_{n-1}$ is the element of $S^{(2)}_{0c}$ which contains the point $\mathbf{Y}_{n-1}$ (see definition [8.54]).

2. VARIATIONS OF THE ELEMENT NODAL MULTIPLIERS:

   – $\{\delta\mathbf{\Lambda}^{e^{(1)}}\}$: the column-vector with $3nne^{(1)}$ components grouping together the triplets of components of the vector $\delta\boldsymbol{\lambda}$ at the nodes of element $e^{(1)}$.

Recall the convention from section 8.3 according to which each component in question is relative to the fixed orthonormal basis $(\mathbf{e}_1, \mathbf{e}_2, \mathbf{e}_3)$.

The element column-vectors are extracted from the global column-vectors $\{\delta\mathbf{U}\}$ and $\{\delta\mathbf{\Lambda}\}$ with the help of the same localization matrices as in [8.35]:

$$\begin{aligned}
\{\delta\mathbf{U}^{(1),e^{(1)}}\} &= [\mathbb{L}^{e^{(1)}}]\{\delta\mathbf{U}^{(1)}\} \\
\{\delta\mathbf{U}^{(2),e^{(2)}}\} &= [\mathbb{L}^{e^{(2)}}]\{\delta\mathbf{U}^{(2)}\} \qquad \{\delta\mathbf{U}^{(2),e^{(2)}_{n-1}}\} = [\mathbb{L}^{e^{(2)}_{n-1}}]\{\delta\mathbf{U}^{(2)}\} \\
\{\delta\mathbf{\Lambda}^{e^{(1)}}\} &= [\mathbb{L}^{e^{(1)}}]\{\delta\mathbf{\Lambda}\}
\end{aligned} \qquad [9.33]$$

*Variations of the fields **U** and λ*

The three-component column-vectors representing the vectors $\delta\mathbf{U}^{(1)}$, $\delta\mathbf{U}^{(2)}$ and $\delta\boldsymbol{\lambda}$ at a current point are denoted as follows:

NOTATION.–

1. VARIATIONS OF THE DISPLACEMENT VECTORS:

   – $\{\delta\mathbf{U}^{(1)}\}$ designates the column-vector containing the three components of the variation of the displacement vector $\delta\mathbf{U}^{(1)}$;

   – similarly, $\{\delta\mathbf{U}^{(2)}\}$ designates the column-vector containing the three components of the variation of the displacement vector $\delta\mathbf{U}^{(2)}$;

   – we will also need the variation of the displacement $\delta\mathbf{U}^{(2)}(\mathbf{Y}_{n-1})$ that features in the variation $\delta(\mathbf{V}_T\Delta t)$ of [9.56]. We denote by $\{\delta\mathbf{U}^{(2)}(\mathbf{Y}_{n-1})\}$ the column-vector containing the three components of the variation vector $\delta\mathbf{U}^{(2)}(\mathbf{Y}_{n-1})$.

2. VARIATION OF THE MULTIPLIER VECTOR:

   – $\{\delta\boldsymbol{\lambda}\}$ is the column-vector comprising the three components of vector $\delta\boldsymbol{\lambda}$.

We calculate the preceding vectors by interpolating the element nodal vectors as follows.

• As was done for the displacement, the variation of the displacement at a current point of the element is obtained by interpolating the nodal variations. The variation of the displacement $\delta\mathbf{U}^{(1)}$ is obtained by interpolating the variations of the nodal displacements of element $e^{(1)}$:

$$\boldsymbol{\xi} \mapsto \underset{(3,1)}{\{\delta\mathbf{U}^{(1)}\}} = \underset{(3,3nne^{(1)})}{[\mathcal{N}^{e^{(1)}}(\boldsymbol{\xi})]} \underset{(3nne^{(1)},1)}{\{\delta\mathbf{U}^{(1),e^{(1)}}\}} \qquad [9.34]$$

• Likewise, the variation of the displacement $\delta\mathbf{U}^{(2)}$ is obtained by interpolating the variations of the nodal displacements of element $e^{(2)}$:

$$\boldsymbol{\eta} \mapsto \underset{(3,1)}{\{\delta\mathbf{U}^{(2)}\}} = \underset{(3,3nne^{(2)})}{[\mathcal{N}^{e^{(2)}}(\boldsymbol{\eta})]} \underset{(3nne^{(2)},1)}{\{\delta\mathbf{U}^{(2),e^{(2)}}\}} \qquad [9.35]$$

• The variation of the displacement $\delta\mathbf{U}^{(2)}(\mathbf{Y}_{n-1})$ is obtained by interpolating the variations of the displacement of the nodes of element $e^{(2)}_{n-1}$ (see the definition of $e^{(2)}_{n-1}$ in [8.54]):

$$\boldsymbol{\eta} \mapsto \underset{(3,1)}{\{\delta\mathbf{U}^{(2)}(\mathbf{Y}_{n-1})\}} = \underset{(3,3nne^{(2)}_{n-1})}{[\mathcal{N}^{e^{(2)}_{n-1}}(\boldsymbol{\eta}_{n-1})]} \underset{(3nne^{(2)}_{n-1},1)}{\{\delta\mathbf{U}^{(2),e^{(2)}_{n-1}}\}} \qquad [9.36]$$

where $nne^{(2)}_{n-1}$ is the number of nodes of element $e^{(2)}_{n-1}$.

• The variation of the multiplier at a current point of the element $e^{(1)}$ is interpolated in the same manner as the variations of the displacements, namely, from the nodal variations:

$$\underset{(3,1)}{\{\delta\boldsymbol{\lambda}\}} = \underset{(3,3nne^{(1)})}{[\mathcal{N}^{e^{(1)}}(\boldsymbol{\xi})]} \underset{(3nne^{(1)},1)}{\{\delta\boldsymbol{\Lambda}^{e^{(1)}}\}} \qquad [9.37]$$

It is the Cartesian components of the multiplier vector $\delta\boldsymbol{\lambda}$ (components in the fixed orthonormal basis $(\mathbf{e}_1, \mathbf{e}_2, \mathbf{e}_3)$), which are interpolated.

Obtaining the variations $\delta\lambda_N$ and $\delta\boldsymbol{\lambda}_T$

Since the virtual contact power $\mathcal{P}^*_{\text{contact}}$ includes λ_N and $\boldsymbol{\lambda}_T$ among its arguments, its variation $\delta\mathcal{P}^*_{\text{contact}}$ depends on the variations $\delta\lambda_N$ and $\delta\boldsymbol{\lambda}_T$ of λ_N and $\boldsymbol{\lambda}_T$, respectively. The question that now presents itself is how we can link $\delta\lambda_N$ and $\delta\boldsymbol{\lambda}_T$ to the vector $\delta\boldsymbol{\lambda}$ obtained in [9.37]. We choose to calculate $\delta\lambda_N$ and $\delta\boldsymbol{\lambda}_T$ via the following relation:

$$\boxed{\delta\boldsymbol{\lambda} = \delta\lambda_N\boldsymbol{\nu} - \delta\boldsymbol{\lambda}_T} \qquad \text{with } \delta\boldsymbol{\lambda}_T \perp \boldsymbol{\nu} \tag{9.38}$$

where $\boldsymbol{\nu}$ is the unit normal at the point $\mathbf{y} \in S_c^{(2)}$, calculated from the displacement fields $\mathbf{U}^{(1)}, \mathbf{U}^{(2)}$. In other words:

- $\delta\lambda_N$ is the component of vector $\delta\boldsymbol{\lambda}$ along the normal $\boldsymbol{\nu}$;
- $\delta\boldsymbol{\lambda}_T$ is (the opposite of) the projection of $\delta\boldsymbol{\lambda}$ onto the plane orthogonal to $\boldsymbol{\nu}$.

The relation [9.38] can be written in an equivalent way:

$$\boxed{\begin{aligned} \delta\lambda_N &= \delta\boldsymbol{\lambda}.\boldsymbol{\nu} \\ \delta\boldsymbol{\lambda}_T &= -(\mathbf{I} - \boldsymbol{\nu}\otimes\boldsymbol{\nu}).\delta\boldsymbol{\lambda} \end{aligned}} \tag{9.39}$$

We can easily prove the following result:

PROPERTY.– [9.40]

Given a normal vector $\boldsymbol{\nu}$, let us denote $E^{\perp\nu}$ the vector space orthogonal to $\boldsymbol{\nu}$ (it is a two-dimensional vector subspace, comparable to $\mathbb{R}^2$). The mapping:

$$\begin{aligned} \mathbb{R}^3 &\rightarrow \mathbb{R}\times E^{\perp\nu} \\ \delta\boldsymbol{\lambda} &\mapsto (\delta\lambda_N, \delta\boldsymbol{\lambda}_T \perp \boldsymbol{\nu}) \text{ given by [9.39]} \end{aligned}$$

is *a bijection.*

Given the normal field $\boldsymbol{\nu}$, we can easily check that an arbitrary vector field $\delta\boldsymbol{\lambda}$ generates an arbitrary field $\delta\lambda_N$ and an arbitrary field $\delta\boldsymbol{\lambda}_T$ in the set of planes orthogonal to the normals $\boldsymbol{\nu}$. Observe that the variation $\delta\lambda_N$ is general, without any condition on its sign being assumed, we do not need to consider the $\delta\lambda_N$ such that λ_N is ≥ 0.

The normal vector $\boldsymbol{\nu}$ in [9.38]–[9.39] is calculated by means of the real displacement fields $\mathbf{U}^{(1)}, \mathbf{U}^{(2)}$. The displacement variations $\delta\mathbf{U}^{(1)}$ and $\delta\mathbf{U}^{(2)}$ generate a variation $\delta\boldsymbol{\nu}$, but $\delta\boldsymbol{\nu}$ does *not* play a role in [9.38]–[9.39]. To insist that $\boldsymbol{\nu}$ comes from the fields $\mathbf{U}^{(1)}, \mathbf{U}^{(2)}$ and not from $\mathbf{U}^{(1)} + \delta\mathbf{U}^{(1)}, \mathbf{U}^{(2)} + \delta\mathbf{U}^{(2)}$, we say that $\boldsymbol{\nu}$ is the normal *before the variation*. The normal $\boldsymbol{\nu}$ and the displacements $\mathbf{U}^{(1)}, \mathbf{U}^{(2)}$ are the fields *before the variation.*

REMARKS.–

1. In numerical computations, the normal $\boldsymbol{\nu}$ which features in [9.39] comes from the displacement fields $\mathbf{U}^{(1)}, \mathbf{U}^{(2)}$ that were *recently acquired, that is to say, from the previous numerical iteration.*

   Recall that we are at iteration i of the step n and that we suppose that the solution $\mathbf{U}^{i-1}, \boldsymbol{\Lambda}^{i-1}$ for the previous iteration $i-1$ is known. The tangent matrix in the expression [9.14] that we are about to establish is calculated with the solution $\mathbf{U}^{i-1}, \boldsymbol{\Lambda}^{i-1}$. Each ingredient involved in the computation is one arising from the iteration $i-1$, the relation [9.39] can be written, in fact, as

   $$\delta\lambda_N = \delta\boldsymbol{\lambda}.\boldsymbol{\nu}^{i-1} \qquad \delta\boldsymbol{\lambda}_T = -(\mathbf{I} - \boldsymbol{\nu}^{i-1}\otimes\boldsymbol{\nu}^{i-1}).\delta\boldsymbol{\lambda}$$

2. The procedure of calculating $[\mathbf{K}_{\text{contact}}]$ is a purely algebraic one in which we need to have $\{\delta\mathbf{\Lambda}\}$ ("small" and) *arbitrary*, but we do not need $\{\delta\mathbf{\Lambda}\}$ to make sense physically. We do not need to assume that the components of $\{\delta\mathbf{\Lambda}\}$ are positive. In addition, let us examine for example the question of orthogonality. We deduce from the relations $\boldsymbol{\lambda} = \lambda_N\boldsymbol{\nu} - \boldsymbol{\lambda}_T$ and $\delta\boldsymbol{\lambda} = \delta\lambda_N\boldsymbol{\nu} - \delta\boldsymbol{\lambda}_T$ that $\boldsymbol{\lambda} + \delta\boldsymbol{\lambda} = (\lambda_N + \delta\lambda_N)\boldsymbol{\nu} - (\boldsymbol{\lambda}_T + \delta\boldsymbol{\lambda}_T)$. The vector $\boldsymbol{\lambda}_T + \delta\boldsymbol{\lambda}_T$ is orthogonal to the normal $\boldsymbol{\nu}$ before variation but is not to the new normal $\boldsymbol{\nu} + \delta\boldsymbol{\nu}$ after variation. This is not of any importance, since we do not require that $\boldsymbol{\lambda}_T + \delta\boldsymbol{\lambda}_T$ is orthogonal to $\boldsymbol{\nu} + \delta\boldsymbol{\nu}$. ■

Arbitrary character and independence of the fields $\delta\mathbf{U}^{(1)}$, $\delta\mathbf{U}^{(2)}$, $\delta\lambda_N$, $\delta\boldsymbol{\lambda}_T \perp \boldsymbol{\nu}$

We have assumed that the column-vectors $\{\delta\mathbf{U}\}$ and $\{\delta\mathbf{\Lambda}\}$ are *arbitrary and independent*. Let us see if these properties are carried over to the vector fields $\delta\mathbf{U}^{(1)}$, $\delta\mathbf{U}^{(2)}$, $\delta\lambda_N$, $\delta\boldsymbol{\lambda}_T$.

• Arbitrary character of the fields $\delta\mathbf{U}^{(1)}$, $\delta\mathbf{U}^{(2)}$, $\delta\lambda_N$ and $\delta\boldsymbol{\lambda}_T$.

The column-vector $\{\delta\mathbf{U}\}$ generates the vector fields $\delta\mathbf{U}^{(1)}$, $\delta\mathbf{U}^{(2)}$ in bodies 1 and 2 via the following chain:

$$\{\delta\mathbf{U}\} \Rightarrow \left| \begin{matrix} \{\delta\mathbf{U}^{(1)}\} \\ \{\delta\mathbf{U}^{(2)}\} \end{matrix} \right. \Rightarrow \left| \begin{matrix} \{\delta\mathbf{U}^{(1),e^{(1)}}\} \\ \{\delta\mathbf{U}^{(2),e^{(2)}}\} \end{matrix} \right. \text{ by extraction [9.33]}$$

$$\Rightarrow \left| \begin{matrix} \delta\mathbf{U}^{(1)} \\ \delta\mathbf{U}^{(2)} \end{matrix} \right. \text{ by interpolation [9.34], [9.35]}$$

On the other hand, the column-vector $\{\delta\mathbf{\Lambda}\}$ generates the vector fields $\delta\boldsymbol{\lambda}$ on $S_{0c}^{(1)}$ via the following chain:

$$\{\delta\mathbf{\Lambda}\} \Rightarrow \{\delta\mathbf{\Lambda}^{e^{(1)}}\} \text{ by extraction [9.33]} \Rightarrow \delta\boldsymbol{\lambda} \text{ by interpolation [9.37]}$$

The two previous chains show that if the column-vectors $\{\delta\mathbf{U}\}$ and $\{\delta\mathbf{\Lambda}\}$ are arbitrary, then the element column-vectors $\{\delta\mathbf{U}^{(1),e^{(1)}}\}$, $\{\delta\mathbf{U}^{(2),e^{(2)}}\}$, $\{\delta\mathbf{\Lambda}^{e^{(1)}}\}$ are too. The arbitrary character is lost somewhat after the interpolation operation, we say that the fields $\delta\mathbf{U}^{(1)}$, $\delta\mathbf{U}^{(2)}$, $\delta\boldsymbol{\lambda}$ are arbitrary up to interpolation. We have seen that if the field $\delta\boldsymbol{\lambda}$ is arbitrary, then the fields $\delta\lambda_N$, $\delta\boldsymbol{\lambda}_T \perp \boldsymbol{\nu}$ are too.

• Independence of the fields $\delta\lambda_N$ and $\delta\boldsymbol{\lambda}_T$ vis-Ãă-vis the displacement field variations $\delta\mathbf{U}^{(1)}$, $\delta\mathbf{U}^{(2)}$.

The same two chains that we have just presented above show that if the column-vectors $\{\delta\mathbf{U}\}$ and $\{\delta\mathbf{\Lambda}\}$ are independent, then the vector fields $\delta\mathbf{U}^{(1)}$, $\delta\mathbf{U}^{(2)}$ and the vector field $\delta\boldsymbol{\lambda}$ are independent. Now, $\delta\lambda_N$ and $\delta\boldsymbol{\lambda}_T$ depend on $\delta\boldsymbol{\lambda}$ via the relation [9.39], then $\delta\mathbf{U}^{(1)}$, $\delta\mathbf{U}^{(2)}$ cannot depend on $\delta\lambda_N$ and $\delta\boldsymbol{\lambda}_T$.

We can also reason conversely. The pair $\delta\lambda_N$ and $\delta\boldsymbol{\lambda}_T$ cannot depend on $\delta\mathbf{U}^{(1)}$, $\delta\mathbf{U}^{(2)}$, since otherwise the vector $\delta\boldsymbol{\lambda} = \delta\lambda_N\boldsymbol{\nu} - \delta\boldsymbol{\lambda}_T$ would depend on $\delta\mathbf{U}^{(1)}$, $\delta\mathbf{U}^{(2)}$ (we cannot combine $\delta\lambda_N$ and $\delta\boldsymbol{\lambda}_T$ to remove the dependence), which would be a contradiction with the observation above.

9.6.2. *Calculation of the variation $\delta\mathcal{P}^*_{contact}$*

Let us continue on now to the second step of the calculation scheme [9.31]. We consider the variations $\delta\mathbf{U}^{(1)}$, $\delta\mathbf{U}^{(2)}$, $\delta\lambda_N$, $\delta\boldsymbol{\lambda}_T \perp \boldsymbol{\nu}$ obtained previously, and we will calculate the corresponding variation $\delta\mathcal{P}^*_{\text{contact}}$ of $\mathcal{P}^*_{\text{contact}}$ which we recall is of the form:

$$\mathcal{P}^*_{\text{contact}} = \int_{S_{0c}^{(1)}} f(\underbrace{\mathbf{U}^{(1)}, \mathbf{U}^{(2)}}_{\text{containing } g, \boldsymbol{\nu}, \mathbf{V}_T}, \lambda_N, \boldsymbol{\lambda}_T; \mathbf{U}^{(1)*}, \mathbf{U}^{(2)*}, \boldsymbol{\lambda}^*) dS_0$$

where f denotes the functional in [8.1].

The variation calculation will be carried out assuming that the variations $\delta\mathbf{U}^{(1)}$, $\delta\mathbf{U}^{(2)}$, $\delta\lambda_N$, $\delta\boldsymbol{\lambda}_T$ are *arbitrary and independent*.

- When we get $\mathcal{P}^*_{\text{contact}}$, we will take $\delta\boldsymbol{\lambda}_T \perp \boldsymbol{\nu}$.
- When we discretize $\mathcal{P}^*_{\text{contact}}$, we will limit ourselves to the fields $\delta\mathbf{U}^{(1)}$, $\delta\mathbf{U}^{(2)}$, $\delta\lambda_N$, $\delta\boldsymbol{\lambda}_T \perp \boldsymbol{\nu}$ defined by the interpolations.

Under the variations $\delta\mathbf{U}^{(1)}$, $\delta\mathbf{U}^{(2)}$, $\delta\lambda_N$, $\delta\boldsymbol{\lambda}_T$, the virtual contact power $\mathcal{P}^*_{\text{contact}}$ undergoes a change. *The difference* which arises from this is denoted $\Delta\mathcal{P}^*_{\text{contact}}$. According to the theory of calculus of variations, the variations $\delta\mathbf{U}^{(1)}$, $\delta\mathbf{U}^{(2)}$, $\delta\lambda_N$, $\delta\boldsymbol{\lambda}_T$ are considered as the basic infinitesimals (i.e. of the first order), and *the variation* $\delta\mathcal{P}^*_{\text{contact}}$ corresponding to the previous variations is by definition the part of the difference $\Delta\mathcal{P}^*_{\text{contact}}$, which is of the first order like $\delta\mathbf{U}^{(1)}$, $\delta\mathbf{U}^{(2)}$, $\delta\lambda_N$, $\delta\boldsymbol{\lambda}_T$.

We have

$$\Delta\mathcal{P}^*_{\text{contact}} = \Delta \int_{S_{0c}^{(1)}} f dS_0 = \int_{S_{0c}^{(1)}} \Delta f dS_0$$

where the difference Δf is

$$\begin{aligned}\Delta f = {} & f(\mathbf{U}^{(1)} + \delta\mathbf{U}^{(1)}, \mathbf{U}^{(2)} + \delta\mathbf{U}^{(2)}, \lambda_N + \delta\lambda_N, \boldsymbol{\lambda}_T + \delta\boldsymbol{\lambda}_T; \mathbf{U}^{(1)*}, \mathbf{U}^{(2)*} + \delta\mathbf{U}^{(2)*}, \boldsymbol{\lambda}^*) \\ & - f(\mathbf{U}^{(1)}, \mathbf{U}^{(2)}, \lambda_N, \boldsymbol{\lambda}_T; \mathbf{U}^{(1)*}, \mathbf{U}^{(2)*}, \boldsymbol{\lambda}^*)\end{aligned}$$

From this, we deduce the variation of $\mathcal{P}^*_{\text{contact}}$:

$$\boxed{\delta\mathcal{P}^*_{\text{contact}} = \delta \int_{S_{0c}^{(1)}} f dS_0 = \int_{S_{0c}^{(1)}} \delta f dS_0}$$

The variation $\delta\mathcal{P}^*_{\text{contact}}$ is the sum of the variations on the elements $e^{(1)} \subset S_{0c}^{(1)}$:

$$\delta\mathcal{P}^*_{\text{contact}} = \sum_{e^{(1)} \subset S_{0c}^{(1)}} (\delta\mathcal{P}^*_{\text{contact}})^{e^{(1)}} \qquad [9.41]$$

9.6.3. *Calculation of the variation* $(\delta\mathcal{P}^*_{\text{contact}})^{e^{(1)}}$

According to relation [9.41], the calculation reduces to that of the variation $(\delta\mathcal{P}^*_{\text{contact}})^{e^{(1)}}$ on each element $e^{(1)} \subset S_{0c}^{(1)}$:

$$\boxed{(\delta\mathcal{P}^*_{\text{contact}})^{e^{(1)}} = \int_{e^{(1)}} \delta f dS_0}$$

We deduce from [8.3]:

$$\left(\delta\mathcal{P}^*_{\text{contact}}\right)^{e^{(1)}} = -\int_{e^{(1)}} \Big\{ -\delta\langle\hat{\lambda}_N\rangle(\mathbf{U}^{(1)*} - \mathbf{U}^{(2)*}).\boldsymbol{\nu} - \langle\hat{\lambda}_N\rangle\,\delta\left[(\mathbf{U}^{(1)*} - \mathbf{U}^{(2)*}).\boldsymbol{\nu}\right]$$

$$-\delta\langle 1 - \frac{\mu\langle\hat{\lambda}_N\rangle}{||\hat{\boldsymbol{\lambda}}_T||}\rangle(\mathbf{U}^{(1)*} - \mathbf{U}^{(2)*}).\hat{\boldsymbol{\lambda}}_T + \left(1 - \langle 1 - \frac{\mu\langle\hat{\lambda}_N\rangle}{||\hat{\boldsymbol{\lambda}}_T||}\rangle\right)\,\delta\left[(\mathbf{U}^{(1)*} - \mathbf{U}^{(2)*}).\hat{\boldsymbol{\lambda}}_T\right]\Big\}dS_0$$

$$-\int_{e^{(1)}} \left\{-\frac{1}{\epsilon_N}(\delta\lambda_N - \delta\langle\hat{\lambda}_N\rangle)\mathbf{m} + \frac{1}{\epsilon_T}\left[\delta\boldsymbol{\lambda}_T + \delta\langle 1 - \frac{\mu\langle\hat{\lambda}_N\rangle}{||\hat{\boldsymbol{\lambda}}_T||}\rangle\hat{\boldsymbol{\lambda}}_T - \left(1 - \langle 1 - \frac{\mu\langle\hat{\lambda}_N\rangle}{||\hat{\boldsymbol{\lambda}}_T||}\rangle\right)\,\delta\hat{\boldsymbol{\lambda}}_T\right]\right\}.\boldsymbol{\lambda}^* dS_0 \quad [9.42]$$

We have placed a minus sign before the integrals to conform with the minus sign in [9.32].

The field $\mathbf{m}$ did not undergo any variation since it is *independent* of the fields $\mathbf{U}^{(1)}, \mathbf{U}^{(2)}, \lambda_N, \boldsymbol{\lambda}_T$.

Let us explain the variations in [9.42]. The variations $\delta\hat{\lambda}_N, \delta\hat{\boldsymbol{\lambda}}_T$ are straightforward:

$$\begin{aligned}\delta\hat{\lambda}_N &= \delta\lambda_N + \epsilon_N\delta g\\ \delta\hat{\boldsymbol{\lambda}}_T &= \delta\boldsymbol{\lambda}_T + \epsilon_T\delta\mathbf{V}_T\end{aligned} \quad [9.43]$$

The virtual velocity $\mathbf{U}^{(2)*}(\mathbf{Y})$ requires special attention, since it undergoes a variation, which is by no means obvious. To convince ourselves of this, recall that $\{\mathbf{U}^{(2)*}(\mathbf{Y})\}$ is obtained by the interpolation [8.13] of the vector of element nodal quantities $\{\mathbf{U}^{(2)*,e^{(2)}}\}$, and that the last is itself extracted from the global nodal vector $\{\mathbf{U}^{(2)*}\}$ thanks to the relation [8.35]:

$$\{\mathbf{U}^{(2)*}(\mathbf{Y})\} \underset{[8.13]}{=} [\mathbb{N}^{e^{(2)}}(\boldsymbol{\eta})]\{\mathbf{U}^{(2)*,e^{(2)}}\} \underset{[8.35]}{=} [\mathbb{N}^{e^{(2)}}(\boldsymbol{\eta})][\mathbb{L}^{e^{(2)}}]\{\mathbf{U}^{(2)*}\} \quad [9.44]$$

The variations $\delta\mathbf{U}^{(1)}$, $\delta\mathbf{U}^{(2)}$ of the displacements do not affect the global nodal vector $\{\mathbf{U}^{(2)*}\}$ of the virtual velocities. On the contrary, they make the point $\mathbf{Y}$ vary, and hence its pre-image $\boldsymbol{\eta}$ as well, as can be seen in Figure A.1 in the Appendix. Consequently, the virtual velocity $\mathbf{U}^{(2)*}(\mathbf{Y}(\mathbf{X}, t))$, evaluated at the point $\mathbf{Y}$, undergoes a variation equal to $\delta\mathbf{U}^{(2)*} = \mathbf{U}^{(2)*}_{,\alpha}\delta\eta^\alpha$!

Therefore, the terms containing $\mathbf{U}^{(2)*}$ in [9.42] undergo the following variations:

$$\begin{aligned}\delta[(\mathbf{U}^{(1)*} - \mathbf{U}^{(2)*}).\boldsymbol{\nu}] &= (\mathbf{U}^{(1)*} - \mathbf{U}^{(2)*}).\delta\boldsymbol{\nu} - \mathbf{U}^{(2)*}_{,\alpha}.\boldsymbol{\nu}\,\delta\eta^\alpha\\ \delta[(\mathbf{U}^{(1)*} - \mathbf{U}^{(2)*}).\hat{\boldsymbol{\lambda}}_T] &= (\mathbf{U}^{(1)*} - \mathbf{U}^{(2)*}).\delta\hat{\boldsymbol{\lambda}}_T - \mathbf{U}^{(2)*}_{,\alpha}.\hat{\boldsymbol{\lambda}}_T\,\delta\eta^\alpha\end{aligned} \quad [9.45]$$

REMARKS.–

(i) The variation of $\mathbf{U}^{(2)*}(\mathbf{Y}(\mathbf{X}, t))$ does not contradict the fact mentioned in [9.24], namely, that $\{\mathbf{U}^*\}$ is independent of $\{\mathbf{U}\}$. Indeed, $\{\mathbf{U}^*\}$ is the column-vector of *nodal* virtual velocities of the structure and it does not vary with $\{\mathbf{U}\}$ (see definition [8.34]).

(ii) As the variations $\delta\mathbf{U}^{(1)}$, $\delta\mathbf{U}^{(2)}$ of the displacements are infinitesimals of the first order, we can always assume that, during variations, the point $\mathbf{Y}$ changes but remains within the interior of the same element $e^{(2)} \subset S^{(2)}_{0c}$. The point $\boldsymbol{\eta}$ in [9.44] varies, but the element $e^{(2)}$ remains the same.

This reasoning works well if the point $\mathbf{Y}$ is found strictly within the interior of the element $e^{(2)}$. If $\mathbf{Y}$ is located on the boundary of the element $e^{(2)}$, it may come out slightly from $e^{(2)}$, but we can continue to use the previous results by imagining that $e^{(2)}$ has been enlarged slightly. ■

It remains to calculate the variations $\delta\langle\hat{\lambda}_N\rangle$ and $\delta\langle 1 - \frac{\mu\langle\hat{\lambda}_N\rangle}{||\hat{\boldsymbol{\lambda}}_T||}\rangle$. To do this, let us suppose that:

HYPOTHESIS.– [9.46]

At the considered point $\mathbf{X} \in e^{(1)}$:

1. $\hat{\lambda}_N \equiv \lambda_N + \epsilon_N g \neq 0$ ($\hat{\lambda}_N$ can be $\lessgtr 0$ but not $= 0$). Numerically, this hypothesis is essentially satisfied, since it is rare to have $\hat{\lambda}_N = 0$ exactly (whereas it is often that we have $\hat{\boldsymbol{\lambda}}_T = \mathbf{0}$).

2. $||\hat{\boldsymbol{\lambda}}_T|| \neq \mu\langle\hat{\lambda}_N\rangle$ (we can have $\lessgtr$ but not $=$). Numerically, this hypothesis is also satisfied. If $\mu = 0$, this hypothesis is written as $\hat{\boldsymbol{\lambda}}_T \neq \mathbf{0}$ at the point $\mathbf{X} \in e^{(1)}$ considered.

This hypothesis is necessary to use the following result: if $x \neq 0$, then $\delta\langle x\rangle = H(x)\,\delta x$, where $H(x)$ is the Heaviside function defined by (see Figure 9.1):

$$H(x) = \begin{cases} 1 \text{ if } x \geq 0 \\ 0 \text{ if } x < 0 \end{cases}$$

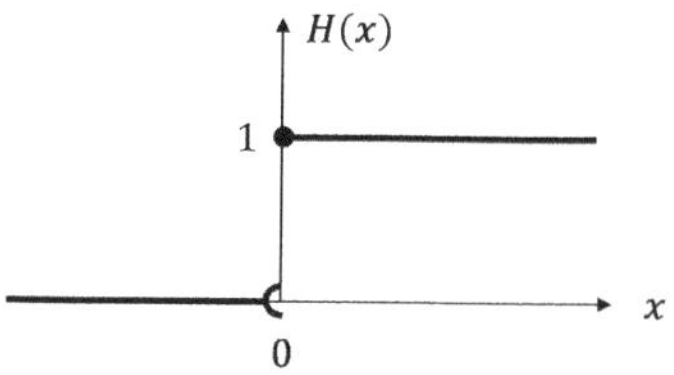

Figure 9.1. *Heaviside function*

From the hypothesis $[9.46]_1$, we have

$$\delta\langle\hat{\lambda}_N\rangle = H(\hat{\lambda}_N)\delta\hat{\lambda}_N \qquad [9.47]$$

Following the hypothesis $[9.46]_2$, we have

$$\delta\langle 1 - \frac{\mu\langle\hat{\lambda}_N\rangle}{||\hat{\boldsymbol{\lambda}}_T||}\rangle = -H(1 - \frac{\mu\langle\hat{\lambda}_N\rangle}{||\hat{\boldsymbol{\lambda}}_T||})\delta\frac{\mu\langle\hat{\lambda}_N\rangle}{||\hat{\boldsymbol{\lambda}}_T||}$$

where

$$\begin{aligned} \delta\frac{\mu\langle\hat{\lambda}_N\rangle}{||\hat{\boldsymbol{\lambda}}_T||} &= \mu\frac{\delta\langle\hat{\lambda}_N\rangle||\hat{\boldsymbol{\lambda}}_T|| - \langle\hat{\lambda}_N\rangle\delta||\hat{\boldsymbol{\lambda}}_T||}{||\hat{\boldsymbol{\lambda}}_T||^2} && \text{by using } \delta\left(\frac{u}{v}\right) = \frac{v\delta u - u\delta v}{v^2} \\ &= \mu\frac{H(\hat{\lambda}_N)}{||\hat{\boldsymbol{\lambda}}_T||}\delta\hat{\lambda}_N - \frac{\mu\langle\hat{\lambda}_N\rangle}{||\hat{\boldsymbol{\lambda}}_T||}\frac{\hat{\boldsymbol{\lambda}}_T.\delta\hat{\boldsymbol{\lambda}}_T}{||\hat{\boldsymbol{\lambda}}_T||^2} && \text{knowing that } \hat{\boldsymbol{\lambda}}_T \neq \mathbf{0} \\ & && \text{and that if } \mathbf{x} \neq \mathbf{0} \text{ then } \delta||\mathbf{x}|| = \frac{\mathbf{x}}{||\mathbf{x}||}\delta\mathbf{x} \end{aligned}$$

Hence

$$\delta\langle 1 - \frac{\mu\langle\hat{\lambda}_N\rangle}{||\hat{\boldsymbol{\lambda}}_T||}\rangle = -H(1 - \frac{\mu\langle\hat{\lambda}_N\rangle}{||\hat{\boldsymbol{\lambda}}_T||})\frac{\mu H(\hat{\lambda}_N)}{||\hat{\boldsymbol{\lambda}}_T||}\delta\hat{\lambda}_N + H(1 - \frac{\mu\langle\hat{\lambda}_N\rangle}{||\hat{\boldsymbol{\lambda}}_T||})\frac{\mu\langle\hat{\lambda}_N\rangle}{||\hat{\boldsymbol{\lambda}}_T||}\frac{\hat{\boldsymbol{\lambda}}_T.\delta\hat{\boldsymbol{\lambda}}_T}{||\hat{\boldsymbol{\lambda}}_T||^2} \qquad [9.48]$$

As some rather long expression reappear frequently, we will give them abbreviated names:

Notation.– We define the following *dimensionless* coefficients.

The first coefficient has already be mentioned in [8.23]:

$$c_{slst} \equiv 1 - \langle 1 - \frac{\mu\langle\hat{\lambda}_N\rangle}{||\hat{\boldsymbol{\lambda}}_T||}\rangle = \begin{cases} 0 & \text{if } \hat{\lambda}_N < 0 \quad \text{(algorithmic gap)} \\ \frac{\mu\hat{\lambda}_N}{||\hat{\boldsymbol{\lambda}}_T||} & \text{if } \hat{\lambda}_N \geq 0 \text{ and } ||\hat{\boldsymbol{\lambda}}_T|| \geq \mu\hat{\lambda}_N \quad \text{(algorithmic contact with slip)} \\ 1 & \text{if } \hat{\lambda}_N \geq 0 \text{ and } ||\hat{\boldsymbol{\lambda}}_T|| < \mu\hat{\lambda}_N \quad \text{(algorithmic contact with stick)} \end{cases} \qquad [9.49]$$

The notation c_{slst} recalls that the coefficient is nonzero only in the algorithmic slip and stick cases.

In fact, inside the conditions of the algorithmic contact with slip case – namely $\hat{\lambda}_N \geq 0$ and $||\hat{\boldsymbol{\lambda}}_T|| \geq \mu\hat{\lambda}_N$ – the equalities do not occur on $S^{(1)}_{0c}$ due to the hypothesis [9.46]. That being said, we leave these equalities in the sequel.

Furthermore, noting that

$$H(1 - \frac{\mu\langle\hat{\lambda}_N\rangle}{||\hat{\boldsymbol{\lambda}}_T||}) = \begin{cases} 1 & \text{if } \hat{\lambda}_N < 0 \quad \text{(algorithmic gap)} \\ 1 & \text{if } \hat{\lambda}_N \geq 0 \text{ and } ||\hat{\boldsymbol{\lambda}}_T|| \geq \mu\hat{\lambda}_N \quad \text{(algorithmic contact with slip)} \\ 0 & \text{if } \hat{\lambda}_N \geq 0 \text{ and } ||\hat{\boldsymbol{\lambda}}_T|| < \mu\hat{\lambda}_N \quad \text{(algorithmic contact with stick)} \end{cases}$$

we can determine the values of the two coefficients defined as follows:

$$c_{sl} \equiv H(1 - \frac{\mu\langle\hat{\lambda}_N\rangle}{||\hat{\boldsymbol{\lambda}}_T||})\frac{\mu\langle\hat{\lambda}_N\rangle}{||\hat{\boldsymbol{\lambda}}_T||} = \begin{cases} 0 & \text{(algorithmic gap)} \\ \frac{\mu\hat{\lambda}_N}{||\hat{\boldsymbol{\lambda}}_T||} & \text{(algorithmic contact with slip)} \\ 0 & \text{(algorithmic contact with stick)} \end{cases}$$

$$d_{sl} \equiv H(1 - \frac{\mu\langle\hat{\lambda}_N\rangle}{||\hat{\boldsymbol{\lambda}}_T||})H(\hat{\lambda}_N) = \begin{cases} 0 & \text{(algorithmic gap)} \\ 1 & \text{(algorithmic contact with slip)} \\ 0 & \text{(algorithmic contact with stick)} \end{cases} \qquad [9.50]$$

Note that $c_{sl} = \frac{\mu\hat{\lambda}_N}{||\hat{\boldsymbol{\lambda}}_T||}d_{sl}$. The notation c_{sl} recalls that the coefficient is nonzero only in the algorithmic slip case. The notation d_{sl} is analogous to the previous one, where the letter d recalls the δ of the Kronecker symbol (0 or 1).

With the notation [9.50], the expression [9.48] can be written in a condensed way:

$$\delta\langle 1 - \frac{\mu\langle\hat{\lambda}_N\rangle}{||\hat{\boldsymbol{\lambda}}_T||}\rangle = -d_{sl}\frac{\mu\delta\hat{\lambda}_N}{||\hat{\boldsymbol{\lambda}}_T||} + c_{sl}\frac{\hat{\boldsymbol{\lambda}}_T.\delta\hat{\boldsymbol{\lambda}}_T}{||\hat{\boldsymbol{\lambda}}_T||^2} \qquad [9.51]$$

Inserting [9.45], [9.47] and [9.51] into [9.42], we arrive at

$$
\begin{aligned}
\left(\delta \mathcal{P}^{*}_{\text{contact}}\right)^{e^{(1)}} = &- \int_{e^{(1)}} \Big\{ -H(\hat{\lambda}_N)(\mathbf{U}^{(1)*} - \mathbf{U}^{(2)*}).\boldsymbol{\nu}\delta\hat{\lambda}_N - \langle \hat{\lambda}_N \rangle \left((\mathbf{U}^{(1)*} - \mathbf{U}^{(2)*}).\delta\boldsymbol{\nu} - \mathbf{U}^{(2)*}_{,\alpha}.\boldsymbol{\nu}\delta\eta^{\alpha} \right) \\
&+ d_{sl}\, \mu (\mathbf{U}^{(1)*} - \mathbf{U}^{(2)*}). \frac{\hat{\boldsymbol{\lambda}}_l}{\|\hat{\boldsymbol{\lambda}}_T\|} \delta\hat{\lambda}_N - c_{sl}(\mathbf{U}^{(1)*} - \mathbf{U}^{(2)*}).\hat{\boldsymbol{\lambda}}_T \frac{\hat{\boldsymbol{\lambda}}_T.\delta\hat{\boldsymbol{\lambda}}_T}{\|\hat{\boldsymbol{\lambda}}_T\|^2} \\
&+ c_{slst}(\mathbf{U}^{(1)*} - \mathbf{U}^{(2)*}).\delta\hat{\boldsymbol{\lambda}}_T - c_{slst}\mathbf{U}^{(2)*}_{,\alpha}.\hat{\boldsymbol{\lambda}}_T \delta\eta^{\alpha} \Big\} dS_0 \\
&- \int_{e^{(1)}} \Big\{ -\frac{1}{\epsilon_N}(\delta\lambda_N - H(\hat{\lambda}_N)\delta\hat{\lambda}_N)\mathbf{m} \\
&+ \frac{1}{\epsilon_T}\left[\delta\boldsymbol{\lambda}_T - d_{sl}\,\mu \frac{\hat{\boldsymbol{\lambda}}_T}{\|\hat{\boldsymbol{\lambda}}_T\|}\delta\hat{\lambda}_N + c_{sl}\hat{\boldsymbol{\lambda}}_T \frac{\hat{\boldsymbol{\lambda}}_T.\delta\hat{\boldsymbol{\lambda}}_T}{\|\hat{\boldsymbol{\lambda}}_T\|^2} - c_{slst}\delta\hat{\boldsymbol{\lambda}}_T \right] \Big\} .\boldsymbol{\lambda}^{*} dS_0
\end{aligned}
\quad [9.52]
$$

To abbreviate the expressions further, we will use the following notation:

NOTATION.–

$$
\boxed{\boldsymbol{\ell}_T \equiv \frac{\hat{\boldsymbol{\lambda}}_T}{\|\hat{\boldsymbol{\lambda}}_T\|}} \quad \text{(unit vector)} \quad \text{and} \quad \boxed{\mathbf{P} \equiv c_{slst}\,\mathbf{I} - c_{sl}\,\boldsymbol{\ell}_T \otimes \boldsymbol{\ell}_T} \quad [9.53]
$$

The letter **P** recalls that when $c_{slst} = c_{sl}$, the tensor **P** is proportional to the projection operator onto the plane perpendicular to $\boldsymbol{\ell}_T$.

With this notation and now by taking into account [9.43], we can break the integrand in [9.52] into four blocks to facilitate the discretization in the sequel (despite its appearance the integrand is still a *scalar* expression):

$$
\left(\delta \mathcal{P}^{*}_{\text{contact}}\right)^{e^{(1)}} = - \int_{e^{(1)}}
\begin{array}{c|l|l|}
 & \delta g, \delta\boldsymbol{\nu}, \delta\mathbf{V}_T, \delta\eta^{\alpha} & \delta\lambda_N, \delta\boldsymbol{\lambda}_T \\
\hline
\mathbf{U}^{(1)*} - \mathbf{U}^{(2)*}, \mathbf{U}^{(2)*}_{,\alpha} &
\begin{array}{l}
-H(\hat{\lambda}_N)\epsilon_N(\mathbf{U}^{(1)*} - \mathbf{U}^{(2)*}).\boldsymbol{\nu}\delta g \\
-\langle \hat{\lambda}_N \rangle \left((\mathbf{U}^{(1)*} - \mathbf{U}^{(2)*}).\delta\boldsymbol{\nu} - \mathbf{U}^{(2)*}_{,\alpha}.\boldsymbol{\nu}\delta\eta^{\alpha} \right) \\
+d_{sl}\,\mu\epsilon_N(\mathbf{U}^{(1)*} - \mathbf{U}^{(2)*}).\boldsymbol{\ell}_T\delta g \\
+\epsilon_T(\mathbf{U}^{(1)*} - \mathbf{U}^{(2)*}).\mathbf{P}.\delta\mathbf{V}_T \\
-c_{slst}\,\mathbf{U}^{(2)*}_{,\alpha}.\hat{\boldsymbol{\lambda}}_T\delta\eta^{\alpha}
\end{array} &
\begin{array}{l}
-H(\hat{\lambda}_N)(\mathbf{U}^{(1)*} - \mathbf{U}^{(2)*}).\boldsymbol{\nu}\delta\lambda_N \\
\\
+d_{sl}\,\mu(\mathbf{U}^{(1)*} - \mathbf{U}^{(2)*}).\boldsymbol{\ell}_T\delta\lambda_N \\
+(\mathbf{U}^{(1)*} - \mathbf{U}^{(2)*}).\mathbf{P}.\delta\boldsymbol{\lambda}_T \\
\\
\end{array} \\
\hline
\boldsymbol{\lambda}^{*} &
\begin{array}{l}
+H(\hat{\lambda}_N)\boldsymbol{\lambda}^{*}.\mathbf{m}\delta g \\
-d_{sl}\,\mu\frac{\epsilon_N}{\epsilon_T}\boldsymbol{\lambda}^{*}.\boldsymbol{\ell}_T\delta g \\
-\boldsymbol{\lambda}^{*}.\mathbf{P}.\delta\mathbf{V}_T
\end{array} &
\begin{array}{l}
-\frac{1}{\epsilon_N}(1 - H(\hat{\lambda}_N))\boldsymbol{\lambda}^{*}.\mathbf{m}\delta\lambda_N \\
-d_{sl}\,\mu\frac{1}{\epsilon_T}\boldsymbol{\lambda}^{*}.\boldsymbol{\ell}_T\delta\lambda_N \\
-\boldsymbol{\lambda}^{*}.\frac{1}{\epsilon_T}(\mathbf{P} - \mathbf{I}).\delta\boldsymbol{\lambda}_T
\end{array} \\
\hline
\end{array}
\; dS_0 \quad [9.54]
$$

For example, the block (2,1) (bottom-left block) contains the terms corresponding on one part to $\boldsymbol{\lambda}^{*}$ and on the other part to $\delta g, \delta\boldsymbol{\nu}, \delta\mathbf{V}_T, \delta\eta^{\alpha}$.

In the block (1,1), it is necessary to group together the terms $(\mathbf{U}^{(1)*} - \mathbf{U}^{(2)*}).\delta\boldsymbol{\nu} - \mathbf{U}^{(2)*}_{,\alpha}.\boldsymbol{\nu}\delta\eta^{\alpha}$, since these will later yield the *symmetric* matrix $[\mathbf{k}^{e,2}_{\text{contact}}]$ in [9.70].

In [9.54], the variations $\delta\eta^{\alpha}$, δg and $\delta\boldsymbol{\nu}$ are given by the relations [A.5], [A.7] and [A.9], established in the Appendix:

$$
\begin{aligned}
\delta\eta^{\alpha} &= \bar{a}^{\alpha\beta}\mathbf{a}_{\beta}.\left(\delta\mathbf{U}^{(1)}(\mathbf{X}) - \delta\mathbf{U}^{(2)}(\mathbf{Y}(\mathbf{X},t))\right) - g\bar{a}^{\alpha\beta}\boldsymbol{\nu}.\delta\mathbf{U}^{(2)}_{,\beta}(\boldsymbol{\eta}) \\
\delta g &= -\left(\delta\mathbf{U}^{(1)}(\mathbf{X}) - \delta\mathbf{U}^{(2)}(\mathbf{Y}(\mathbf{X},t))\right).\boldsymbol{\nu} \\
\delta\boldsymbol{\nu} &= -\left(b_{\alpha\gamma}\bar{a}^{\gamma\beta}\mathbf{a}^{\alpha}\otimes\mathbf{a}_{\beta}\right).\left(\delta\mathbf{U}^{(1)}(\mathbf{X}) - \delta\mathbf{U}^{(2)}(\mathbf{Y}(\mathbf{X},t))\right) - (\bar{a}^{\alpha\beta}\mathbf{a}_{\beta}\otimes\boldsymbol{\nu}).\delta\mathbf{U}^{(2)}_{,\alpha}(\boldsymbol{\eta})
\end{aligned}
\qquad [9.55]
$$

As for the slip variation, it is given by [A.14]:

$$
\delta(\mathbf{V}_T\Delta t) = \mathbf{W}^1.\delta\mathbf{U}^{(1)}(\mathbf{X}) + \mathbf{W}^2.\delta\mathbf{U}^{(2)}(\mathbf{Y}) - \delta\mathbf{U}^{(2)}(\mathbf{Y}_{n-1}) - \mathbf{Z}^{\alpha}.\delta\mathbf{U}^{(2)}_{,\alpha}(\boldsymbol{\eta}) \qquad [9.56]
$$

where $\mathbf{W}^{\alpha}$ and $\mathbf{Z}^{\alpha}, \alpha \in \{1,2\}$ are *dimensionless* second-order tensors defined in [A.15]:

$$
\begin{aligned}
\mathbf{W}^1 &\equiv \bar{a}^{\alpha\beta}\mathbf{a}_{\alpha}\otimes\mathbf{a}_{\beta} \\
\mathbf{W}^2 &\equiv \mathbf{I} - \mathbf{W}^1 \\
\mathbf{Z}^{\alpha} &\equiv g\bar{a}^{\alpha\beta}\mathbf{a}_{\beta}\otimes\boldsymbol{\nu}
\end{aligned}
\qquad [9.57]
$$

($\mathbf{W}^1$ and $\mathbf{W}^2$ are *symmetric*, but not $\mathbf{Z}^1$ and $\mathbf{Z}^2$).

By replacing in [9.54] $\delta\lambda_N$ and $\delta\boldsymbol{\lambda}_T$ by their expressions [9.39] as a function of the vector $\delta\boldsymbol{\lambda}$, we obtain

$$
\left(\delta\mathcal{P}^{*}_{\text{contact}}\right)^{e^{(1)}} = -\int_{e^{(1)}}
\begin{array}{c|l|l|}
 & \delta g, \delta\boldsymbol{\nu}, \delta\mathbf{V}_T, \delta\eta^{\alpha} & \delta\boldsymbol{\lambda} \\
\hline
\mathbf{U}^{(1)*} - \mathbf{U}^{(2)*}, \mathbf{U}^{(2)*}_{,\alpha} &
\begin{array}{l}
-H(\hat{\lambda}_N)\epsilon_N(\mathbf{U}^{(1)*} - \mathbf{U}^{(2)*}).\boldsymbol{\nu}\delta g \\
-\langle\hat{\lambda}_N\rangle\left((\mathbf{U}^{(1)*} - \mathbf{U}^{(2)*}).\delta\boldsymbol{\nu} - \mathbf{U}^{(2)*}_{,\alpha}.\boldsymbol{\nu}\delta\eta^{\alpha}\right) \\
+d_{sl}\,\mu\epsilon_N(\mathbf{U}^{(1)*} - \mathbf{U}^{(2)*}).\boldsymbol{\ell}_T\delta g \\
+\epsilon_T(\mathbf{U}^{(1)*} - \mathbf{U}^{(2)*}).\mathbf{P}.\delta\mathbf{V}_T \\
-c_{slst}\,\mathbf{U}^{(2)*}_{,\alpha}.\hat{\boldsymbol{\lambda}}_T\delta\eta^{\alpha}
\end{array}
&
\begin{array}{l}
-H(\hat{\lambda}_N)(\mathbf{U}^{(1)*} - \mathbf{U}^{(2)*}).\boldsymbol{\nu}\,(\boldsymbol{\nu}.\delta\boldsymbol{\lambda}) \\
\\
+d_{sl}\,\mu(\mathbf{U}^{(1)*} - \mathbf{U}^{(2)*}).\boldsymbol{\ell}_T\,(\boldsymbol{\nu}.\delta\boldsymbol{\lambda}) \\
-(\mathbf{U}^{(1)*} - \mathbf{U}^{(2)*}).\mathbf{P}.(\mathbf{I} - \boldsymbol{\nu}\otimes\boldsymbol{\nu}).\delta\boldsymbol{\lambda} \\
\\
\end{array} \\
\hline
\boldsymbol{\lambda}^{*} &
\begin{array}{l}
+H(\hat{\lambda}_N)\boldsymbol{\lambda}^{*}.\mathbf{m}\delta g \\
-d_{sl}\,\mu\dfrac{\epsilon_N}{\epsilon_T}\boldsymbol{\lambda}^{*}.\boldsymbol{\ell}_T\delta g \\
-\boldsymbol{\lambda}^{*}.\mathbf{P}.\delta\mathbf{V}_T
\end{array}
&
\begin{array}{l}
-\dfrac{1}{\epsilon_N}(1 - H(\hat{\lambda}_N))\boldsymbol{\lambda}^{*}.\mathbf{m}\,(\boldsymbol{\nu}.\delta\boldsymbol{\lambda}) \\
-d_{sl}\,\mu\dfrac{1}{\epsilon_T}\boldsymbol{\lambda}^{*}.\boldsymbol{\ell}_T\,(\boldsymbol{\nu}.\delta\boldsymbol{\lambda}) \\
+\boldsymbol{\lambda}^{*}.\dfrac{1}{\epsilon_T}(\mathbf{P} - \mathbf{I}).(\mathbf{I} - \boldsymbol{\nu}\otimes\boldsymbol{\nu}).\delta\boldsymbol{\lambda}
\end{array} \\
\hline
\end{array}
\; dS_0 \qquad [9.58]
$$

We thus pass:

1. from the expression [9.54] for the variation $\left(\delta\mathcal{P}^{*}_{\text{contact}}\right)^{e^{(1)}}$, functional of $\delta\mathbf{U}^{(1)}$, $\delta\mathbf{U}^{(2)}$, $\delta\lambda_N$, $\delta\boldsymbol{\lambda}_T$;

2. to a new expression [9.58], functional of $\delta\mathbf{U}^{(1)}$, $\delta\mathbf{U}^{(2)}$, $\delta\boldsymbol{\lambda}$.

In the variation [9.54], the arbitrary variations $\delta\lambda_N$ and $\delta\boldsymbol{\lambda}_T$ play separately, while in the variation [9.58], they are grouped together in the vector $\delta\boldsymbol{\lambda}$.

Thanks to the bijection [9.40], we have the following equivalence:

THEOREM.– [9.59]

The variation [9.54] holds $\forall\delta\mathbf{U}^{(1)}, \forall\delta\mathbf{U}^{(2)}, \forall\delta\lambda_N, \forall\delta\boldsymbol{\lambda}_T\perp\boldsymbol{\nu} \quad \Leftrightarrow$
the variation [9.58] holds $\forall\delta\mathbf{U}^{(1)}, \forall\delta\mathbf{U}^{(2)}, \forall\delta\boldsymbol{\lambda}$.

The equivalence [9.59] is useful: it makes it possible to reduce to the variation [9.58] containing the vector $\delta\boldsymbol{\lambda}$, *easier to interpolate* than $\delta\lambda_N$ and $\delta\boldsymbol{\lambda}_T$ considered separately.

9.6.4. *Discretization of the variation* $(\delta\mathcal{P}^*_{contact})^{e^{(1)}}$ – *Element contact tangent matrix* $[K^e_{contact}]$

Passing now onto the third and final step of the computational scheme [9.31].

By interpolation, we express the continuous quantities which appear in the variation [9.58] as functions of the nodal quantities:

the vectors $\mathbf{U}^{(i)*}, i \in \{1,2\}$,	as functions of	$\{\mathbf{U}^{(i)*,e^{(i)}}\}$	by the interpolations [8.9], [8.13];
the vector λ^*	as a function of	$\{\mathbf{\Lambda}^{*,e^{(1)}}\}$	by the interpolation [8.18].

And:

the vectors $\delta\mathbf{U}^{(i)}, i \in \{1,2\}$,	as functions of	$\{\delta\mathbf{U}^{(i),e^{(i)}}\}$	by the interpolations [9.34], [9.35];
the vector $\delta\mathbf{U}^{(2)}(\mathbf{Y}_{n-1})$	as a function of	$\{\delta\mathbf{U}^{(2),e^{(2)}_{n-1}}\}$	by the interpolation [9.36];
the vector $\delta\lambda$	as a function of	$\{\delta\mathbf{\Lambda}^{e^{(1)}}\}$	by the interpolation [9.37].

By means of the previous interpolations, we can discretize the variation $(\delta\mathcal{P}^*_{\text{contact}})^{e^{(1)}}$ relative to the element $e^{(1)} \subset S^{(1)}_{0c}$ – relation [9.58] – that is to say to express $(\delta\mathcal{P}^*_{\text{contact}})^{e^{(1)}}$ as a function of the nodal quantities, then put $(\delta\mathcal{P}^*_{\text{contact}})^{e^{(1)}}$ into the form:

$$(\delta\mathcal{P}^*_{\text{contact}})^{e^{(1)}}_{\text{discretized}} = -\int_{e^{(1)}} \begin{Bmatrix} \{\mathbf{U}^{(1)*,e^{(1)}}\} \\ \{\mathbf{U}^{(2)*,e^{(2)}}\} \\ \{\mathbf{\Lambda}^{*,e^{(1)}}\} \end{Bmatrix}^T [\mathbf{K}^e_{\text{contact}}] \begin{Bmatrix} \{\delta\mathbf{U}^{(1),e^{(1)}}\} \\ \{\delta\mathbf{U}^{(2),e^{(2)}}\} \\ \{\delta\mathbf{U}^{(2),e^{(2)}_{n-1}}\} \\ \{\delta\mathbf{\Lambda}^{e^{(1)}}\} \end{Bmatrix} dS_0 \qquad [9.60]$$

In the previous expression, the variation $\{\delta\mathbf{U}^{(2),e^{(2)}_{n-1}}\}$ is necessary for the tangential contact.

Recall that we have considered the arbitrary column-vectors $\{\delta\mathbf{U}\}$, $\{\delta\mathbf{\Lambda}\}$ (while remaining "small") and we must obtain an expression for the variation $\delta\mathcal{P}^*_{\text{contact}}$, which is valid for all such vectors. The expression [9.60] has been obtained by discretizing the variation [9.58]. As the latter is valid $\forall\delta\mathbf{U}^{(1)}$, $\forall\delta\mathbf{U}^{(2)}$, $\forall\delta\lambda$, we can replace $\delta\mathbf{U}^{(1)}$, $\delta\mathbf{U}^{(2)}$, $\delta\lambda$ with the *particular* expressions which are the interpolations in the elements:

1. We have replaced the fields $\delta\mathbf{U}^{(1)}$, $\delta\mathbf{U}^{(2)}$ with the piecewise-defined fields [9.34] and [9.35], namely, $\{\delta\mathbf{U}^{(1)}\} = [\mathcal{N}^{e^{(1)}}(\boldsymbol{\xi})]\{\delta\mathbf{U}^{(1),e^{(1)}}\}$ and $\{\delta\mathbf{U}^{(2)}\} = [\mathcal{N}^{e^{(2)}}(\boldsymbol{\eta})]\{\delta\mathbf{U}^{(2),e^{(2)}}\}$. This is a classical operation in the finite element method.

2. We have replaced the field $\delta\lambda$ with the piecewise-defined fields [9.37]: $\{\delta\lambda\} = [\mathcal{N}^{e^{(1)}}(\boldsymbol{\xi})]\{\delta\mathbf{\Lambda}^{e^{(1)}}\}$.

The expression [9.60] obtained is valid regardless of the column-vectors $\{\delta\mathbf{U}^{(1),e^{(1)}}\}$, $\{\delta\mathbf{U}^{(2),e^{(2)}}\}$, $\{\delta\mathbf{U}^{(2),e^{(2)}_{n-1}}\}$, $\{\delta\mathbf{\Lambda}^{e^{(1)}}\}$, which are all extracted from the column-vectors $\{\delta\mathbf{U}\}$, $\{\delta\mathbf{\Lambda}\}$. After assembly we will obtain, as desired, an expression for the variation $\delta\mathcal{P}^*_{\text{contact}}$ which is valid for the arbitrary column-vectors $\{\delta\mathbf{U}\}$, $\{\delta\mathbf{\Lambda}\}$ considered. Thanks to this fact, the tangent matrix $[\mathbf{K}_{\text{contact}}]$ that will be obtained later will be *complete*.

The matrix $[\mathbf{K}^e_{\text{contact}}]$ identified in [9.60] is a *rectangular* matrix.

DEFINITION.– The *rectangular* matrix $[\mathbf{K}^e_{\text{contact}}]$ is called *the element contact tangent matrix*.

Recall that, according to the notation [8.25], the index e groups together the three elements $e^{(1)}, e^{(2)}$ and $e^{(2)}_{n-1}$.

REMARK.– The remark made about the element contact vectors [8.21]–[8.22] also applies here. In classical finite element theory, a matrix is said to be an *element* one – for example, the element mass matrix – when it is relative or corresponds to *an entire* given *element*. Here, the term "element" used for the matrix $[\mathbf{K}^e_{\text{contact}}]$ does not have quite the same meaning and should be understood as an extension of the classical term. We will understand this remark more when we discuss "the assembly" of such matrix in a short while. ■

The explicit expression of $[\mathbf{K}^e_{\text{contact}}]$ will be given later on, and for the moment we will focus solely on its form. Let us decompose $[\mathbf{K}^e_{\text{contact}}]$ into four blocks, in an analogous way to the decomposition [9.11] for the global contact matrix $[\mathbf{K}_{\text{contact}}]$:

$$[\mathbf{K}^e_{\text{contact}}] = \left[\begin{array}{c|c} \mathbf{K}^e_{UU} & \mathbf{K}^e_{U\Lambda} \\ \hline \mathbf{K}^e_{\Lambda U} & \mathbf{K}^e_{\Lambda\Lambda} \end{array}\right] \qquad [9.61]$$

The following expression displays the nodal variables relative to each block of the matrix $[\mathbf{K}^e_{\text{contact}}]$:

$[\mathbf{K}^e_{\text{contact}}] =$

	$<\mathbf{U}^{(1),e^{(1)}}>$ $<\mathbf{U}^{(2),e^{(2)}}>$ $<\mathbf{U}^{(2),e^{(2)}_{n-1}}>$	$<\mathbf{\Lambda}^{e^{(1)}}>$
$\{\mathbf{U}^{(1),e^{(1)}}\}$ $\{\mathbf{U}^{(2),e^{(2)}}\}$	$[\mathbf{K}^e_{UU}]$ = block (1,1) = block $\{\mathbf{U}^e\}, <\mathbf{U}>^e$	$[\mathbf{K}^e_{U\Lambda}]$ = block (1,2) = block $\{\mathbf{U}^e\}, <\mathbf{\Lambda}^{e^{(1)}}>$
$\{\mathbf{\Lambda}^{e^{(1)}}\}$	$[\mathbf{K}^e_{\Lambda U}]$ = block (2,1) = block $\{\mathbf{\Lambda}^{e^{(1)}}\}, <\mathbf{U}>^e$	$[\mathbf{K}^e_{\Lambda\Lambda}]$ = block (2,2) = block $\{\mathbf{\Lambda}^{e^{(1)}}\}, <\mathbf{\Lambda}^{e^{(1)}}>$

[9.62]

To be more precise, that is with the names and the sizes of the sub-blocks, similar to the decomposition [9.12]:

$$[\mathbf{K}^e_{\text{contact}}] = \begin{array}{c} \\ \{\mathbf{U}^{(1),e^{(1)}}\} \\ \{\mathbf{U}^{(2),e^{(2)}}\} \\ \{\mathbf{\Lambda}^{e^{(1)}}\} \end{array} \begin{array}{c} <\mathbf{U}^{(1),e^{(1)}}><\mathbf{U}^{(2),e^{(2)}}><\mathbf{U}^{(2),e^{(2)}_{n-1}}><\mathbf{\Lambda}^{e^{(1)}}> \\ \left[\begin{array}{ccc|c} \mathbf{K}_{U^{e(1)}U^{e(1)}} & \mathbf{K}_{U^{e(1)}U^{e(2)}} & \mathbf{K}_{U^{e(1)}U^{e(2)}_{n-1}} & \mathbf{K}_{U^{e(1)}\Lambda^{e(1)}} \\ \hline \mathbf{K}_{U^{e(2)}U^{e(1)}} & \mathbf{K}_{U^{e(2)}U^{e(2)}} & \mathbf{K}_{U^{e(2)}U^{e(2)}_{n-1}} & \mathbf{K}_{U^{e(2)}\Lambda^{e(1)}} \\ \hline \mathbf{K}_{\Lambda^{e(1)}U^{e(1)}} & \mathbf{K}_{\Lambda^{e(1)}U^{e(2)}} & \mathbf{K}_{\Lambda^{e(1)}U^{e(2)}_{n-1}} & \mathbf{K}_{\Lambda^{e(1)}\Lambda^{e(1)}} \end{array}\right] \end{array} \qquad [9.63]$$

$$= \begin{array}{c} \\ 3nne^{(1)} \\ 3nne^{(2)} \\ 3nne^{(1)} \end{array} \begin{array}{c} \begin{array}{cccc} 3nne^{(1)} & 3nne^{(2)} & 3nne^{(2)}_{n-1} & 3nne^{(1)} \end{array} \\ \left[\begin{array}{ccc|c} \cdots & \cdots & \cdots & \cdots \\ \hline \cdots & \cdots & \cdots & \cdots \\ \hline \cdots & \cdots & \cdots & \cdots \end{array}\right] \end{array}$$

The expressions of the four blocks $[\mathbf{K}^e_{UU}]$, $[\mathbf{K}^e_{U\Lambda}]$, $[\mathbf{K}^e_{\Lambda U}]$ and $[\mathbf{K}^e_{\Lambda\Lambda}]$ will be made explicit later on.

The situation is the same as in section 8.7 where we studied the discretization of the element virtual contact power $\left(\mathcal{P}^*_{\text{contact}}\right)^{e^{(1)}}$. The integration in [9.60] should be done with respect to the variable $\mathbf{X} \in e^{(1)}$ or, which amounts to the same effect, with respect to the variable $\boldsymbol{\xi}$ belonging to the parent element of $e^{(1)}$. The quantities in the integrand are linked to the point $\boldsymbol{\xi}$ by the same dependency chain described in Tables 8.1 and 8.2.

Since the column-vectors $\{\mathbf{U}^{(1)*,e^{(1)}}\}$, $\{\delta\mathbf{U}^{(1),e^{(1)}}\}$, $\{\boldsymbol{\Lambda}^{*,e^{(1)}}\}$ and $\{\delta\boldsymbol{\Lambda}^{e^{(1)}}\}$ in [9.60] are independent of the integration variable $\boldsymbol{\xi}$, we can bring them outside of the integral. Conversely, we must leave the column-vectors $\{\mathbf{U}^{(2)*,e^{(2)}}\}$, $\{\delta\mathbf{U}^{(2),e^{(2)}}\}$ (respectively, $\{\delta\mathbf{U}^{(2),e^{(2)}_{n-1}}\}$) related to the element $e^{(2)}$ (respectively, $e^{(2)}_{n-1}$) inside the integral, since the elements $e^{(2)}$, $e^{(2)}_{n-1}$ *depend on the point* $\boldsymbol{\xi}$.

In [9.63], we therefore have two types of submatrices which have to be treated differently:

1. The submatrices such as $\mathbf{K}_{U^{e(1)}\Lambda^{e(1)}}$ which can be separated from the nodal vectors $\{\mathbf{U}^{(1)*,e^{(1)}}\}$ and $\{\delta\boldsymbol{\Lambda}^{e^{(1)}}\}$. These lead to element matrices which are defined classically in the finite element method, and hence do not pose any particular problems.

2. The submatrices such as $\mathbf{K}_{U^{e(2)}\Lambda^{e(1)}}$ which remain attached to the nodal vectors $\{\mathbf{U}^{(2)*,e^{(2)}}\}$ and $\{\delta\boldsymbol{\Lambda}^{e^{(1)}}\}$. These require special attention since they are slightly more complicated.

The presentation is cumbersome because of this distinction. In the sequel, for the sake of uniformity of the exposition, we decide to proceed as we did in section 8.7, namely, to apply to all the submatrices the same specific treatment which is normally reserved for the type-two submatrices only. The reader can check that for type-one submatrices, this treatment is indeed equivalent to the classical treatment found in the finite element method.

9.6.5. *Discretization of the variation* $\delta\mathcal{P}^*_{contact}$ *– Contact tangent matrix* $[K_{contact}]$

By assembly, we are going to obtain the discrete expression for $\delta\mathcal{P}^*_{\text{contact}}$ as a function of the variations $\{\delta\mathbf{U}\}, \{\delta\boldsymbol{\Lambda}\}$. By substituting [9.60] into [9.41], we obtain

$$\begin{aligned}\left(\delta\mathcal{P}^*_{\text{contact}}\right)_{\text{discretized}} &= \sum_{e^{(1)}\subset S^{(1)}_{0c}} \left(\delta\mathcal{P}^*_{\text{contact}}\right)^{e^{(1)}}_{\text{discretized}} \\ &= -\sum_{e^{(1)}\subset S^{(1)}_{0c}} \int_{e^{(1)}} \left\{\begin{matrix}\{\mathbf{U}^{(1)*,e^{(1)}}\}\\ \{\mathbf{U}^{(2)*,e^{(2)}}\}\\ \{\boldsymbol{\Lambda}^{*,e^{(1)}}\}\end{matrix}\right\}^T [\mathbf{K}^e_{\text{contact}}] \left\{\begin{matrix}\{\delta\mathbf{U}^{(1),e^{(1)}}\}\\ \{\delta\mathbf{U}^{(2),e^{(2)}}\}\\ \{\delta\mathbf{U}^{(2),e^{(2)}_{n-1}}\}\\ \{\delta\boldsymbol{\Lambda}^{e^{(1)}}\}\end{matrix}\right\} dS_0\end{aligned} \qquad [9.64]$$

We know that:

- the element virtual vectors are linked to the virtual global vectors by [8.35];

- the element variation vectors are linked to the global variation vectors by [9.33].

The assembly of the matrices will be performed symbolically thanks to the localization matrices that appear in these relations. The integrand in [9.64] becomes

$$\left\{\begin{matrix}\{\mathbf{U}^{(1)*,e^{(1)}}\}\\ \{\mathbf{U}^{(2)*,e^{(2)}}\}\\ \{\boldsymbol{\Lambda}^{*,e^{(1)}}\}\end{matrix}\right\}^T [\mathbf{K}^e_{\text{contact}}] \left\{\begin{matrix}\{\delta\mathbf{U}^{(1),e^{(1)}}\}\\ \{\delta\mathbf{U}^{(2),e^{(2)}}\}\\ \{\delta\mathbf{U}^{(2),e^{(2)}_{n-1}}\}\\ \{\delta\boldsymbol{\Lambda}^{e^{(1)}}\}\end{matrix}\right\} = \left\{\begin{matrix}[\mathbb{L}^{e^{(1)}}]\{\mathbf{U}^{(1)*}\}\\ [\mathbb{L}^{e^{(2)}}]\{\mathbf{U}^{(2)*}\}\\ [\mathbb{L}^{e^{(1)}}]\{\boldsymbol{\Lambda}^*\}\end{matrix}\right\}^T [\mathbf{K}^e_{\text{contact}}] \left\{\begin{matrix}[\mathbb{L}^{e^{(1)}}]\{\delta\mathbf{U}^{(1)}\}\\ [\mathbb{L}^{e^{(2)}}]\{\delta\mathbf{U}^{(2)}\}\\ [\mathbb{L}^{e^{(2)}_{n-1}}]\{\delta\mathbf{U}^{(2)}\}\\ [\mathbb{L}^{e^{(1)}}]\{\delta\boldsymbol{\Lambda}\}\end{matrix}\right\}$$

$$= \left\{\begin{array}{c} \{\mathbf{U}^{(1)*}\} \\ \{\mathbf{U}^{(2)*}\} \\ \{\boldsymbol{\Lambda}^{*}\} \end{array}\right\}^{T} \begin{bmatrix} [\mathbb{L}^{e^{(1)}}] & & \\ & [\mathbb{L}^{e^{(2)}}] & \\ & & [\mathbb{L}^{e^{(1)}}] \end{bmatrix}^{T} [\mathbf{K}^{e}_{\text{contact}}] \begin{bmatrix} [\mathbb{L}^{e^{(1)}}] & & \\ & [\mathbb{L}^{e^{(2)}}] & \\ & [\mathbb{L}^{e^{(2)}}_{n-1}] & \\ & & [\mathbb{L}^{e^{(1)}}] \end{bmatrix} \left\{\begin{array}{c} \{\delta\mathbf{U}^{(1)}\} \\ \{\delta\mathbf{U}^{(2)}\} \\ \{\delta\boldsymbol{\Lambda}\} \end{array}\right\}$$

With the notation [9.63], the last RHS becomes

$$\left\{\begin{array}{c} \{\mathbf{U}^{(1)*}\} \\ \{\mathbf{U}^{(2)*}\} \\ \{\boldsymbol{\Lambda}^{*}\} \end{array}\right\}^{T} \left[\begin{array}{c|c|c} \mathbb{L}^{e^{(1)}T}\mathbf{K}_{U^{e(1)}U^{e(1)}}\mathbb{L}^{e^{(1)}} & \begin{array}{l}\mathbb{L}^{e^{(1)}T}\mathbf{K}_{U^{e(1)}U^{e(2)}}\mathbb{L}^{e^{(2)}} + \\ \mathbb{L}^{e^{(1)}T}\mathbf{K}_{U^{e(1)}U^{e(2)}_{n-1}}\mathbb{L}^{e^{(2)}}_{n-1}\end{array} & \mathbb{L}^{e^{(1)}T}\mathbf{K}_{U^{e(1)}\Lambda^{e(1)}}\mathbb{L}^{e^{(1)}} \\ \hline \mathbb{L}^{e^{(2)}T}\mathbf{K}_{U^{e(2)}U^{e(1)}}\mathbb{L}^{e^{(1)}} & \begin{array}{l}\mathbb{L}^{e^{(2)}T}\mathbf{K}_{U^{e(2)}U^{e(2)}}\mathbb{L}^{e^{(2)}} + \\ \mathbb{L}^{e^{(2)}T}\mathbf{K}_{U^{e(2)}U^{e(2)}_{n-1}}\mathbb{L}^{e^{(2)}}_{n-1}\end{array} & \mathbb{L}^{e^{(2)}T}\mathbf{K}_{U^{e(2)}\Lambda^{e(1)}}\mathbb{L}^{e^{(1)}} \\ \hline \mathbb{L}^{e^{(1)}T}\mathbf{K}_{\Lambda^{e(1)}U^{e(1)}}\mathbb{L}^{e^{(1)}} & \begin{array}{l}\mathbb{L}^{e^{(1)}T}\mathbf{K}_{\Lambda^{e(1)}U^{e(2)}}\mathbb{L}^{e^{(2)}} + \\ \mathbb{L}^{e^{(1)}T}\mathbf{K}_{\Lambda^{e(1)}U^{e(2)}_{n-1}}\mathbb{L}^{e^{(2)}}_{n-1}\end{array} & \mathbb{L}^{e^{(1)}T}\mathbf{K}_{\Lambda^{e(1)}\Lambda^{e(1)}}\mathbb{L}^{e^{(1)}} \end{array}\right] \left\{\begin{array}{c} \{\delta\mathbf{U}^{(1)}\} \\ \{\delta\mathbf{U}^{(2)}\} \\ \{\delta\boldsymbol{\Lambda}\} \end{array}\right\}$$

The discretized expression [9.64] of the variation $\delta\mathcal{P}^{*}_{\text{contact}}$ can be written therefore as

$$\left(\delta\mathcal{P}^{*}_{\text{contact}}\right)_{\text{discretized}} = -\left\{\begin{array}{c} \{\mathbf{U}^{(1)*}\} \\ \{\mathbf{U}^{(2)*}\} \\ \{\boldsymbol{\Lambda}^{*}\} \end{array}\right\}^{T}$$

$$\sum_{e^{(1)}\subset S^{(1)}_{0c}} \int_{e^{(1)}} \left[\begin{array}{c|c|c} \mathbb{L}^{e^{(1)}T}\mathbf{K}_{U^{e(1)}U^{e(1)}}\mathbb{L}^{e^{(1)}} & \begin{array}{l}\mathbb{L}^{e^{(1)}T}\mathbf{K}_{U^{e(1)}U^{e(2)}}\mathbb{L}^{e^{(2)}} + \\ \mathbb{L}^{e^{(1)}T}\mathbf{K}_{U^{e(1)}U^{e(2)}_{n-1}}\mathbb{L}^{e^{(2)}}_{n-1}\end{array} & \mathbb{L}^{e^{(1)}T}\mathbf{K}_{U^{e(1)}\Lambda^{e(1)}}\mathbb{L}^{e^{(1)}} \\ \hline \mathbb{L}^{e^{(2)}T}\mathbf{K}_{U^{e(2)}U^{e(1)}}\mathbb{L}^{e^{(1)}} & \begin{array}{l}\mathbb{L}^{e^{(2)}T}\mathbf{K}_{U^{e(2)}U^{e(2)}}\mathbb{L}^{e^{(2)}} + \\ \mathbb{L}^{e^{(2)}T}\mathbf{K}_{U^{e(2)}U^{e(2)}_{n-1}}\mathbb{L}^{e^{(2)}}_{n-1}\end{array} & \mathbb{L}^{e^{(2)}T}\mathbf{K}_{U^{e(2)}\Lambda^{e(1)}}\mathbb{L}^{e^{(1)}} \\ \hline \mathbb{L}^{e^{(1)}T}\mathbf{K}_{\Lambda^{e(1)}U^{e(1)}}\mathbb{L}^{e^{(1)}} & \begin{array}{l}\mathbb{L}^{e^{(1)}T}\mathbf{K}_{\Lambda^{e(1)}U^{e(2)}}\mathbb{L}^{e^{(2)}} + \\ \mathbb{L}^{e^{(1)}T}\mathbf{K}_{\Lambda^{e(1)}U^{e(2)}_{n-1}}\mathbb{L}^{e^{(2)}}_{n-1}\end{array} & \mathbb{L}^{e^{(1)}T}\mathbf{K}_{\Lambda^{e(1)}\Lambda^{e(1)}}\mathbb{L}^{e^{(1)}} \end{array}\right] dS_0 \left\{\begin{array}{c} \{\delta\mathbf{U}^{(1)}\} \\ \{\delta\mathbf{U}^{(2)}\} \\ \{\delta\boldsymbol{\Lambda}\} \end{array}\right\}$$

which is of the form

$$\left(\delta\mathcal{P}^{*}_{\text{contact}}\right)_{\text{discretized}} = -\left\{\begin{array}{c} \{\mathbf{U}^{(1)*}\} \\ \{\mathbf{U}^{(2)*}\} \\ \{\boldsymbol{\Lambda}^{*}\} \end{array}\right\}^{T} [\mathbf{K}_{\text{contact}}] \left\{\begin{array}{c} \{\delta\mathbf{U}^{(1)}\} \\ \{\delta\mathbf{U}^{(2)}\} \\ \{\delta\boldsymbol{\Lambda}\} \end{array}\right\} \qquad [9.65]$$

where the matrix $[\mathbf{K}_{\text{contact}}]$ thus identified – equal to the matrix $[\mathbf{B}_{\text{contact}}]$ in [9.32] – is the contact tangent matrix $[\mathbf{K}_{\text{contact}}]$ [9.11] that we have been looking for:

$$[\mathbf{K}_{\text{contact}}] = \begin{array}{c} \\ \{\mathbf{U}^{(1)}\} \\ \\ \{\mathbf{U}^{(2)}\} \\ \\ \{\boldsymbol{\Lambda}\} \end{array} \sum_{e^{(1)}\subset S^{(1)}_{0c}} \int_{e^{(1)}} \begin{array}{c} \begin{array}{ccc} \quad < \mathbf{U}^{(1)} > \quad & \quad < \mathbf{U}^{(2)} > \quad & \quad < \boldsymbol{\Lambda} > \quad \end{array} \\ \left[\begin{array}{c|c|c} \mathbb{L}^{e^{(1)}T}\mathbf{K}_{U^{e(1)}U^{e(1)}}\mathbb{L}^{e^{(1)}} & \begin{array}{l}\mathbb{L}^{e^{(1)}T}\mathbf{K}_{U^{e(1)}U^{e(2)}}\mathbb{L}^{e^{(2)}} + \\ \mathbb{L}^{e^{(1)}T}\mathbf{K}_{U^{e(1)}U^{e(2)}_{n-1}}\mathbb{L}^{e^{(2)}}_{n-1}\end{array} & \mathbb{L}^{e^{(1)}T}\mathbf{K}_{U^{e(1)}\Lambda^{e(1)}}\mathbb{L}^{e^{(1)}} \\ \hline \mathbb{L}^{e^{(2)}T}\mathbf{K}_{U^{e(2)}U^{e(1)}}\mathbb{L}^{e^{(1)}} & \begin{array}{l}\mathbb{L}^{e^{(2)}T}\mathbf{K}_{U^{e(2)}U^{e(2)}}\mathbb{L}^{e^{(2)}} + \\ \mathbb{L}^{e^{(2)}T}\mathbf{K}_{U^{e(2)}U^{e(2)}_{n-1}}\mathbb{L}^{e^{(2)}}_{n-1}\end{array} & \mathbb{L}^{e^{(2)}T}\mathbf{K}_{U^{e(2)}\Lambda^{e(1)}}\mathbb{L}^{e^{(1)}} \\ \hline \mathbb{L}^{e^{(1)}T}\mathbf{K}_{\Lambda^{e(1)}U^{e(1)}}\mathbb{L}^{e^{(1)}} & \begin{array}{l}\mathbb{L}^{e^{(1)}T}\mathbf{K}_{\Lambda^{e(1)}U^{e(2)}}\mathbb{L}^{e^{(2)}} + \\ \mathbb{L}^{e^{(1)}T}\mathbf{K}_{\Lambda^{e(1)}U^{e(2)}_{n-1}}\mathbb{L}^{e^{(2)}}_{n-1}\end{array} & \mathbb{L}^{e^{(1)}T}\mathbf{K}_{\Lambda^{e(1)}\Lambda^{e(1)}}\mathbb{L}^{e^{(1)}} \end{array}\right] \end{array} dS_0$$

[9.66]

Compare the decomposition of the previous matrix to that [9.12].

Although the element matrices $[\mathbf{K}^e_{\text{contact}}]$ are *rectangular*, in the end the assembly of these matrices produce the *square* contact tangent matrix $[\mathbf{K}_{\text{contact}}]$ [9.11], as it should be.

The submatrices in [9.66] are of two types:

1. The first type is formed of submatrices such as

$$\mathbf{K}_{U^{(1)}U^{(1)}} = \sum_{e^{(1)} \subset S^{(1)}_{0c}} \int_{e^{(1)}} \mathbb{L}^{e^{(1)}T} \mathbf{K}_{U^{e(1)}U^{e(1)}} \mathbb{L}^{e^{(1)}} dS_0$$

We can bring the matrix $[\mathbb{L}^{e^{(1)}}]^T$ outside of the integral over $e^{(1)}$, so that

$$\mathbf{K}_{U^{(1)}U^{(1)}} = \sum_{e^{(1)} \subset S^{(1)}_{0c}} \mathbb{L}^{e^{(1)}T} \left[\int_{e^{(1)}} \mathbf{K}_{U^{e(1)}U^{e(1)}} dS_0 \right] \mathbb{L}^{e^{(1)}}$$

The left and right products of the integral with the matrices $[\mathbb{L}^{e^{(1)}}]^T$ and $[\mathbb{L}^{e^{(1)}}]$ have the effect of expanding this integral into a matrix of the same size as $\mathbf{K}_{U^{(1)}U^{(1)}}$. Therefore, the submatrix $\mathbf{K}_{U^{(1)}U^{(1)}}$ is *the assembly* of the element matrices $\int_{e^{(1)}} \mathbf{K}_{U^{e(1)}U^{e(1)}} dS_0$, that is the assembly in the usual finite element sense.

2. The second type is formed from the submatrices such as

$$\mathbf{K}_{U^{(2)}U^{(1)}} = \sum_{e^{(1)} \subset S^{(1)}_{0c}} \int_{e^{(1)}} \mathbb{L}^{e^{(2)}T} \mathbf{K}_{U^{e(2)}U^{e(1)}} \mathbb{L}^{e^{(1)}} dS_0$$

For this type of matrix, the situation is a little bit more complicated. As we have seen in the dependency chain in Table 8.1, the element $e^{(2)}$ *varies* with the point $\boldsymbol{\xi}$ and, consequently, we can *not* bring the matrix $[\mathbb{L}^{e^{(2)}}]^T$ outside of the integral over $e^{(1)}$. The submatrix $\mathbf{K}_{U^{(2)}U^{(1)}}$ is *not* the assembly of the element matrices $\int_{e^{(1)}} \mathbf{K}_{U^{e(2)}U^{e(1)}} dS_0$.

The submatrix $\sum_{e^{(1)} \subset S^{(1)}_{0c}} \int_{e^{(1)}} \mathbb{L}^{e^{(1)}T} \mathbf{K}_{U^{e(1)}U^{e^{(2)}_{n-1}}} \mathbb{L}^{e^{(2)}_{n-1}} dS_0$ in [9.66] is also of the second type, because the element $e^{(2)}_{n-1}$ *varies* with the point $\boldsymbol{\xi}$.

This subtlety has been mentioned by Laursen [77] who utilized a different contact formulation to ours.

To better see the difference between the two types of matrices, we will express their numerical calculation in detail. We use the numerical Gaussian integration method [8.43] to decompose the integral over an element $e^{(1)} \subset S^{(1)}_{0c}$ into a linear combination of the integrand calculated at the different integration points $\boldsymbol{\xi}$ of the element $e^{(1)}$.

The submatrix $\mathbf{K}_{U^{(1)}U^{(1)}}$ is obtained by the classical assembly as follows:

LOOP OVER THE ELEMENTS $e^{(1)}$ OF THE SURFACE $S^{(1)}_{0c}$

LOOP OVER THE INTEGRATION POINTS ξ OF EACH ELEMENT $e^{(1)}$

Calculate the matrix $\mathbf{K}_{U^{e(1)}U^{e(1)}}\, j(\xi)$ at the point ξ.

Add the result into a matrix.

END OF LOOP

This loop give the integral $\int_{e^{(1)}} \mathbf{K}_{U^{e(1)}U^{e(1)}} dS_0$.

Assemble the obtained matrix in $[\mathbf{K}_{U^{(1)}U^{(1)}}]$.

This assembly is performed symbolically by the addition of the term:

$$\mathbb{L}^{e^{(1)}T} \left[\int_{e^{(1)}} \mathbf{K}_{U^{e(1)}U^{e(1)}} dS_0 \right] \mathbb{L}^{e^{(1)}}.$$

END OF LOOP

where we recall that $j(\xi)$ is the product of the area Jacobian at the point ξ with the integral weight, defined in [8.44]. On the other hand, the submatrix $\mathbf{K}_{U^{(2)}U^{(1)}}$ is obtained by the following *nonclassical assembly*:

LOOP OVER THE ELEMENTS $e^{(1)}$ OF THE SURFACE $S^{(1)}_{0c}$

LOOP OVER THE INTEGRATION POINTS ξ OF EACH ELEMENT $e^{(1)}$

Calculate the matrix $\mathbf{K}_{U^{e(2)}U^{e(1)}}\, j(\xi)$ at the point ξ.

Assembly this matrix into $[\mathbf{K}_{U^{(2)}U^{(1)}}]$.

This assembly is performed symbolically by the addition of the term

$$\mathbb{L}^{e^{(2)}T} \mathbf{K}_{U^{e(2)}U^{e(1)}} \mathbb{L}^{e^{(1)}}\, j(\xi).$$

END OF LOOP

END OF LOOP

For $\mathbf{K}_{U^{(1)}U^{(1)}}$, we finish calculating the integral over $e^{(1)}$, then we assemble it. For $\mathbf{K}_{U^{(2)}U^{(1)}}$, we calculate the integrand at each integration point ξ and assembly it as we go along.

9.6.6. *Explicit expression for the element contact tangent matrix* $[K^e_{contact}]$

We are going to give the explicit expression for the element contact tangent matrix $[\mathbf{K}^e_{\text{contact}}]$ [9.62] by expressing the four blocks $[\mathbf{K}^e_{UU}]$, $[\mathbf{K}^e_{U\Lambda}]$, $[\mathbf{K}^e_{\Lambda U}]$ and $[\mathbf{K}^e_{\Lambda\Lambda}]$.

The expressions established so far are valid for an arbitrary vector field $\mathbf{m}$ satisfying $\mathbf{m} \not\perp \boldsymbol{\nu}$ at every point. From now on, let us take

$$\boxed{\mathbf{m} = \boldsymbol{\nu}} \qquad [9.67]$$

This choice, which is permissible since $\boldsymbol{\nu}$ is indeed non-orthogonal to itself, is also the simplest choice.

THEOREM.– The *rectangular* matrix $[\mathbf{K}^e_{UU}]$ is relative to the nodal displacements $\{\mathbf{U}^{(1),e^{(1)}}\}$ on $e^{(1)} \subset S^{(1)}_{0c}$, $\{\mathbf{U}^{(2),e^{(2)}}\}$ on $e^{(2)} \subset S^{(2)}_{0c}$ and $\{\mathbf{U}^{(2),e^{(2)}_{n-1}}\}$ on $e^{(2)}_{n-1} \subset S^{(2)}_{0c}$. It has dimension $(3nne^{(1)} + 3nne^{(2)}) \times (3nne^{(1)} + 3nne^{(2)} + 3nne^{(2)}_{n-1})$ and its expression is

$$\boxed{[\mathbf{K}^e_{UU}] = H(\hat{\lambda}_N)\epsilon_N[\mathbf{k}^{e,1}_{\text{contact}}] + \langle \hat{\lambda}_N \rangle [\mathbf{k}^{e,2}_{\text{contact}}] - d_{sl}\,\mu\epsilon_N[\mathbf{k}^{e,3}_{\text{contact}}] + \epsilon_T[\mathbf{k}^{e,4}_{\text{contact}}] + c_{slst}\,[\mathbf{k}^{e,5}_{\text{contact}}]} \qquad [9.68]$$

where the matrices $[\mathbf{k}^{e,1}_{\text{contact}}]$ to $[\mathbf{k}^{e,5}_{\text{contact}}]$ are listed below.

• The matrices $[\mathbf{k}^{e,1}_{\text{contact}}]$ and $[\mathbf{k}^{e,2}_{\text{contact}}]$ are square ones of order $3nne^{(1)} + 3nne^{(2)}$ and are symmetric:

$$[\mathbf{k}^{e,1}_{\text{contact}}] = \begin{array}{c} \\ \{\mathbf{U}^{(1),e^{(1)}}\} \\ \\ \{\mathbf{U}^{(2),e^{(2)}}\} \end{array} \begin{array}{c} \begin{array}{cc} \qquad <\mathbf{U}^{(1),e^{(1)}}> \qquad\qquad & \qquad <\mathbf{U}^{(2),e^{(2)}}> \end{array} \\ \left[\begin{array}{c|c} \underset{(3nne^{(1)},3)}{[\mathcal{N}^{e^{(1)}}]^T}\ \underset{(3,3)}{\{\boldsymbol{\nu}\}<\boldsymbol{\nu}>}\ \underset{(3,3nne^{(1)})}{[\mathcal{N}^{e^{(1)}}]} & -[\mathcal{N}^{e^{(1)}}]^T\{\boldsymbol{\nu}\}<\boldsymbol{\nu}>[\mathcal{N}^{e^{(2)}}] \\ \hline -\underset{(3nne^{(2)},3)}{[\mathcal{N}^{e^{(2)}}]^T}\ \underset{(3,3)}{\{\boldsymbol{\nu}\}<\boldsymbol{\nu}>}\ \underset{(3,3nne^{(1)})}{[\mathcal{N}^{e^{(1)}}]} & [\mathcal{N}^{e^{(2)}}]^T\{\boldsymbol{\nu}\}<\boldsymbol{\nu}>[\mathcal{N}^{e^{(2)}}] \end{array}\right] \end{array} \qquad [9.69]$$

The dimensions of some typical matrices are recalled just below these matrices. Hence for example, the matrix $[\mathcal{N}^{e^{(1)}}]$ *has dimension* $3 \times 3nne^{(1)}$, *and the matrix* $\{\boldsymbol{\nu}\} < \boldsymbol{\nu} >$ *has dimension* 3×3.

$$[\mathbf{k}^{e,2}_{\text{contact}}] = \begin{array}{c} \{\mathbf{U}^{(1),e^{(1)}}\} \\ \\ \\ \{\mathbf{U}^{(2),e^{(2)}}\} \end{array} \begin{array}{c} \begin{array}{cc} \qquad <\mathbf{U}^{(1),e^{(1)}}> \qquad\qquad & \qquad <\mathbf{U}^{(2),e^{(2)}}> \end{array} \\ \left[\begin{array}{c|c} \underset{(3nne^{(1)},3)}{[\mathcal{N}^{e^{(1)}}]^T}\ \underset{(3,3)}{[b_{\alpha\gamma}\bar{a}^{\gamma\beta}\mathbf{a}^\alpha \otimes \mathbf{a}_\beta]}\ \underset{(3,3nne^{(1)})}{[\mathcal{N}^{e^{(1)}}]} & \begin{array}{l} -[\mathcal{N}^{e^{(1)}}]^T[b_{\alpha\gamma}\bar{a}^{\gamma\beta}\mathbf{a}^\alpha \otimes \mathbf{a}_\beta][\mathcal{N}^{e^{(2)}}] \\ +[\mathcal{N}^{e^{(1)}}]^T[\bar{a}^{\alpha\beta}\mathbf{a}_\beta \otimes \boldsymbol{\nu}][\mathcal{N}^{e^{(2)}}_{,\alpha}] \end{array} \\ \hline \begin{array}{l} -[\mathcal{N}^{e^{(2)}}]^T[b_{\alpha\gamma}\bar{a}^{\gamma\beta}\mathbf{a}^\alpha \otimes \mathbf{a}_\beta][\mathcal{N}^{e^{(1)}}] \\ +\underset{(3nne^{(2)},3)}{[\mathcal{N}^{e^{(2)}}_{,\alpha}]^T}\ \underset{(3,3)}{[\bar{a}^{\alpha\beta}\boldsymbol{\nu} \otimes \mathbf{a}_\beta]}\ \underset{(3,3nne^{(1)})}{[\mathcal{N}^{e^{(1)}}]^T} \end{array} & \begin{array}{l} [\mathcal{N}^{e^{(2)}}]^T[b_{\alpha\gamma}\bar{a}^{\gamma\beta}\mathbf{a}^\alpha \otimes \mathbf{a}_\beta][\mathcal{N}^{e^{(2)}}] \\ -[\mathcal{N}^{e^{(2)}}]^T[\bar{a}^{\alpha\beta}\mathbf{a}_\beta \otimes \boldsymbol{\nu}][\mathcal{N}^{e^{(2)}}_{,\alpha}] \\ -[\mathcal{N}^{e^{(2)}}_{,\alpha}]^T[\bar{a}^{\alpha\beta}\boldsymbol{\nu} \otimes \mathbf{a}_\beta][\mathcal{N}^{e^{(2)}}] \\ -[\mathcal{N}^{e^{(2)}}_{,\alpha}]^T[g\bar{a}^{\alpha\beta}\boldsymbol{\nu} \otimes \boldsymbol{\nu}][\mathcal{N}^{e^{(2)}}_{,\beta}] \end{array} \end{array}\right] \end{array}$$

[9.70]

In [9.70], the notation $[\mathbf{T}]$ denotes, according to the notational convention from section 8.3, the 3×3 matrix representing the second-order tensor $\mathbf{T}$ in the Cartesian basis $(\mathbf{e}_1, \mathbf{e}_2, \mathbf{e}_3)$. Hence, for example, if $\mathbf{T} = T_\alpha{}^\beta \mathbf{a}^\alpha \otimes \mathbf{a}_\beta = T^{\alpha\beta}\mathbf{a}_\alpha \otimes \mathbf{a}_\beta$, the (p, q) component of the matrix $[\mathbf{T}]$ is

$$T_{pq} = \mathbf{e}_p.\mathbf{T}.\mathbf{e}_q = T_\alpha{}^\beta(\mathbf{a}^\alpha)_p(\mathbf{a}_\beta)_q = T^{\alpha\beta}(\mathbf{a}_\alpha)_p(\mathbf{a}_\beta)_q$$

where $(\mathbf{a}^\alpha)_p$ is the pth component of $\mathbf{a}^\alpha$ in the fixed basis.

From [2.31], the tensor $b_{\alpha\gamma}\bar{a}^{\gamma\beta}\mathbf{a}^\alpha \otimes \mathbf{a}_\beta = \bar{a}^{\alpha\gamma}b_{\gamma\beta}\mathbf{a}_\alpha \otimes \mathbf{a}^\beta$ is symmetric. Consequently, the matrix $[\mathbf{k}^{e,2}_{\text{contact}}]$ is too.

• The matrix $[\mathbf{k}^{e,3}_{\text{contact}}]$ in [9.68] is square, with the same dimension as $[\mathbf{k}^{1}_{\text{contact}}]$ albeit non-symmetric:

$$[\mathbf{k}^{e,3}_{\text{contact}}] = \begin{array}{c} \\ \{\mathbf{U}^{(1),e^{(1)}}\} \\ \{\mathbf{U}^{(2),e^{(2)}}\} \end{array} \begin{array}{c} \begin{array}{cc} \qquad <\mathbf{U}^{(1),e^{(1)}}> \qquad\qquad & \qquad <\mathbf{U}^{(2),e^{(2)}}> \end{array} \\ \left[\begin{array}{c|c} [\mathcal{N}^{e^{(1)}}]^T\{\boldsymbol{\ell}_T\}<\boldsymbol{\nu}>[\mathcal{N}^{e^{(1)}}] & -[\mathcal{N}^{e^{(1)}}]^T\{\boldsymbol{\ell}_T\}<\boldsymbol{\nu}>[\mathcal{N}^{e^{(2)}}] \\ \hline -[\mathcal{N}^{e^{(2)}}]^T\{\boldsymbol{\ell}_T\}<\boldsymbol{\nu}>[\mathcal{N}^{e^{(1)}}] & [\mathcal{N}^{e^{(2)}}]^T\{\boldsymbol{\ell}_T\}<\boldsymbol{\nu}>[\mathcal{N}^{e^{(2)}}] \end{array}\right] \end{array} \qquad [9.71]$$

• The matrix $[\mathbf{k}^{e,4}_{\text{contact}}]$ is *rectangular* of dimension $(3nne^{(1)} + 3nne^{(2)}) \times (3nne^{(1)} + 3nne^{(2)} + 3nne^{(2)}_{n-1})$. It is within this matrix that the point $\boldsymbol{\eta}_{n-1}$ plays a role.

$[\mathbf{k}^{e,4}_{\text{contact}}] =$

$$\begin{array}{c} \{\mathbf{U}^{(1),e^{(1)}}\} \\ \\ \\ \{\mathbf{U}^{(2),e^{(2)}}\} \end{array} \begin{array}{c} \begin{array}{ccc} <\mathbf{U}^{(1),e^{(1)}}> \qquad\qquad & \qquad <\mathbf{U}^{(2),e^{(2)}}> \qquad\qquad & \qquad <\mathbf{U}^{(2),e^{(2)}_{n-1}}> \end{array} \\ \left[\begin{array}{c|c|c} [\mathcal{N}^{e^{(1)}}]^T\underset{(3,3)}{[\mathbf{P}]}\underset{(3,3)}{[\mathbf{W}^1]}[\mathcal{N}^{e^{(1)}}] & \begin{array}{l} [\mathcal{N}^{e^{(1)}}]^T\underset{(3,3)}{[\mathbf{P}]}\underset{(3,3)}{[\mathbf{W}^2]}[\mathcal{N}^{e^{(2)}}] \\ -[\mathcal{N}^{e^{(1)}}]^T[\mathbf{P}]\underset{(3,3)}{[\mathbf{Z}^\alpha]}[\mathcal{N}^{e^{(2)}}_{,\alpha}] \end{array} & -[\mathcal{N}^{e^{(1)}}]^T[\mathbf{P}][\mathcal{N}^{e^{(2)}_{n-1}}(\boldsymbol{\eta}_{n-1})] \\ \hline -[\mathcal{N}^{e^{(2)}}]^T[\mathbf{P}][\mathbf{W}^1][\mathcal{N}^{e^{(1)}}] & \begin{array}{l} -[\mathcal{N}^{e^{(2)}}]^T[\mathbf{P}][\mathbf{W}^2][\mathcal{N}^{e^{(2)}}] \\ +[\mathcal{N}^{e^{(2)}}]^T[\mathbf{P}][\mathbf{Z}^\alpha][\mathcal{N}^{e^{(2)}}_{,\alpha}] \end{array} & [\mathcal{N}^{e^{(2)}}]^T[\mathbf{P}][\mathcal{N}^{e^{(2)}_{n-1}}(\boldsymbol{\eta}_{n-1})] \end{array}\right] \end{array}$$

[9.72]

where $[\mathbf{P}], [\mathbf{W}^\alpha], [\mathbf{Z}^\alpha]$ denotes the 3×3 matrices representing the tensors $\mathbf{P}, \mathbf{W}^\alpha, \mathbf{Z}^\alpha$ defined in [9.53] and [9.57], in the fixed basis $(\mathbf{e}_1, \mathbf{e}_2, \mathbf{e}_3)$: $P_{ij} = \mathbf{e}_i.\mathbf{P}.\mathbf{e}_j$, $W^1_{ij} = \mathbf{e}_i.\mathbf{W}^1.\mathbf{e}_j = \bar{a}^{\alpha\beta}(2\mathbf{a}_\alpha - \mathbf{a}_\alpha(\eta_{n-1}, t))_i(\mathbf{a}_\beta)_j$, $Z^\alpha_{ij} = \mathbf{e}_i.\mathbf{Z}^\alpha.\mathbf{e}_j = g\bar{a}^{\alpha\beta}(2\mathbf{a}_\beta - \mathbf{a}_\beta(\eta_{n-1}, t))_i \nu_j$.

• The last matrix in [9.68], $[\mathbf{k}^{e,5}_{\text{contact}}]$, is square, with the same dimension as $[\mathbf{k}^1_{\text{contact}}]$ albeit non-symmetric:

$$[\mathbf{k}^{e,5}_{\text{contact}}] =
\begin{array}{c} \\ \{\mathbf{U}^{(1),e^{(1)}}\} \\ \{\mathbf{U}^{(2),e^{(2)}}\} \end{array}
\begin{array}{c|c}
< \mathbf{U}^{(1),e^{(1)}} > & < \mathbf{U}^{(2),e^{(2)}} > \\
\underset{(3nne^{(1)},3nne^{(1)})}{[\mathbf{0}]} & \underset{(3nne^{(1)},3nne^{(2)})}{[\mathbf{0}]} \\
\hline
-\underset{(3nne^{(2)},3)}{[\mathcal{N}^{e^{(2)}}_{,\alpha}]^T} \underset{(3,1)}{\{\hat{\boldsymbol{\lambda}}_T\}} \underset{(1,3)}{< \bar{a}^{\alpha\beta}\mathbf{a}_\beta >} \underset{(3,3nne^{(1)})}{[\mathcal{N}^{e^{(1)}}]} & [\mathcal{N}^{e^{(2)}}_{,\alpha}]^T\{\hat{\boldsymbol{\lambda}}_T\} \left(< \bar{a}^{\alpha\beta}\mathbf{a}_\beta > [\mathcal{N}^{e^{(2)}}] + g < \bar{a}^{\alpha\beta}\boldsymbol{\nu} > [\mathcal{N}^{e^{(2)}}_{,\beta}]\right)
\end{array} \quad [9.73]$$

PROOF. The matrix $[\mathbf{K}^e_{UU}]$ comes from the discretization of the terms of the $(1, 1)$ block of the expression [9.58]:

$$\begin{cases}
-H(\hat{\lambda}_N)\epsilon_N(\mathbf{U}^{(1)*} - \mathbf{U}^{(2)*}).\boldsymbol{\nu}\delta g \\
-\langle\hat{\lambda}_N\rangle\left((\mathbf{U}^{(1)*} - \mathbf{U}^{(2)*}).\delta\boldsymbol{\nu} - \mathbf{U}^{(2)*}_{,\alpha}.\boldsymbol{\nu}\delta\eta^\alpha\right) \\
+d_{sl}\,\mu\epsilon_N(\mathbf{U}^{(1)*} - \mathbf{U}^{(2)*}).\boldsymbol{\ell}_T\delta g \\
+\epsilon_T(\mathbf{U}^{(1)*} - \mathbf{U}^{(2)*}).\mathbf{P}.\delta\mathbf{V}_T \\
-c_{slst}\,\mathbf{U}^{(2)*}_{,\alpha}.\hat{\boldsymbol{\lambda}}_T\delta\eta^\alpha
\end{cases}$$

We will show that the five preceding lines generate the five matrices $[\mathbf{k}^{e,1}_{\text{contact}}]$ to $[\mathbf{k}^{e,5}_{\text{contact}}]$, respectively.

1. From the expression [9.55] of δg, we have

$$(\mathbf{U}^{(1)*} - \mathbf{U}^{(2)*}).\boldsymbol{\nu}\delta g = -(\mathbf{U}^{(1)*} - \mathbf{U}^{(2)*}).(\boldsymbol{\nu}\otimes\boldsymbol{\nu}).\left(\delta\mathbf{U}^{(1)} - \delta\mathbf{U}^{(2)}\right)$$

From the interpolations [8.9], [8.13] of $\mathbf{U}^{(1)*}, \mathbf{U}^{(2)*}$, [9.34] and [9.35] of $\delta\mathbf{U}^{(1)}, \delta\mathbf{U}^{(2)}$, the previous expression becomes

$$(\mathbf{U}^{(1)*} - \mathbf{U}^{(2)*}).\boldsymbol{\nu}\delta g = -\left\{\begin{array}{c}\{\mathbf{U}^{(1)*,e^{(1)}}\} \\ \{\mathbf{U}^{(2)*,e^{(2)}}\}\end{array}\right\}^T [\mathbf{k}^{e,1}_{\text{contact}}] \left\{\begin{array}{c}\{\delta\mathbf{U}^{(1),e^{(1)}}\} \\ \{\delta\mathbf{U}^{(2),e^{(2)}}\}\end{array}\right\}$$

where the matrix $[\mathbf{k}^{e,1}_{\text{contact}}]$ is defined by [9.69].

If we identify this expression with [9.60], then we get the first term of $[\mathbf{K}^e_{UU}]$ in [9.68].

2. From the expression [9.55] of $\delta\boldsymbol{\nu}, \delta\eta^\alpha$, we have

$$\begin{aligned}
&\delta\left[(\mathbf{U}^{(1)*} - \mathbf{U}^{(2)*}).\boldsymbol{\nu}\right] = \left(\mathbf{U}^{(1)*}(\xi) - \mathbf{U}^{(2)*}(\eta)\right).\delta\boldsymbol{\nu} - \mathbf{U}^{(2)*}_{,\alpha}.\boldsymbol{\nu}\delta\eta^\alpha \\
&= -\left(\mathbf{U}^{(1)*} - \mathbf{U}^{(2)*}\right).\left(b_{\alpha\gamma}\bar{a}^{\gamma\beta}\mathbf{a}^\alpha\otimes\mathbf{a}_\beta\right).\left(\delta\mathbf{U}^{(1)} - \delta\mathbf{U}^{(2)}\right) \\
&- \left(\mathbf{U}^{(1)*} - \mathbf{U}^{(2)*}\right).\left(\bar{a}^{\alpha\beta}\mathbf{a}_\beta\otimes\boldsymbol{\nu}\right).\delta\mathbf{U}^{(2)}_{,\alpha} \\
&- \mathbf{U}^{(2)*}_{,\alpha}.\left(\bar{a}^{\alpha\beta}\boldsymbol{\nu}\otimes\mathbf{a}_\beta\right).\left(\delta\mathbf{U}^{(1)} - \delta\mathbf{U}^{(2)}\right) \\
&+ g\bar{a}^{\alpha\beta}\mathbf{U}^{(2)*}_{,\alpha}.(\boldsymbol{\nu}\otimes\boldsymbol{\nu}).\delta\mathbf{U}^{(2)}_{,\beta}
\end{aligned}$$

From the interpolations [8.9], [8.13] of $\mathbf{U}^{(1)*}, \mathbf{U}^{(2)*}$, [9.34] and [9.35] of $\delta\mathbf{U}^{(1)}, \delta\mathbf{U}^{(2)}$, the previous expression becomes

$$\delta\left[(\mathbf{U}^{(1)*} - \mathbf{U}^{(2)*}).\boldsymbol{\nu}\right] = -\left\{ \begin{matrix} \{\mathbf{U}^{(1)*,e^{(1)}}\} \\ \{\mathbf{U}^{(2)*,e^{(2)}}\} \end{matrix} \right\}^T [\mathbf{k}^{e,2}_{\text{contact}}] \left\{ \begin{matrix} \{\delta\mathbf{U}^{(1),e^{(1)}}\} \\ \{\delta\mathbf{U}^{(2),e^{(2)}}\} \end{matrix} \right\}$$

where the matrix $[\mathbf{k}^{e,2}_{\text{contact}}]$ is defined by [9.70], from which we get the second term of $[\mathbf{K}^e_{UU}]$ in [9.68].

3. By reasoning analogously in what follows, we get

$$\begin{aligned}(\mathbf{U}^{(1)*} - \mathbf{U}^{(2)*}).\boldsymbol{\ell}_T\delta g &= -(\mathbf{U}^{(1)*} - \mathbf{U}^{(2)*}).(\boldsymbol{\ell}_T \otimes \boldsymbol{\nu}).\left(\delta\mathbf{U}^{(1)} - \delta\mathbf{U}^{(2)}\right) \\ &= -\left\{ \begin{matrix} \{\mathbf{U}^{(1)*,e^{(1)}}\} \\ \{\mathbf{U}^{(2)*,e^{(2)}}\} \end{matrix} \right\}^T [\mathbf{k}^{e,3}_{\text{contact}}] \left\{ \begin{matrix} \{\delta\mathbf{U}^{(1),e^{(1)}}\} \\ \{\delta\mathbf{U}^{(2),e^{(2)}}\} \end{matrix} \right\}\end{aligned}$$

where the matrix $[\mathbf{k}^{e,3}_{\text{contact}}]$ is defined by [9.71], from which we get the third term of $[\mathbf{K}^e_{UU}]$ in [9.68].

4. From the expression [9.56] of $\delta(\mathbf{V}_T\Delta t)$, we have

$$(\mathbf{U}^{(1)*} - \mathbf{U}^{(2)*}).\mathbf{P}.\delta\mathbf{V}_T = (\mathbf{U}^{(1)*} - \mathbf{U}^{(2)*}).\mathbf{P}.$$
$$\left(\mathbf{W}^1.\delta\mathbf{U}^{(1)}(\mathbf{X}) + \mathbf{W}^2.\delta\mathbf{U}^{(2)}(\mathbf{Y}) - \delta\mathbf{U}^{(2)}(\mathbf{Y}_{n-1}) - \mathbf{Z}^\alpha.\delta\mathbf{U}^{(2)}_{,\alpha}(\boldsymbol{\eta})\right)$$

With the same interpolations as those above, this expression becomes

$$(\mathbf{U}^{(1)*} - \mathbf{U}^{(2)*}).\mathbf{P}.\delta\mathbf{V}_T = \left\{ \begin{matrix} \{\mathbf{U}^{(1)*,e^{(1)}}\} \\ \{\mathbf{U}^{(2)*,e^{(2)}}\} \end{matrix} \right\}^T \begin{bmatrix} [\mathcal{N}^{e^{(1)}}]^T \\ -[\mathcal{N}^{e^{(2)}}]^T \end{bmatrix} [\mathbf{P}]$$
$$\left[[\mathbf{W}^1][\mathcal{N}^{e^{(1)}}], [\mathbf{W}^2][\mathcal{N}^{e^{(2)}}] - [\mathbf{Z}^\alpha][\mathcal{N}^{e^{(2)}}_{,\alpha}], -[\mathcal{N}^{e^{(2)}_{n-1}}(\boldsymbol{\eta}_{n-1})]\right] \left\{ \begin{matrix} \{\delta\mathbf{U}^{(1),e^{(1)}}\} \\ \{\delta\mathbf{U}^{(2),e^{(2)}}\} \\ \{\delta\mathbf{U}^{(2),e^{(2)}_{n-1}}\} \end{matrix} \right\}$$

Or, equivalently:

$$(\mathbf{U}^{(1)*} - \mathbf{U}^{(2)*}).\mathbf{P}.\delta\mathbf{V}_T = \left\{ \begin{matrix} \{\mathbf{U}^{(1)*,e^{(1)}}\} \\ \{\mathbf{U}^{(2)*,e^{(2)}}\} \end{matrix} \right\}^T [\mathbf{k}^{e,4}_{\text{contact}}] \left\{ \begin{matrix} \{\delta\mathbf{U}^{(1),e^{(1)}}\} \\ \{\delta\mathbf{U}^{(2),e^{(2)}}\} \\ \{\delta\mathbf{U}^{(2),e^{(2)}_{n-1}}\} \end{matrix} \right\}$$

where the matrix $[\mathbf{k}^{e,4}_{\text{contact}}]$ is defined by [9.72], from which we get the fourth term of $[\mathbf{K}^e_{UU}]$ in [9.68].

5. Finally, we have

$$\begin{aligned}\mathbf{U}^{(2)*}_{,\alpha}.\hat{\boldsymbol{\lambda}}_T\delta\eta^\alpha &= \mathbf{U}^{(2)*}_{,\alpha}.\hat{\boldsymbol{\lambda}}_T\bar{a}^{\alpha\beta}\mathbf{a}_\beta.\left(\delta\mathbf{U}^{(1)} - \delta\mathbf{U}^{(2)}\right) - \mathbf{U}^{(2)*}_{,\alpha}.\hat{\boldsymbol{\lambda}}_T g\bar{a}^{\alpha\beta}\boldsymbol{\nu}.\delta\mathbf{U}^{(2)}_{,\beta} \\ &= -\left\{ \begin{matrix} \{\mathbf{U}^{(1)*,e^{(1)}}\} \\ \{\mathbf{U}^{(2)*,e^{(2)}}\} \end{matrix} \right\}^T [\mathbf{k}^{e,5}_{\text{contact}}] \left\{ \begin{matrix} \{\delta\mathbf{U}^{(1),e^{(1)}}\} \\ \{\delta\mathbf{U}^{(2),e^{(2)}}\} \end{matrix} \right\}\end{aligned}$$

where the matrix $[\mathbf{k}^{e,5}_{\text{contact}}]$ is defined by [9.73]. From which we get the fifth and final term of $[\mathbf{K}^e_{UU}]$ in [9.68]. ■

REMARK. Since $[\mathbf{K}^e_{UU}]$ involves only the displacements of the nodes of the elements $e^{(1)} \subset S^{(1)}_{0c}$ and $e^{(2)}, e^{(2)}_{n-1} \subset S^{(2)}_{0c}$, *the assembled matrix* $[\boldsymbol{K}_{UU}]$ *involves only the displacements of the nodes on* $S^{(1)}_{0c}$ *and* $S^{(2)}_{0c}$, the terms relative to the nodes outside of $S^{(1)}_{0c} \cup S^{(2)}_{0c}$ *vanish.*

In programming therefore, we do not calculate $[\mathbf{K}_{UU}]$ in its entirety, but only the matrix denoted $[\mathbf{k}_{UU}]$ defined as being the part of $[\mathbf{K}_{UU}]$ relative to the nodes of $S^{(1)}_{0c} \cup S^{(2)}_{0c}$, with dimension $(3nnsc1 + 3nnsc2, 3nnsc1 + 3nnsc2)$ which is small in relation to $(3nn, 3nn)$. The matrix $[\mathbf{k}_{UU}]$ is assembled in $[\mathbf{K}_T]$. ■

THEOREM.– The matrix $[\mathbf{K}^e_{U\Lambda}]$ in [9.61], which reflects the coupling between the displacements and the multipliers, is rectangular and of dimension $(3nne^{(1)} + 3nne^{(2)}) \times 3nne^{(1)}$:

$$[\mathbf{K}^e_{U\Lambda}] = \begin{bmatrix} \underset{(3nne^{(1)},3)}{[\mathcal{N}^{e^{(1)}}]^T} \\ \underset{(3nne^{(2)},3)}{-[\mathcal{N}^{e^{(2)}}]^T} \end{bmatrix} \left[-H(\hat{\lambda}_N) \underset{(3,1)}{\{\boldsymbol{\nu}\}} \underset{(1,3)}{<\boldsymbol{\nu}>} + d_{sl}\, \mu \{\boldsymbol{\ell}_T\} <\boldsymbol{\nu}> - [\mathbf{P}][\mathbf{I} - \boldsymbol{\nu} \otimes \boldsymbol{\nu}] \right] \underset{(3,3nne^{(1)})}{[\mathcal{N}^{e^{(1)}}]} \qquad [9.74]$$

PROOF. The matrix $[\mathbf{K}^e_{U\Lambda}]$ comes from the discretization of the terms in the $(1, 2)$ block from the expression [9.58]:

$$\begin{cases} -H(\hat{\lambda}_N)(\mathbf{U}^{(1)*} - \mathbf{U}^{(2)*}).\boldsymbol{\nu}\; (\boldsymbol{\nu}.\delta\boldsymbol{\lambda}) \\ +d_{sl}\, \mu(\mathbf{U}^{(1)*} - \mathbf{U}^{(2)*}).\boldsymbol{\ell}_T\; (\boldsymbol{\nu}.\delta\boldsymbol{\lambda}) \\ -(\mathbf{U}^{(1)*} - \mathbf{U}^{(2)*}).\mathbf{P}.(\mathbf{I} - \boldsymbol{\nu} \otimes \boldsymbol{\nu}).\delta\boldsymbol{\lambda} \end{cases}$$

By using the interpolations [8.9], [8.13] of $\mathbf{U}^{(1)*}, \mathbf{U}^{(2)*}$ and [9.37] of $\delta\boldsymbol{\lambda}$, we obtain [9.74]. ■

THEOREM.– Likewise, the matrix $[\mathbf{K}^e_{\Lambda U}]$ in [9.61] that reflects the coupling between the displacements and the multipliers is rectangular of dimension $3nne^{(1)} \times (3nne^{(1)} + 3nne^{(2)})$:

$$\begin{aligned} &[\mathbf{K}^e_{\Lambda U}] = \\ &\underset{(3nne^{(1)},3)}{[\mathcal{N}^{e^{(1)}}]^T} \Bigg(\left[-H(\hat{\lambda}_N) \underset{(3,1)}{\{\boldsymbol{\nu}\}} \underset{(1,3)}{<\boldsymbol{\nu}>} + d_{sl}\, \mu \frac{\epsilon_N}{\epsilon_T} \{\boldsymbol{\ell}_T\} <\boldsymbol{\nu}> \right] \left[\underset{(3,3nne^{(1)})}{[\mathcal{N}^{e^{(1)}}]}, -\underset{(3,3nne^{(2)})}{[\mathcal{N}^{e^{(2)}}]}, \underset{(3,3nne^{(2)}_{n-1})}{[\mathbf{0}]} \right] \\ &-[\mathbf{P}] \left[\underset{(3,3)}{[\mathbf{W}^1]} \underset{(3,3nne^{(1)})}{[\mathcal{N}^{e^{(1)}}]}, \underset{(3,3)}{[\mathbf{W}^2]} \underset{(3,3nne^{(2)})}{[\mathcal{N}^{e^{(2)}}]} - [\mathbf{Z}^\alpha][\mathcal{N}^{e^{(2)}}_{,\alpha}], \underset{(3,3nne^{(2)}_{n-1})}{-[\mathcal{N}^{e^{(2)}_{n-1}}(\boldsymbol{\eta}_{n-1})]} \right] \Bigg) \end{aligned} \qquad [9.75]$$

PROOF. The matrix $[\mathbf{K}^e_{\Lambda U}]$ comes from the discretization of the terms in the $(2, 1)$ block of expression [9.58] (recall that we have set $\mathbf{m} = \boldsymbol{\nu}$):

$$\begin{cases} +H(\hat{\lambda}_N)\boldsymbol{\lambda}^*.\boldsymbol{\nu}\delta g \\ -d_{sl}\, \mu \dfrac{\epsilon_N}{\epsilon_T} \boldsymbol{\lambda}^*.\boldsymbol{\ell}_T \delta g \\ -\boldsymbol{\lambda}^*.\mathbf{P}.\delta\mathbf{V}_T \end{cases}$$

1. From the expression [9.55] of δg, we have

$$\begin{aligned} \boldsymbol{\lambda}^*.\boldsymbol{\nu}\delta g &= -\boldsymbol{\lambda}^*.(\boldsymbol{\nu} \otimes \boldsymbol{\nu}).\left(\delta\mathbf{U}^{(1)} - \delta\mathbf{U}^{(2)}\right) \\ \boldsymbol{\lambda}^*.\boldsymbol{\ell}_T\delta g &= -\boldsymbol{\lambda}^*.(\boldsymbol{\ell}_T \otimes \boldsymbol{\nu}).\left(\delta\mathbf{U}^{(1)} - \delta\mathbf{U}^{(2)}\right) \end{aligned}$$

Invoking the interpolations [8.18] of $\boldsymbol{\lambda}^*$, [9.34] and [9.35] of $\delta\mathbf{U}^{(1)}, \delta\mathbf{U}^{(2)}$, these expressions become

$$\begin{aligned} \boldsymbol{\lambda}^*.\boldsymbol{\nu}\delta g &= -<\boldsymbol{\lambda}^*>^{e^{(1)}} [\mathcal{N}^{e^{(1)}}]^T\{\boldsymbol{\nu}\} <\boldsymbol{\nu}> \left[[\mathcal{N}^{e^{(1)}}], -[\mathcal{N}^{e^{(2)}}]\right] \left\{ \begin{array}{c} \{\delta\mathbf{U}^{(1),e^{(1)}}\} \\ \{\delta\mathbf{U}^{(2),e^{(2)}}\} \end{array} \right\} \\ \boldsymbol{\lambda}^*.\boldsymbol{\ell}_T\delta g &= -<\boldsymbol{\lambda}^*>^{e^{(1)}} [\mathcal{N}^{e^{(1)}}]^T\{\boldsymbol{\ell}_T\} <\boldsymbol{\nu}> \left[[\mathcal{N}^{e^{(1)}}], -[\mathcal{N}^{e^{(2)}}]\right] \left\{ \begin{array}{c} \{\delta\mathbf{U}^{(1),e^{(1)}}\} \\ \{\delta\mathbf{U}^{(2),e^{(2)}}\} \end{array} \right\} \end{aligned}$$

2. From the expression [9.56] of $\delta(\mathbf{V}_T\Delta t)$, we have

$$\boldsymbol{\lambda}^*.\mathbf{P}.\delta\mathbf{V}_T = \boldsymbol{\lambda}^*.\mathbf{P}.\left(\mathbf{W}^1.\delta\mathbf{U}^{(1)}(\mathbf{X}) + \mathbf{W}^2.\delta\mathbf{U}^{(2)}(\mathbf{Y}) - \delta\mathbf{U}^{(2)}(\mathbf{Y}_{n-1}) - \mathbf{Z}^\alpha.\delta\mathbf{U}^{(2)}_{,\alpha}(\boldsymbol{\eta})\right)$$

With the same interpolations as before, we obtain

$$\boldsymbol{\lambda}^*.\mathbf{P}.\delta\mathbf{V}_T =<\boldsymbol{\lambda}^*>^{e^{(1)}} [\mathcal{N}^{e^{(1)}}]^T[\mathbf{P}]$$

$$\left[[\mathbf{W}^1][\mathcal{N}^{e^{(1)}}], [\mathbf{W}^2][\mathcal{N}^{e^{(2)}}] - [\mathbf{Z}^\alpha][\mathcal{N}^{e^{(2)}}_{,\alpha}], -[\mathcal{N}^{e^{(2)}_{n-1}}(\boldsymbol{\eta}_{n-1})]\right] \left\{ \begin{array}{c} \{\delta\mathbf{U}^{(1),e^{(1)}}\} \\ \{\delta\mathbf{U}^{(2),e^{(2)}}\} \\ \{\delta\mathbf{U}^{(2),e^{(2)}_{n-1}}\} \end{array} \right\}$$

3. By gathering the results found above and identifying the whole with [9.60], we arrive at [9.75]. ■

THEOREM.– Finally, the matrix $[\mathbf{K}^e_{\Lambda\Lambda}]$ in [9.61] is relative to the nodal multipliers $\{\boldsymbol{\Lambda}^{e^{(1)}}\}$, it is square and of order $3nne^{(1)}$:

$$[\mathbf{K}^e_{\Lambda\Lambda}] = \underset{(3nne^{(1)},3)}{[\mathcal{N}^{e^{(1)}}]^T} \left[-\frac{1}{\epsilon_N}(1 - H(\hat{\lambda}_N))\underset{(3,1)}{\{\boldsymbol{\nu}\}} \underset{(1,3)}{<\boldsymbol{\nu}>} - d_{sl}\frac{\mu}{\epsilon_T}\{\boldsymbol{\ell}_T\} <\boldsymbol{\nu}> + \frac{1}{\epsilon_T}[\mathbf{P} - \mathbf{I}][\mathbf{I} - \boldsymbol{\nu}\otimes\boldsymbol{\nu}] \right] \underset{(3,3nne^{(1)})}{[\mathcal{N}^{e^{(1)}}]} \quad [9.76]$$

PROOF. The matrix $[\mathbf{K}^e_{\Lambda\Lambda}]$ comes from the discretization of the terms in the (2, 2) block of the expression [9.58]:

$$\left\{ \begin{array}{l} -\dfrac{1}{\epsilon_N}(1 - H(\hat{\lambda}_N))\boldsymbol{\lambda}^*.\boldsymbol{\nu}\,(\boldsymbol{\nu}.\delta\boldsymbol{\lambda}) \\ -d_{sl}\,\mu\dfrac{1}{\epsilon_T}\boldsymbol{\lambda}^*.\boldsymbol{\ell}_T\,(\boldsymbol{\nu}.\delta\boldsymbol{\lambda}) \\ +\boldsymbol{\lambda}^*.\dfrac{1}{\epsilon_T}\,(\mathbf{P} - \ \mathbf{I}).(\mathbf{I} - \boldsymbol{\nu}\otimes\boldsymbol{\nu}).\delta\boldsymbol{\lambda} \end{array} \right.$$

By using the interpolations [8.18] of $\boldsymbol{\lambda}^*$ and [9.37] of $\delta\boldsymbol{\lambda}$, we easily obtain [9.76]. ■

From the programming point of view, it is more efficient to distinguish between three exclusive cases: (i) without algorithmic contact (algorithmic gap), (ii) algorithmic contact with slip and (iii) algorithmic contact with stick. We will therefore deduce from the general expressions [9.68]–[9.76] the expression for the element contact tangent matrix specific to these particular cases, by remembering that the matrices are written in each point of integration.

9.6.7. $[K^e_{contact}]$ in the case of the algorithmic gap at the considered integration point

Suppose that there is an *algorithmic gap* at the integration point considered: $\hat{\lambda}_N < 0$.
From [9.49], [9.50] and [9.53], here $c_{slst} = c_{sl} = d_{sl} = 0$, from which $\boxed{[\mathbf{P}] = [\mathbf{0}]}$.
The element contact tangent matrix becomes therefore the *square symmetric* matrix:

$$\boxed{\begin{array}{l} [\mathbf{K}^e_{\text{contact}}]_{gap} = \\ \begin{array}{cc} & \begin{array}{cc} <\mathbf{U}^{(1),e^{(1)}}><\mathbf{U}^{(2),e^{(2)}}> & <\mathbf{\Lambda}^{e^{(1)}}> \end{array} \\ \begin{array}{c} \{\mathbf{U}^{(1),e^{(1)}}\} \\ \{\mathbf{U}^{(2),e^{(2)}}\} \\ \{\mathbf{\Lambda}^{e^{(1)}}\} \end{array} & \left[\begin{array}{c|c} [\mathbf{0}] & \\ \hline & -[\mathcal{N}^{e^{(1)}}]^T \left[\dfrac{1}{\epsilon_N}\{\nu\}<\nu> + \dfrac{1}{\epsilon_T}[\mathbf{I} - \nu \otimes \nu]\right] [\mathcal{N}^{e^{(1)}}] \end{array}\right] \end{array} \end{array}} \quad [9.77]$$

We observe that $[\mathbf{K}^e_{\text{contact}}]_{gap}$ does not depend on the coefficient of friction μ, which is plausible.

The matrix $[\mathbf{K}^e_{\text{contact}}]_{gap}$ is no longer rectangular but *square*, since the matrices which involve the nodal displacements $\{\mathbf{U}^{(2),e^{(2)}_{n-1}}\}$, namely, $[\mathbf{k}^{e,4}_{\text{contact}}]$ and $[\mathbf{K}^e_{\Lambda U}]$, expressions [9.72] and [9.75], are zero here. There are no more degrees of freedom $\{\mathbf{U}^{(2),e^{(2)}_{n-1}}\}$ in the matrix $[\mathbf{K}^e_{\text{contact}}]_{gap}$.

Unlike in [9.62], the index e here in $[\mathbf{K}^e_{\text{contact}}]_{gap}$ only groups $e^{(1)}$ and $e^{(2)}$, there is no longer $e^{(2)}_{n-1}$.

The block $[\mathbf{K}^e_{\Lambda\Lambda}]_{gap}$ is a symmetric matrix. Moreover, knowing that $\mathbf{I} - \nu \otimes \nu = \mathbf{a}_\alpha \otimes \mathbf{a}^\alpha$, we have

$$\frac{1}{\epsilon_N}\{\nu\}<\nu> + \frac{1}{\epsilon_T}[\mathbf{I} - \nu \otimes \nu] = \frac{1}{\epsilon_T}\left[\mathbf{a}_\alpha \otimes \mathbf{a}^\alpha + \frac{\epsilon_N}{\epsilon_T}\nu \otimes \nu\right]$$

The block $[\mathbf{K}^e_{\Lambda\Lambda}]_{gap}$ is therefore *negative-definite* and invertible.

9.6.8. $[K^e_{contact}]$ in the case of algorithmic contact with slip at the considered integration point

Suppose there is *algorithmic* contact with slip at the integration point considered: $\hat{\lambda}_N \geq 0$ and $||\hat{\lambda}_T|| \geq \mu\hat{\lambda}_N$.

From [9.49], [9.50] and [9.53], here:

$$c_{slst} = c_{sl} = \frac{\mu\hat{\lambda}_N}{||\hat{\lambda}_T||}, d_{sl} = 1 \quad \Rightarrow \quad \boxed{[\mathbf{P}] = \frac{\mu\hat{\lambda}_N}{||\hat{\lambda}_T||}\Big([\mathbf{I}] - \{\boldsymbol{\ell}_T\}<\boldsymbol{\ell}_T>\Big)}, \text{with } 0 \leq \frac{\mu\hat{\lambda}_N}{||\hat{\lambda}_T||} \leq 1$$

Since $||\hat{\lambda}_T|| \geq \mu\hat{\lambda}_N \geq 0$ and $\hat{\lambda}_N \neq 0$ from the hypothesis [9.46], we have $||\hat{\lambda}_T|| \neq 0$ and the vector $\boldsymbol{\ell}_T \equiv \dfrac{\hat{\lambda}_T}{||\hat{\lambda}_T||}$ as well as the vector $\mathbf{P}$ are well defined.

The element contact tangent matrix is of the rectangular form [9.62]:

$$\boxed{[\mathbf{K}^e_{\text{contact}}]_{slip} = \begin{array}{cc} & \begin{array}{cc} <\mathbf{U}^{(1),e^{(1)}}><\mathbf{U}^{(2),e^{(2)}}><\mathbf{U}^{(2),e^{(2)}_{n-1}}> & <\mathbf{\Lambda}^{e^{(1)}}> \end{array} \\ \begin{array}{c} \{\mathbf{U}^{(1),e^{(1)}}\} \\ \{\mathbf{U}^{(2),e^{(2)}}\} \\ \{\mathbf{\Lambda}^{e^{(1)}}\} \end{array} & \left[\begin{array}{c|c} [\mathbf{K}^e_{UU}]_{slip} & [\mathbf{K}^e_{U\Lambda}]_{slip} \\ \hline [\mathbf{K}^e_{\Lambda U}]_{slip} & [\mathbf{K}^e_{\Lambda\Lambda}]_{slip} \end{array}\right] \end{array}} \quad [9.78]$$

with

$$[\mathbf{K}^e_{UU}]_{slip} = \epsilon_N[\mathbf{k}^{e,1}_{\text{contact}}] + \hat{\lambda}_N[\mathbf{k}^{e,2}_{\text{contact}}] - \mu\epsilon_N[\mathbf{k}^{e,3}_{\text{contact}}] + \epsilon_T[\mathbf{k}^{e,4}_{\text{contact}}] + \frac{\mu\hat{\lambda}_N}{||\hat{\boldsymbol{\lambda}}_T||}[\mathbf{k}^{e,5}_{\text{contact}}] \qquad [9.79]$$

In this expression, the matrices $[\mathbf{k}^{e,3}_{\text{contact}}]$ and $\frac{1}{||\hat{\boldsymbol{\lambda}}_T||}[\mathbf{k}^{e,5}_{\text{contact}}]$ contain $\{\boldsymbol{\ell}_T\} \equiv \frac{\{\hat{\boldsymbol{\lambda}}_T\}}{||\hat{\boldsymbol{\lambda}}_T||}$. The matrix $[\mathbf{k}^{e,4}_{\text{contact}}]$ is rectangular and contains $\{\boldsymbol{\ell}_T\} < \boldsymbol{\ell}_T >= \frac{\{\hat{\boldsymbol{\lambda}}_T\} < \hat{\boldsymbol{\lambda}}_T >}{||\hat{\boldsymbol{\lambda}}_T||^2}$.

$$[\mathbf{K}^e_{U\Lambda}]_{slip} - \begin{bmatrix} [\mathcal{N}^{e(1)}]^T \\ -[\mathcal{N}^{e(2)}]^T \end{bmatrix} \left[-\{\boldsymbol{\nu}\} < \boldsymbol{\nu} > +\mu\{\boldsymbol{\ell}_T\} < \boldsymbol{\nu} > -[\mathbf{P}]\,[\mathbf{I} - \boldsymbol{\nu}\otimes\boldsymbol{\nu}]\right] [\mathcal{N}^{e(1)}] \qquad [9.80]$$

$$\begin{aligned}[\mathbf{K}^e_{\Lambda U}]_{slip} = [\mathcal{N}^{e(1)}]^T \Bigg(&\left[\mu\frac{\epsilon_N}{\epsilon_T}\{\boldsymbol{\ell}_T\} < \boldsymbol{\nu} > -\{\boldsymbol{\nu}\} < \boldsymbol{\nu} >\right] \left[[\mathcal{N}^{e(1)}], -[\mathcal{N}^{e(2)}], \underset{(3,3nne^{(2)}_{n-1})}{[\mathbf{0}]}\right] \\ &-[\mathbf{P}] \left[[\mathbf{W}^1][\mathcal{N}^{e(1)}], [\mathbf{W}^2][\mathcal{N}^{e(2)}] - [\mathbf{Z}^\alpha][\mathcal{N}^{e(2)}_{,\alpha}], \underset{(3,3nne^{(2)}_{n-1})}{-[\mathcal{N}^{e^{(2)}_{n-1}}(\boldsymbol{\eta}_{n-1})]}\right]\Bigg)\end{aligned} \qquad [9.81]$$

$$[\mathbf{K}^e_{\Lambda\Lambda}]_{slip} = [\mathcal{N}^{e(1)}]^T \left[-\frac{\mu}{\epsilon_N}\{\boldsymbol{\ell}_T\} < \boldsymbol{\nu} > +\frac{1}{\epsilon_T}[\mathbf{P} - \mathbf{I}]\,[\mathbf{I} - \boldsymbol{\nu}\otimes\boldsymbol{\nu}]\right] [\mathcal{N}^{e(1)}] \qquad [9.82]$$

Even without the columns corresponding to $\{\mathbf{U}^{(2),e^{(2)}_{n-1}}\}$, the matrix $[\mathbf{K}^e_{\text{contact}}]_{slip}$ (which then becomes square) is *not symmetric*, except if $\mu = 0$.

- Let us examine the case where $\hat{\boldsymbol{\lambda}}_T = \mathbf{0}$ at the considered point.

The case with algorithmic contact with slip is the only one of the three contact cases where the expressions for the contact matrices contain $\hat{\boldsymbol{\lambda}}_T$ in the denominator and are no longer valid when $\hat{\boldsymbol{\lambda}}_T = \mathbf{0}$.

Recall the comments from section 8.7.1 when $\hat{\boldsymbol{\lambda}}_T = \mathbf{0}$, notably the fact that, from the hypotheses $\hat{\lambda}_N \geq 0$ and $||\hat{\boldsymbol{\lambda}}_T|| \geq \mu\hat{\lambda}_N$, the equality $\hat{\boldsymbol{\lambda}}_T = \mathbf{0}$ imply that $\hat{\lambda}_N = 0$.

We shall admit that, in the case of algorithmic contact with slip at the considered integration point and $\hat{\boldsymbol{\lambda}}_T = \mathbf{0}$, the expressions for the contact matrices can be deduced from those above by formally setting (see [8.31]):

$$\frac{\mu\hat{\lambda}_N}{||\hat{\boldsymbol{\lambda}}_T||} = 1 \qquad \text{and} \qquad \boldsymbol{\ell}_T \equiv \frac{\hat{\boldsymbol{\lambda}}_T}{||\hat{\boldsymbol{\lambda}}_T||} = \mathbf{0} \qquad [9.83]$$

We thus obtain $[\mathbf{P}] = [\mathbf{I}]$, then the contact tangent matrices:

$$[\mathbf{K}^e_{UU}]_{slip} = \epsilon_N[\mathbf{k}^{e,1}_{\text{contact}}] + \epsilon_T[\mathbf{k}^{e,4}_{\text{contact}}]$$

$$[\mathbf{K}^e_{U\Lambda}]_{slip} = -\begin{bmatrix} [\mathcal{N}^{e(1)}]^T \\ -[\mathcal{N}^{e(2)}]^T \end{bmatrix} [\mathcal{N}^{e(1)}]$$

$$[\mathbf{K}^e_{\Lambda U}]_{slip} = -[\mathcal{N}^{e^{(1)}}]^T\Bigg(\{\boldsymbol{\nu}\} < \boldsymbol{\nu} > \Bigg[[\mathcal{N}^{e^{(1)}}], -[\mathcal{N}^{e^{(2)}}], \underset{(3,3nne^{(2)}_{n-1})}{[\mathbf{0}]}\Bigg]$$

$$+ \Bigg[[\mathbf{W}^1][\mathcal{N}^{e^{(1)}}], [\mathbf{W}^2][\mathcal{N}^{e^{(2)}}] - [\mathbf{Z}^\alpha][\mathcal{N}^{e^{(2)}}_{,\alpha}], \underset{(3,3nne^{(2)}_{n-1})}{-[\mathcal{N}^{e^{(2)}_{n-1}}(\boldsymbol{\eta}_{n-1})]}\Bigg]\Bigg)$$

$$[\mathbf{K}^e_{\Lambda\Lambda}]_{slip} = [\mathbf{0}]$$

We observe the same thing in section 8.7.1 on the contact force and the contact residual vectors, namely, that the following two contact states:

1. the algorithmic contact with slip at the considered integration point and $\hat{\boldsymbol{\lambda}}_T = \mathbf{0}$ (point **O** in Figure 8.5);
2. the algorithmic contact with stick at the integration point considered and $\hat{\lambda}_N \to 0$ (point **A** in the same figure).

They lead to the same expressions for the contact matrices. In other words, the above-found matrices are identical to the matrix [9.84] with $\hat{\lambda}_N \to 0$ (so $\hat{\boldsymbol{\lambda}}_T \to \mathbf{0}$).

9.6.9. $[K^e_{contact}]$ *in the case of algorithmic contact with* **stick** *at the considered integration point*

Suppose that there is algorithmic contact with stick at the integration point considered: $\hat{\lambda}_N \geq 0$ and $\|\hat{\boldsymbol{\lambda}}_T\| < \mu\hat{\lambda}_N$.

From [9.49], [9.50] and [9.53], here $c_{slst} = 1, c_{sl} = 0, d_{sl} = 0$, d'oÃź $\boxed{[\mathbf{P}] = [\mathbf{I}]}$.

The element contact tangent matrix is given by

$$[\mathbf{K}^e_{\text{contact}}]_{stick} =$$

$$\begin{array}{c|c} & < \mathbf{U}^{(1),e^{(1)}} >< \mathbf{U}^{(2),e^{(2)}} >< \mathbf{U}^{(2),e^{(2)}_{n-1}} > \qquad\qquad < \boldsymbol{\Lambda}^{e^{(1)}} > \\ \begin{array}{c}\{\mathbf{U}^{(1),e^{(1)}}\}\\ \{\mathbf{U}^{(2),e^{(2)}}\}\\ \\ \{\boldsymbol{\Lambda}^{e^{(1)}}\}\end{array} & \left[\begin{array}{c|c} \epsilon_N[\mathbf{k}^{e,1}_{\text{contact}}] + \hat{\lambda}_N[\mathbf{k}^{e,2}_{\text{contact}}] + \epsilon_T[\mathbf{k}^{e,4}_{\text{contact}}] + [\mathbf{k}^{e,5}_{\text{contact}}] & -\begin{bmatrix}[\mathcal{N}^{e^{(1)}}]^T\\ -[\mathcal{N}^{e^{(2)}}]^T\end{bmatrix}[\mathcal{N}^{e^{(1)}}] \\ \hline \begin{array}{l} -[\mathcal{N}^{e^{(1)}}]^T\Big(\{\boldsymbol{\nu}\} < \boldsymbol{\nu} > \Big[[\mathcal{N}^{e^{(1)}}], -[\mathcal{N}^{e^{(2)}}], 3\overset{3nne^{(2)}_{n-1}}{[\mathbf{0}]}\Big] \\ + \Big[[\mathbf{W}^1][\mathcal{N}^{e^{(1)}}], \mathbf{W}^2][\mathcal{N}^{e^{(2)}}] - [\mathbf{Z}^\alpha][\mathcal{N}^{e^{(2)}}_{,\alpha}], \\ -[\mathcal{N}^{e^{(2)}_{n-1}}(\boldsymbol{\eta}_{n-1})]\Big]\Big) \end{array} & [\mathbf{0}] \end{array}\right] \end{array}$$

[9.84]

We observe that $[\mathbf{K}^e_{\text{contact}}]_{stick}$ does not depend on the coefficient of friction μ! That said, the case of algorithmic contact with stick is subject to the condition $\|\hat{\boldsymbol{\lambda}}_T\| < \mu\hat{\lambda}_N$ which depends on μ.

In the algorithmic stick case, it is possible to have $\hat{\boldsymbol{\lambda}}_T = \mathbf{0}$ (we can satisfy $\|\hat{\boldsymbol{\lambda}}_T = \mathbf{0}\| < \mu\hat{\lambda}_N$ if $\hat{\lambda}_N$ is > 0) and then the matrix $[\mathbf{k}^{e,5}_{\text{contact}}]$ in [9.84] vanishes according to [9.73].

Recall that the matrix $[\mathbf{k}^{e,4}_{\text{contact}}]$ in the block $[\mathbf{K}^e_{UU}]$ is rectangular.

Since the matrix $[\mathbf{K}^e_{\text{contact}}]_{stick}$ is rectangular, symmetry is irrelevant here. That being said, you may wonder if in some cases the matrix becomes square and, in such a case, if it can further be symmetric. Suppose for example that we are *at the convergence* (that is to say, at the last iteration of a step) and that the algorithmic stick case at the integration point considered leads to *contact with effective stick* (and no longer algorithmic stick): $g = 0$ and $\mathbf{Y} = \mathbf{Y}_{n-1}$. In this case, there is no energy dissipation at the contact point and we hope to have a symmetric matrix. Let us see if this is really the case.

1. The consequences of $\mathbf{Y} = \mathbf{Y}_{n-1}$ are the following:

$$\mathbf{Y} = \mathbf{Y}_{n-1} \;\Rightarrow \begin{cases} .\; e^{(2)} = e^{(2)}_{n-1} \\ .\; \boldsymbol{\eta} = \boldsymbol{\eta}_{n-1} \end{cases}$$
$$\Rightarrow \begin{cases} .\; \lfloor \mathcal{N}^{e^{(2)}} \rfloor = \lfloor \mathcal{N}^{e^{(2)}_{n-1}}(\boldsymbol{\eta}_{n-1}) \rfloor \\ .\; \text{we can group together the columns corresponding to } e^{(2)} \text{ and } e^{(2)}_{n-1} \end{cases}$$

Doing this, the matrices $[\mathbf{k}^{e,4}_{\text{contact}}]$ and $[\mathbf{K}^e_{\text{contact}}]_{stick}$ simplify a bit and moreover they become square and we can ask the question on whether they are symmetric or not.

2. The consequences of $g = 0$ and $\boldsymbol{\eta} = \boldsymbol{\eta}_{n-1}$ are the following:

$$\begin{cases} .\; g = 0 \Rightarrow \bar{a}^{\alpha\beta} = a^{\alpha\beta} \\ .\; \boldsymbol{\eta} = \boldsymbol{\eta}_{n-1} \end{cases} \Rightarrow \begin{cases} .\; \mathbf{W}^1 = a^{\alpha\beta}\mathbf{a}_\alpha \otimes \mathbf{a}_\beta = \mathbf{I} - \boldsymbol{\nu} \otimes \boldsymbol{\nu} \\ .\; \mathbf{Z}^\alpha = \mathbf{0} \end{cases}$$

$$\Rightarrow \begin{cases} .\; \text{the matrix } [\mathbf{k}^{e,4}_{\text{contact}}] \text{ is symmetric} \\ .\; \text{the blocks } [\mathbf{K}^e_{U\Lambda}] \text{ and } [\mathbf{K}^e_{\Lambda U}] \text{ are symmetric, transposed of each other.} \end{cases}$$

There remains the matrix $[\mathbf{k}^{e,5}_{\text{contact}}]$ which is not always symmetric! Consequently, in the case considered, the matrix $[\mathbf{K}^e_{\text{contact}}]_{stick}$ is *not* symmetric.

9.6.10. *Symmetry of the contact tangent matrix* $[K_{contact}]$

We have just examined the symmetry of the element contact tangent matrix $[\mathbf{K}^e_{\text{contact}}]$ in the three exclusive cases: algorithmic gap; algorithmic contact with slip; and algorithmic contact with stick. In the algorithmic gap case, the matrix $[\mathbf{K}^e_{\text{contact}}]$ is symmetric, yet in the two other cases this is not certain. Let us focus now on the symmetry of the contact tangent matrix $[\mathbf{K}_{\text{contact}}]$ of the entire system which, from [9.66], is obtained by assembling the submatrices of $[\mathbf{K}^e_{\text{contact}}]$ (see the definition of these submatrices in [9.63]).

The particular case when $\mu = 0$ will be studied in detail in section 9.8. There, we will see that in the absence of friction, the contact tangent matrix $[\mathbf{K}^e_{\text{contact}}]$ is *symmetric* at every integration point whether there is gap or contact. Consequently, the assembled tangent matrix $[\mathbf{K}_{\text{contact}}]$ is also symmetric regardless of the contact nature (gap or no gap).

In the presence of friction, the terms arising from the friction make the matrix $[\mathbf{K}^e_{\text{contact}}]$ non-symmetric. Generally, even if $[\mathbf{K}^e_{\text{contact}}]$ is symmetric at every integration point, it can be non-symmetric at another point, so that the assembled tangent matrix $[\mathbf{K}_{\text{contact}}]$ is *non-symmetric* in general.

In comparison with the frictionless case, practically the tangent matrix $[\mathbf{K}_{\text{contact}}]$ requires twice the memory space for its storage. A calculation with friction therefore is more costly in terms of memory and in computation time than a calculation without friction.

To reduce this cost, we could either use the symmetrization technique of Laursen [83], which allows us to work with symmetric matrices even in frictional cases, or to use iterative solving methods for linear systems, much faster not only for symmetric matrices but also for non-symmetric ones. In the scope of this book, we have not exploited any of these possibilities; in the friction case, we content ourselves with solving the non-symmetric system using Gauss's triangularization method, knowing that this can generate significant computational costs.

Lastly, even if the contact tangent matrix $[\mathbf{K}_{\text{contact}}]$ is symmetric, the global tangent matrix [9.10], as for it, can be non-symmetric because of other sources of non-symmetry like, for example, certain follower forces applied on the system.

9.7. Particular case of two non-contacting bodies

Let us examine the particularly simple case of two bodies which are not in contact at a given step n.

HYPOTHESES.– [9.85]

Suppose that at the numerical iteration $i-1$ and at *every* integration point of every element $e^{(1)} \subset S^{(1)}_{0c}$:

1. there is an algorithmic gap ($\hat{\lambda}_N < 0$);
2. furthermore, the contact constraints are zero: $\lambda_N = 0$ and $\lambda_T = \mathbf{0}$.

In these conditions, the element contact force vectors and the element residual contact vector, at the iteration $i-1$ and at every integration point $S^{(1)}_{0c}$, are given by [8.26] and [8.27]:

$$\begin{aligned} \left\{\mathbf{\Phi}^{e^{(1)}}_{\text{contact}}(\mathbf{U}^e, \mathbf{\Lambda}^{e^{(1)}})\right\} &= \{\mathbf{0}\} \\ \left\{\mathbf{\Phi}^{e^{(2)}}_{\text{contact}}(\mathbf{U}^e, \mathbf{\Lambda}^{e^{(1)}})\right\} &= \{\mathbf{0}\} \end{aligned} \qquad [9.86]$$

$$\left\{\mathbf{R}^{e^{(1)}}_{\Lambda}(\mathbf{U}^e, \mathbf{\Lambda}^{e^{(1)}})\right\} \underset{[8.27]}{=} [\mathcal{N}^{e^{(1)}}(\xi)]^T \left(-\frac{\lambda_N}{\epsilon_N}\{\mathbf{m}\} + \frac{1}{\epsilon_T}\{\lambda_T\} \right) \underset{[9.85]_2}{=} \{\mathbf{0}\}$$

The element contact tangent matrix is the *square symmetric* matrix given by [9.77]:

$$[\mathbf{K}^e_{\text{contact}}]_{gap} = \begin{array}{c} \{\mathbf{U}^{(1),e^{(1)}}\} \\ \{\mathbf{U}^{(2),e^{(2)}}\} \\ \{\mathbf{\Lambda}^{e^{(1)}}\} \end{array} \overset{\begin{array}{cc} <\mathbf{U}^{(1),e^{(1)}}><\mathbf{U}^{(2),e^{(2)}}> & <\mathbf{\Lambda}^{e^{(1)}}> \end{array}}{\left[\begin{array}{c|c} [\mathbf{0}] & \\ \hline & -[\mathcal{N}^{e^{(1)}}]^T \left[\frac{1}{\epsilon_N}\{\nu\}<\nu> + \frac{1}{\epsilon_T}[\mathbf{I} - \nu \otimes \nu] \right] [\mathcal{N}^{e^{(1)}}] \end{array} \right]}$$

After assembly, the contact force vectors and the residual contact vector vanish:

$$\{\mathbf{\Phi}_{\text{contact}}\} = \{\mathbf{0}\} \qquad \text{and} \qquad \{\mathbf{R}_{\Lambda}\} = \{\mathbf{0}\}$$

while the contact tangent matrix has the same form as the element one:

$$[\mathbf{K}_{\text{contact}}] = \begin{array}{c} \{\mathbf{U}^{(1)}\} \\ \{\mathbf{U}^{(2)}\} \\ \{\mathbf{\Lambda}\} \end{array} \overset{\begin{array}{cc} <\mathbf{U}^{(1)}><\mathbf{U}^{(2)}> & <\mathbf{\Lambda}> \end{array}}{\left[\begin{array}{c|c} [\mathbf{0}] & \\ \hline & \begin{array}{l} \mathbf{K}_{\Lambda\Lambda} = -\sum_{e^{(1)} \subset S^{(1)}_{0c}} \int_{e^{(1)}} \mathbb{L}^{e^{(1)}T} [\mathcal{N}^{e^{(1)}}]^T \\ \left[\frac{1}{\epsilon_N}\{\nu\}<\nu> + \frac{1}{\epsilon_T}[\mathbf{I} - \nu \otimes \nu] \right] [\mathcal{N}^{e^{(1)}}] \mathbb{L}^{e^{(1)}} dS_0 \end{array} \end{array} \right]} \qquad [9.87]$$

At the ith iteration, we have to solve the system [9.14] for the unknowns $\delta\mathbf{U}$ and $\delta\mathbf{\Lambda}$ which here are decoupled:

$$\left[\begin{array}{c|c} \mathbf{K} = \mathbf{K}_{\text{classical without contact}} & \\ \hline & \mathbf{K}_{\Lambda\Lambda} \end{array}\right] \left\{\begin{array}{c} \delta\mathbf{U} \\ \delta\mathbf{\Lambda} \end{array}\right\} = -\left\{\begin{array}{c} \mathbf{R}_{\text{classical without contact}} \\ \mathbf{0} \end{array}\right\} \qquad [9.88]$$

After taking the boundary conditions into account, the matrix $\mathbf{K}_{\text{classical without contact}}$ is invertible. As for the matrix $\mathbf{K}_{\Lambda\Lambda}$, it has the same form as the mass matrix which exists in dynamics, for example, for body 1:

$$\sum_{e^{(1)}\subset\Omega^{(1)}} \int_{e^{(1)}} \mathbb{L}^{e^{(1)}T}[\mathcal{N}^{e^{(1)}}]^T \rho_0^{(1)}[\mathcal{N}^{e^{(1)}}]\mathbb{L}^{e^{(1)}} d\Omega_0$$

Here, the matrix $\dfrac{1}{\epsilon_N}\{\nu\} < \nu > + \dfrac{1}{\epsilon_T}[\mathbf{I} - \nu\otimes\nu]$, of dimension 3×3, takes the place of the density $\rho_0^{(1)}$. It is symmetric and invertible, with its inverse given by $\epsilon_N\{\nu\} < \nu > + \epsilon_T[\mathbf{I} - \nu\otimes\nu]$. Just like the mass matrix, the matrix $\mathbf{K}_{\Lambda\Lambda}$ is also invertible.

The system [9.88] provides the corrections, so that the new displacement and multiplier values for the ith iteration are:

$$\left\{\begin{array}{lcl} \{\delta\mathbf{U}\}^i &=& \{\delta\mathbf{U}\}_{\text{classical without contact}} \\ \{\delta\mathbf{\Lambda}\}^i &=& \mathbf{0} \end{array}\right. \Rightarrow \left\{\begin{array}{l} \{\mathbf{U}\}^i = \{\mathbf{U}\}^{i-1} + \{\delta\mathbf{U}\}^i \\ \{\mathbf{\Lambda}\}^i = \{\mathbf{\Lambda}\}^{i-1} + \{\delta\mathbf{\Lambda}\}^i = \{\mathbf{0}\} \end{array}\right.$$

The contact constraints remain zero for the ith iteration.

As for the displacement $\{\mathbf{U}\}^i$, it may or may not create contact between the two bodies at the ith iteration, which entails that the first hypothesis in [9.85] may no longer or still be satisfied at the ith iteration.

If the hypotheses [9.85] are satisfied at every iteration until convergence is achieved, then we obtain $\boxed{\{\mathbf{\Lambda}\} = \{\mathbf{0}\}}$ for the considered step as expected.

9.8. Particular case of the frictionless problem

The frictionless problem is a particular case in which the relations are notably simplified. If $\mu = 0$, then there are only the following two possible sub-cases:

1. algorithmic gap ($\hat{\lambda}_N < 0$);

2. algorithmic contact with slip ($\hat{\lambda}_N \geq 0$ and $\|\hat{\lambda}_T\| \geq \mu\hat{\lambda}_N$, with the last condition being automatically satisfied since $\mu = 0$).

The case of algorithmic contact with stick is not possible.

By setting $\mu = 0$ in definitions [8.21]–[8.22], we obtain new expressions for the element contact force vectors and for the element residual contact law vector:

$$\boxed{\begin{array}{lcl} \left\{\mathbf{\Phi}^{e^{(1)}}_{\text{contact}}(\mathbf{U}^e, \mathbf{\Lambda}^{e^{(1)}})\right\} &=& [\mathcal{N}^{e^{(1)}}(\boldsymbol{\xi})]^T\{\nu\}\,\langle\hat{\lambda}_N\rangle \\ \left\{\mathbf{\Phi}^{e^{(2)}}_{\text{contact}}(\mathbf{U}^e, \mathbf{\Lambda}^{e^{(1)}})\right\} &=& -[\mathcal{N}^{e^{(2)}}(\boldsymbol{\eta})]^T\{\nu\}\,\langle\hat{\lambda}_N\rangle \end{array}} \quad \text{(recall: } \langle\hat{\lambda}_N\rangle \text{ is a scalar)} \qquad [9.89]$$

$$\boxed{\left\{\mathbf{R}^{e^{(1)}}_{\Lambda}(\mathbf{U}^e, \mathbf{\Lambda}^{e^{(1)}})\right\} = [\mathcal{N}^{e^{(1)}}(\boldsymbol{\xi})]^T \left(\frac{1}{\epsilon_N}\left(\langle\hat{\lambda}_N\rangle - \lambda_N\right)\{\nu\} + \frac{1}{\epsilon_T}\{\boldsymbol{\lambda}_T\}\right)} \qquad [9.90]$$

Unlike in [8.25], the index e of $\mathbf{U}^e$ groups only $e^{(1)}$ and $e^{(2)}$ together, with $e^{(2)}_{n-1}$ no longer appearing since $\mathbf{V}_T$ does not play any role here.

From [9.49], [9.50] and [9.53], here $c_{slst} = 0$ (the stick case being impossible), $c_{sl} = 0$, whence $\boxed{[\mathbf{P}] = [\mathbf{0}]}$.

By setting $\mu = c_{slst} = 0$ and $[\mathbf{P}] = [\mathbf{0}]$ in the expressions [9.68]–[9.76], then we see that several submatrices of the element contact tangent matrix $[\mathbf{K}^e_{\text{contact}}]$ vanish and that $[\mathbf{K}^e_{\text{contact}}]$ reduces to the following *square symmetric* matrix:

$$
\boxed{
\begin{array}{l}
[\mathbf{K}^e_{\text{contact}}] = \\
\begin{array}{c c c}
 & <\mathbf{U}^{(1),e^{(1)}}><\mathbf{U}^{(2),e^{(2)}}> & <\mathbf{\Lambda}^{e^{(1)}}> \\
\begin{array}{c}\{\mathbf{U}^{(1),e^{(1)}}\}\\ \{\mathbf{U}^{(2),e^{(2)}}\}\\ \\ \{\mathbf{\Lambda}^{e^{(1)}}\}\end{array} &
\multicolumn{2}{c}{\left[\begin{array}{c|c}
H(\hat{\lambda}_N)\epsilon_N[\mathbf{k}^{e,1}_{\text{contact}}] + \langle\hat{\lambda}_N\rangle[\mathbf{k}^{e,2}_{\text{contact}}] & -\begin{bmatrix}[\mathcal{N}^{e^{(1)}}]^T\\ -[\mathcal{N}^{e^{(2)}}]^T\end{bmatrix} H(\hat{\lambda}_N)\{\boldsymbol{\nu}\}<\boldsymbol{\nu}>[\mathcal{N}^{e^{(1)}}] \\ \hline
-[\mathcal{N}^{e^{(1)}}]^T H(\hat{\lambda}_N)\{\boldsymbol{\nu}\}<\boldsymbol{\nu}> \left[[\mathcal{N}^{e^{(1)}}], -[\mathcal{N}^{e^{(2)}}]\right] & -[\mathcal{N}^{e^{(1)}}]^T \left[\dfrac{1}{\epsilon_N}(1 - H(\hat{\lambda}_N))\{\boldsymbol{\nu}\}<\boldsymbol{\nu}> + \dfrac{1}{\epsilon_T}[\mathbf{I} - \boldsymbol{\nu}\otimes\boldsymbol{\nu}]\right] [\mathcal{N}^{e^{(1)}}]
\end{array}\right]}
\end{array}
\end{array}
}
\quad [9.91]
$$

The matrix $[\mathbf{K}^e_{\text{contact}}]$ is no longer rectangular but rather is now *square* for the following reasons:

- The rectangular matrix $[\mathbf{k}^{e,4}_{\text{contact}}]$ in [9.72], which involves the displacements $\{\mathbf{U}^{(2),e^{(2)}_{n-1}}\}$, vanishes here.
- In the expression [9.75] of the matrix $[\mathbf{K}^e_{\Lambda U}]$, the $3 \times 3nne^{(2)}_{n-1}$ rectangular block $[\mathbf{0}]$, corresponding to the columns $<\mathbf{U}^{(2)}>^{e^{(2)}_{n-1}}$, also vanishes.

Unlike in [9.62], the index e of $[\mathbf{K}^e_{\text{contact}}]$ groups only $e^{(1)}$ and $e^{(2)}$ together, with $e^{(2)}_{n-1}$ no longer appearing.

The symmetric nature of $[\mathbf{K}^e_{\text{contact}}]$ is understandable physically, given that there is no friction and hence no energy dissipation.

In the absence of friction, there is no point in calculating the slip $\mathbf{V}_T\Delta t$.

9.8.1. *Algorithmic* gap *case at the considered integration point*

Suppose now that we have *algorithmic gap* at the considered integration point ($\hat{\lambda}_N < 0$), so that the expressions [9.89]–[9.91] become:

$$
\boxed{\begin{array}{l}
\left\{\mathbf{\Phi}^{e^{(1)}}_{\text{contact}}(\mathbf{U}^e, \mathbf{\Lambda}^{e^{(1)}})\right\}_{gap} = \{\mathbf{0}\} \\
\left\{\mathbf{\Phi}^{e^{(2)}}_{\text{contact}}(\mathbf{U}^e, \mathbf{\Lambda}^{e^{(1)}})\right\}_{gap} = \{\mathbf{0}\}
\end{array}}
$$

$$
\boxed{\left\{\mathbf{R}^{e^{(1)}}_{\Lambda}(\mathbf{U}^e, \mathbf{\Lambda}^{e^{(1)}})\right\}_{gap} = [\mathcal{N}^{e^{(1)}}(\boldsymbol{\xi})]^T \left(-\frac{\lambda_N}{\epsilon_N}\{\boldsymbol{\nu}\} + \frac{1}{\epsilon_T}\{\boldsymbol{\lambda}_T\}\right)}
$$

$$
\boxed{[\mathbf{K}^e_{\text{contact}}]_{gap} = \left[\begin{array}{c|c}
[\mathbf{0}] & \\ \hline
 & -[\mathcal{N}^{e^{(1)}}]^T \left[\dfrac{1}{\epsilon_N}\{\boldsymbol{\nu}\}<\boldsymbol{\nu}> + \dfrac{1}{\epsilon_T}[\mathbf{I} - \boldsymbol{\nu}\otimes\boldsymbol{\nu}]\right] [\mathcal{N}^{e^{(1)}}]
\end{array}\right]}
\quad [9.92]
$$

Note that in the algorithmic gap case, the expressions found do not depend on the coefficient of friction μ, which makes sense. Besides, they are the same as those found in the homologous case of frictional contact (see [8.26], [8.27] and [9.77]).

The block $[\mathbf{K}^e_{\Lambda\Lambda}]_{gap}$ is a symmetric matrix, *definite negative*, and also invertible.

Let us consider the nth step and suppose, analogously to the contactless case involving two bodies from section 9.7, that:

(i) the hypotheses [9.85] are satisfied at the $(i-1)$th iteration of the calculation, and at *every* integration point of every element $e^{(1)} \subset S^{(1)}_{0c}$;

(ii) the assembled tangent matrix is invertible.

With these conditions, we obtain $\{\boldsymbol{\Lambda}\} = \{\mathbf{0}\}$ at the ith iteration, which means that the contact constraints remain zero. In particular, the tangential contact stress $\boldsymbol{\lambda}_T$ vanishes as expected, since here there is no friction.

In the frictionless contact case, you may think that each term relative to the tangential stress should vanish and may be surprised at the presence of the tensor $\frac{1}{\epsilon_T}[\mathbf{I} - \boldsymbol{\nu} \otimes \boldsymbol{\nu}]$ inside the matrix [9.92]. In fact, we are going to see that this tensor is important since, without it, the tangent matrix is *not* invertible.

At the ith iteration of the calculation, we have to solve the system [9.14] that we write here as

$$\left[\begin{array}{c|c} \mathbf{K} = \mathbf{K}_{\text{classical without contact}} & \\ \hline & \mathbf{K}_{\Lambda\Lambda} \end{array}\right] \left\{\begin{array}{c} \delta\mathbf{U} \\ \delta\boldsymbol{\Lambda} \end{array}\right\} = -\left\{\begin{array}{c} \mathbf{R}^{i-1} \\ \mathbf{R}^{i-1}_{\Lambda} \end{array}\right\} \qquad [9.93]$$

where $\mathbf{K}_{\Lambda\Lambda}$ has the same expression as in [9.87]. The LHS of [9.93] is identical to that in [9.88]. Under appropriate boundary conditions, the matrix $\mathbf{K}_{\text{classical without contact}}$ is invertible. As for $\mathbf{K}_{\Lambda\Lambda}$, it is also invertible. The solution of the previous system provides the corrections $\{\delta\mathbf{U}\}^i$, $\{\delta\boldsymbol{\Lambda}\}^i$, and then the new values of the displacements $\{\mathbf{U}\}^i$ and the multipliers $\{\boldsymbol{\Lambda}\}^i$ at the ith iteration. Usually the tangential contact stress found should vanish given that there is no friction.

The tensor $\frac{1}{\epsilon_N}\{\boldsymbol{\nu}\} < \boldsymbol{\nu} > + \frac{1}{\epsilon_T}[\mathbf{I} - \boldsymbol{\nu} \otimes \boldsymbol{\nu}]$ inside of the matrix [9.92] is invertible and whose inverse is $\epsilon_N\{\boldsymbol{\nu}\} < \boldsymbol{\nu} > + \epsilon_T[\mathbf{I} - \boldsymbol{\nu} \otimes \boldsymbol{\nu}]$. This forms the block $\mathbf{K}_{\Lambda\Lambda}$ of the system [9.93], which is invertible, and this emphasizes the importance of the tensor $\frac{1}{\epsilon_T}[\mathbf{I} - \boldsymbol{\nu} \otimes \boldsymbol{\nu}]$ within the matrix [9.92].

If we had limited ourselves to considering the frictionless contact problem from the start instead, then we would have had a simpler weak form given by [7.21] but without any term related to the tangential contact stress $\boldsymbol{\lambda}_T$ nor to the slip velocity $\mathbf{V}_T$. Then the matrix [9.92] would not contain the $\frac{1}{\epsilon_T}[\mathbf{I} - \boldsymbol{\nu} \otimes \boldsymbol{\nu}]$ term, nor would we have access to the stress $\boldsymbol{\lambda}_T$, which we already know is zero by construction.

The situation here is different; we started from the general contact framework with friction before then considering the frictionless case as a particular one with $\mu = 0$. Under these conditions, it was necessary to check that the tangential contact stress $\boldsymbol{\lambda}_T$ is zero via a calculation.

9.8.2. *Algorithmic contact with* slip *case at the considered integration point*

Let us suppose that we have *algorithmic* contact with slip at the considered integration point ($\hat{\lambda}_N \geq 0$), so that the expressions [9.89]–[9.91] become:

$$\begin{array}{|lcl|}\hline \left\{\mathbf{\Phi}^{e^{(1)}}_{\text{contact}}(\mathbf{U}^e,\mathbf{\Lambda}^{e^{(1)}})\right\}_{slip} & = & [\mathcal{N}^{e^{(1)}}(\xi)]^T\{\nu\}\,\hat{\lambda}_N \\ \left\{\mathbf{\Phi}^{e^{(2)}}_{\text{contact}}(\mathbf{U}^e,\mathbf{\Lambda}^{e^{(1)}})\right\}_{slip} & = & -[\mathcal{N}^{e^{(2)}}(\eta)]^T\{\nu\}\,\hat{\lambda}_N \\ \hline \end{array}$$

$$\boxed{\left\{\mathbf{R}^{e^{(1)}}_{\Lambda}(\mathbf{U}^e,\mathbf{\Lambda}^{e^{(1)}})\right\}_{slip} = [\mathcal{N}^{e^{(1)}}(\xi)]^T\left(g\{\nu\} + \frac{1}{\epsilon_T}\{\lambda_T\}\right)}$$

$$\boxed{[\mathbf{K}^e_{\text{contact}}]_{slip} = \left[\begin{array}{c|c} \epsilon_N[\mathbf{k}^{e,1}_{\text{contact}}] + \hat{\lambda}_N[\mathbf{k}^{e,2}_{\text{contact}}] & -\left[\begin{array}{c} [\mathcal{N}^{e^{(1)}}]^T \\ -[\mathcal{N}^{e^{(2)}}]^T \end{array}\right]\{\nu\} < \nu > [\mathcal{N}^{e^{(1)}}] \\ \hline -[\mathcal{N}^{e^{(1)}}]^T\{\nu\} < \nu > \left[[\mathcal{N}^{e^{(1)}}], -[\mathcal{N}^{e^{(2)}}]\right] & -\frac{1}{\epsilon_T}[\mathcal{N}^{e^{(1)}}]^T[\mathbf{I} - \nu\otimes\nu][\mathcal{N}^{e^{(1)}}] \end{array}\right]}$$

[9.94]

The block $[\mathbf{K}^e_{\Lambda\Lambda}]_{slip}$ is a symmetric and *negative* matrix. It is not negative-definite however, because it is *non-invertible*.

9.9. Solution via the arc-length method

In a nonlinear calculation, when computing a solution numerically, we would usually assume that the prescribed displacements or the loads depend on a certain number of variable factors. So, for example, a prescribed force can depend on a factor evolving from one step to another according to a given rule. The case of a constant force is a special case in which the variable factor is just a constant value.

The easiest way to shift from one step to another is to prescribe the variable factors at each step. Algorithm 9.1 applies using this mode of incrementation. However in nonlinear problems, it often happens that the response curve may have limit points (under load or displacement control) or bifurcation points. In this case, directly incrementing the variable factors does not enable one to cross the limit points nor describe a given bifurcated branch; we must then use the arc-length method [135, 116, 27], which allows for the continuation to be carried out correctly. An example of this scenario is given in Chapter 10 (section 10.4), where we consider a beam subjected to a follower force and contacting another beam.

The reasoning is similar to that for a calculation with no contact. To avoid overburdening the presentation, we will present only the arc-length method in the presence of a load (either dead or follower), that in the presence of a prescribed displacement can be treated analogously. Furthermore, we shall only cover points specific to contact, without going into detail for the rest which is similar to contactless computations.

We place ourselves in the classical framework for the arc-length method:

(i) The prescribed displacement is constant (with the zero prescribed displacement being a special case).

(ii) The load can depend on several factors, of which only one – denoted by λ – varies (there is no risk of confusion with the multipliers λ_N nor λ_T from the contact problem), and the other factors are constant. Hence, the prescribed external force vector $\mathbf{\Phi}$ is of the form:

$$\mathbf{\Phi} = \lambda.\mathbf{\Phi}^{ref}(\mathbf{U}) + \mathbf{\Phi}_{\text{f constant}} + \sum_j \lambda_{pj}\,\mathbf{\Phi}^{\text{pref }j}(\mathbf{U}) \qquad [9.95]$$

where $\mathbf{\Phi}_{\text{f constant}}$ denotes the force vector due to the constant dead loads, $\mathbf{\Phi}^{\text{pref }j}(\mathbf{U})$ the reference force vector of the jth prescribed constant follower force, λ_{pj} the corresponding constant factor, $\mathbf{\Phi}^{ref}(\mathbf{U})$ the reference force vector corresponding to the variable factor λ, this last vector may depend on the displacement $\mathbf{U}$.

At each step n, we have to solve the following system with the unknowns $\mathbf{U}$, $\mathbf{\Lambda}$ and λ:

$$\begin{aligned} &\mathbf{R}(\mathbf{U},\lambda,\mathbf{\Lambda}) = \mathbf{\Psi}(\mathbf{U}) - \mathbf{\Phi}(\mathbf{U},\lambda) - \mathbf{\Phi}_{\text{contact}}(\mathbf{U},\mathbf{\Lambda}) = \mathbf{0} \\ &\mathbf{R}_{\Lambda}(\mathbf{U},\mathbf{\Lambda}) = \mathbf{0} \\ &C(\Delta\mathbf{U},\Delta\lambda) = \|\Delta\mathbf{U}\|^2 + \Delta\lambda^2.C_{ref}^2 - \Delta\ell^2 = 0 \end{aligned} \qquad [9.96]$$

where $\Delta\mathbf{U} \equiv \mathbf{U}_n - \mathbf{U}_{n-1}$, $\Delta\lambda \equiv \lambda_n - \lambda_{n-1}$, C_{ref} and $\Delta\ell$ are chosen and fixed (positive) scalars.

The constraint relation $[9.96]_3$ is the equation of a hypersurface in ($\mathbf{U}$, λ) space. At the ith step, a Taylor series expansion allows for the variations $\delta\mathbf{\Lambda}$ and $\delta\mathbf{U}$ to be expressed as functions of $\delta\lambda$, analogous to what was done in section 9.3:

$$\begin{aligned} &\delta\mathbf{\Lambda} = \delta\mathbf{\Lambda}_R - \delta\lambda.\mathbf{\Lambda}_{\phi ref} \\ &\delta\mathbf{U} = \delta\mathbf{U}_R - \mathbf{K}^{-1}\mathbf{K}_{U\Lambda}\delta\mathbf{\Lambda}_R + \delta\lambda(\mathbf{U}_{\phi ref} + \mathbf{K}^{-1}\mathbf{K}_{U\Lambda}\mathbf{\Lambda}_{\phi ref}) \end{aligned} \qquad [9.97]$$

where we have defined:

$$\begin{aligned} &\delta\mathbf{U}_R = -\mathbf{K}^{-1}\mathbf{R} \\ &\mathbf{U}_{\phi ref} = \mathbf{K}^{-1}\mathbf{\Phi}^{ref} \\ &\delta\mathbf{\Lambda}_R = -\left[\mathbf{K}_{\Lambda\Lambda} - \mathbf{K}_{\Lambda U}.\mathbf{K}^{-1}.\mathbf{K}_{U\Lambda}\right]^{-1}\{\mathbf{R}_{\Lambda} + \mathbf{K}_{\Lambda U}\delta\mathbf{U}_R\} \\ &\mathbf{\Lambda}_{\phi ref} = \left[\mathbf{K}_{\Lambda\Lambda} - \mathbf{K}_{\Lambda U}.\mathbf{K}^{-1}.\mathbf{K}_{U\Lambda}\right]^{-1}\mathbf{K}_{\Lambda U}\mathbf{U}_{\phi ref} \end{aligned} \qquad [9.98]$$

In a contactless calculation, the relation $[9.97]_1$ does not exist, whereas the relation $[9.97]_2$ is replaced by another simpler one:

$$\delta\mathbf{U} = \delta\mathbf{U}_R + \delta\lambda.\mathbf{U}_{\phi ref} \qquad [9.99]$$

It remains to calculate $\delta\lambda$ in order to exploit [9.97]. By comparing $[9.97]_2$ with [9.99], we see that $\delta\lambda$ can be obtained by simply using the same formula as in a contactless calculation, provided that we replace:

$$\begin{aligned} &\delta\mathbf{U}_R \quad \text{by} \quad \delta\mathbf{U}_R - \mathbf{K}^{-1}\mathbf{K}_{U\Lambda}\delta\mathbf{\Lambda}_R \\ &\mathbf{U}_{\phi ref} \quad \text{by} \quad \mathbf{U}_{\phi ref} + \mathbf{K}^{-1}\mathbf{K}_{U\Lambda}\mathbf{\Lambda}_{\phi ref} \end{aligned} \qquad [9.100]$$

The solution algorithm for the case of the static contact problem with the arc-length method is analogous to that with the direct incrementation of the variable factors (see Algorithm 9.1), while being a little more complicated, because this time we also have to manage the arc length $\Delta\lambda$.

In numerical programming, [9.100] allows to reuse the same calculation module for $\delta\lambda$ which is available in the contactless case in order to deal with both the linearized and the spherical arc length.

9.10. Physical units

In this section, we focus on the SI units of the different quantities involved in the contact matrix. We will use the units that were already established in section 8.12, and we will agree that the unit of a dimensionless quantity is 1.

- Let us examine the units of the contact tangent matrix. As we know that

$$\begin{cases} \mathbf{a}_\alpha : m & \Rightarrow [a_{\alpha\beta}] = m^2 \\ \mathbf{a}^\alpha : 1/m & \Rightarrow [a_{\alpha\beta}] = 1/m^2 \\ b_{\alpha\beta} : m & \\ \mathbf{P}, \mathbf{W}^\alpha, \mathbf{Z}^\alpha : 1 & \end{cases}$$

we have the following units:

$$[\mathbf{k}^{e,1}_{\text{contact}}], [\mathbf{k}^{e,3}_{\text{contact}}], [\mathbf{k}^{e,4}_{\text{contact}}] : 1, \qquad [\mathbf{k}^{e,2}_{\text{contact}}] : 1/m, \qquad [\mathbf{k}^{e,5}_{\text{contact}}] : N/m^3$$

From which we get:

$$\underset{N/m^3}{[\mathbf{K}^e_{UU}]} = H(\hat{\lambda}_N) \underset{N/m^3}{\epsilon_N} \underset{1}{[\mathbf{k}^{e,1}_{\text{contact}}]} + \underset{N/m^2}{\langle \hat{\lambda}_N \rangle} \underset{1/m}{[\mathbf{k}^{e,2}_{\text{contact}}]} - d_{sl}\,\mu \underset{N/m^3}{\epsilon_N} \underset{1}{[\mathbf{k}^{e,3}_{\text{contact}}]}$$

$$+\epsilon_T \underset{1}{[\mathbf{k}^{e,4}_{\text{contact}}]} + c_{slst} \underset{N/m^3}{[\mathbf{k}^{e,5}_{\text{contact}}]}$$

On the other hand:

$$\underset{1}{[\mathbf{K}^e_{U\Lambda}]} = \begin{bmatrix} [\mathcal{N}^{e^{(1)}}]^T \\ -[\mathcal{N}^{e^{(2)}}]^T \end{bmatrix} \left[-H(\hat{\lambda}_N)\{\boldsymbol{\nu}\} < \boldsymbol{\nu} > + d_{sl}\,\mu\{\boldsymbol{\ell}_T\} < \boldsymbol{\nu} > - \underset{1}{[\mathbf{P}]}[\mathbf{I} - \boldsymbol{\nu} \otimes \boldsymbol{\nu}] \right] [\mathcal{N}^{e^{(1)}}]$$

$$\underset{1}{[\mathbf{K}^e_{\Lambda U}]} = [\mathcal{N}^{e^{(1)}}]^T \left(\left[-H(\hat{\lambda}_N)\underset{1}{\{\boldsymbol{\nu}\}} < \boldsymbol{\nu} > + d_{sl}\,\mu \frac{\epsilon_N}{\epsilon_T}\{\boldsymbol{\ell}_T\} < \boldsymbol{\nu} > \right] \left[[\mathcal{N}^{e^{(1)}}], -[\mathcal{N}^{e^{(2)}}], [\mathbf{0}] \right] \right.$$
$$\left. -[\mathbf{P}] \left[[\mathbf{W}^1][\mathcal{N}^{e^{(1)}}], [\mathbf{W}^2][\mathcal{N}^{e^{(2)}}] - [\mathbf{Z}^\alpha][\mathcal{N}^{e^{(2)}}_{,\alpha}], -[\mathcal{N}^{e^{(2)}}_{n-1}(\boldsymbol{\eta}_{n-1})] \right] \right)$$

and:

$$\underset{m^3/N!}{[\mathbf{K}^e_{\Lambda\Lambda}]}$$
$$= [\mathcal{N}^{e^{(1)}}]^T \left[- \underset{m^3/N\,!}{\frac{1}{\epsilon_N}} (1 - H(\hat{\lambda}_N))\{\boldsymbol{\nu}\} < \boldsymbol{\nu} > - d_{sl} \frac{\mu}{\epsilon_T}\{\boldsymbol{\ell}_T\} < \boldsymbol{\nu} > + \frac{1}{\epsilon_T}[\mathbf{P} - \mathbf{I}][\mathbf{I} - \boldsymbol{\nu} \otimes \boldsymbol{\nu}] \right] [\mathcal{N}^{e^{(1)}}]$$

To summarize, the units involved in the element contact matrix [9.61] are

$$[\mathbf{K}^e_{\text{contact}}] = \left[\begin{array}{c|c} \underset{N/m^3}{\mathbf{K}^e_{UU}} & \underset{1}{\mathbf{K}^e_{U\Lambda}} \\ \hline \underset{1}{\mathbf{K}^e_{\Lambda U}} & \underset{m^3/N}{\mathbf{K}^e_{\Lambda\Lambda}} \end{array} \right]$$

For the contact tangent matrix $[\mathbf{K}_{\text{contact}}]$ in [9.66], which is the assembly of the element matrices $[\mathbf{K}^e_{\text{contact}}]$, and after recalling that the dimension of dS_0 is m^2, we can see that the units are

$$[\mathbf{K}_{\text{contact}}] = \left[\begin{array}{c|c} \begin{array}{c}\mathbf{K}_{UU}\\ N/m\end{array} & \begin{array}{c}\mathbf{K}_{U\Lambda}\\ m^2\end{array} \\ \hline \begin{array}{c}\mathbf{K}_{\Lambda U}\\ m^2\end{array} & \begin{array}{c}\mathbf{K}_{\Lambda\Lambda}\\ m^5/N\ !\end{array} \end{array}\right]$$

Note that $\mathbf{K}_{UU}$ does indeed have the classical dimension of stiffness.

- The units of the global tangent matrix [9.13] for the entire structure are

$$[\mathbf{K}_T] = \left[\begin{array}{c|c} \begin{array}{c}\mathbf{K}_{classical} + \mathbf{K}_{UU}\\ N/m \qquad N/m\end{array} & \begin{array}{c}\mathbf{K}_{U\Lambda}\\ m^2\end{array} \\ \hline \begin{array}{c}\mathbf{K}_{\Lambda U}\\ m^2\end{array} & \begin{array}{c}\mathbf{K}_{\Lambda\Lambda}\\ m^5/N\end{array} \end{array}\right]$$

The units of the system [9.14] to be solved at the ith iteration are

$$\left[\begin{array}{c|c} \begin{array}{c}\mathbf{K}_{classical} + \mathbf{K}_{UU}\\ N/m\end{array} & \begin{array}{c}\mathbf{K}_{U\Lambda}\\ m^2\end{array} \\ \hline \begin{array}{c}\mathbf{K}_{\Lambda U}\\ m^2\end{array} & \begin{array}{c}\mathbf{K}_{\Lambda\Lambda}\\ m^5/N\end{array} \end{array}\right] \left\{\begin{array}{c}\delta\mathbf{U}\\ m\\ \delta\mathbf{\Lambda}\\ N/m^2\end{array}\right\} = -\left\{\begin{array}{c}\mathbf{R}^{i-1}\\ N\\ \mathbf{R}_{\Lambda}^{i-1}\\ m^3\end{array}\right\} \qquad [9.101]$$

9.11. Summary of the chapter

In this chapter, we have shown how to solve the system of equations [9.1] of the quasi-static contact problem with the help of the Newton–Raphson scheme. A larger part of the chapter has been dedicated to the calculation of the contact tangent matrix in [9.11]:

$$[\mathbf{K}_{\text{contact}}] = \left[\begin{array}{c|c} \mathbf{K}_{UU} & \mathbf{K}_{U\Lambda} \\ \hline \mathbf{K}_{\Lambda U} & \mathbf{K}_{\Lambda\Lambda} \end{array}\right]$$

which is the principal ingredient that you must know at each iteration of the numerical solution.

The matrix $[\mathbf{K}_{\text{contact}}]$ is not symmetric except in the case of frictionless contact. Certain element tangent matrices which enter into the composition of $[\mathbf{K}_{\text{contact}}]$ can be rectangular, since they also involve the displacement at the point $\mathbf{Y}_{n-1}$ (which appears in the discrete expression for the slip $\mathbf{V}_T\Delta t$), but that does not prevent the global tangent matrix $[\mathbf{K}_{\text{contact}}]$ from being a square one after assembly, as it should be.

The numerical solution method presented in this chapter will be validated by the numerical examples presented in Chapter 10.

10

Numerical Examples of Quasi-static Contact

In this chapter, we shall present some representative numerical examples that will allow us to validate the weak contact form proposed in Chapter 7. The examples presented here are *quasi-static*, and we will solve them by applying the computation method described in Chapter 9. The solution to the dynamic problem will then be approached in the following chapters.

The quasi-static frame, although more restrained than the dynamic frame, suffices to test how the formulation performs in a diverse range of situations – finite strains/displacements, contact with/without friction, finite slip, nonmatching meshes, hyperelastic/elastoplastic materials, solid/membrane structures and dead/follower loads – and also to validate the formulation through convergence with respect to the mesh, precision (in particular, the fulfillment of the contact laws), and stability (no locking phenomena nor spurious contact pressure modes).

Each computational method presented in this chapter is addressed using an in-house and nonlinear numerical code written in Fortran.

10.1. Contact patch test

The first example that we consider here is the contact patch test. Initially, as proposed in some previous studies [125, 132], it is considered to be a basic test that every contact formulation should pass for it to prove its robustness.

In the literature, we can find a diverse range of geometries used, but they are all alike and simple enough. Here, we will consider the geometry proposed in some previous studies [28, 34] and shown in Figure 10.1(a). The upper elastic block is subjected to a uniform distributed force $p = 1,000$ N/m^2 on its upper surface, and pressed against a (lower) elastic block situated below it. If contact takes place without friction, then we have to find a constant contact pressure field on the contact surface and which equals p.

For computational solutions, the two blocks are meshed differently on purpose in order to create a *nonmatching* contact interface, which is composed of nodes that do *not* lie opposite to one another. We must check that the finite element solution indeed gives a uniform contact pressure (and hence a uniform displacement) along the contact interface, even though the meshes are nonmatching.

The calculation is performed assuming that the two bodies are made of an isotropic hyperelastic material that obeys the Saint-Venant Kirchhofff law (4.5), with Young's modulus $E = 10^6$ N/m^2 and Poisson's ratio $\nu = 0$. The problem is modeled using plane strain. The blocks are rectangular and have the same dimensions, $a = 5$ m along $\mathbf{x}$ and $b = 1$ m along $\mathbf{y}$.

In principle, the patch test should pass for an arbitrary mesh, but, in practice, we must perform the test using just a few typical meshes. Here, the patch test is conducted using two meshes – the first is regular, whereas the second is distorted – which are composed of eight-node quadrilaterals (Q8) so that the line elements on the contact surface have three nodes (see Figures 10.1(b) and (c)). In all cases, the meshes are nonmatching over the contact surface.

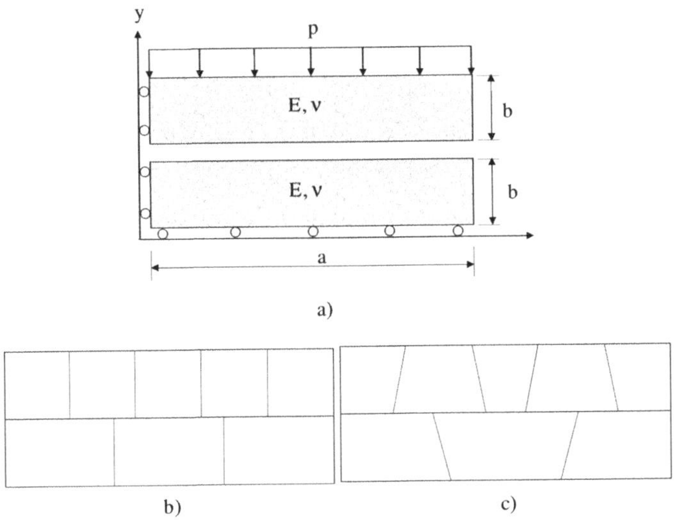

Figure 10.1. *Contact patch test. (a) Problem definition, (b) regular mesh and (c) distorted mesh*

If, in the initial configuration, the two blocks are separated as in Figure 10.1(a), then the upper block would undergo a rigid-body displacement along **y** before touching the lower block and the stiffness matrix of the system would be singular. To avoid this, let us artificially prescribe an initial penetration $g_0 > 0$, equal to $b/10^9$, which is a very small value, meaning that the two blocks are barely touching at the initial instant.

The numerical integration is performed using 3×3 Gaussian points for the Q8 elements, and seven Gaussian points for the line elements on the contact surface. We take the penalty parameter $\epsilon_N = \epsilon_T = 10^{12}$ N/m^3. In order to observe the influence of the contactor/target choice, the calculations will be performed by (i) taking the upper body as the contactor and (ii) taking the lower body as the contactor.

Figure 10.2 shows the deformed meshes and the contact stress distributions along the contact surface. The arrows represent the contact stresses exerted by the target onto the contactor. In all cases, note that the contact stresses are indeed uniform and equal to p up to the fifth to sixth significant digit. The patch test has therefore been verified.

Moreover, each node of the contact surface has the same vertical displacement equal to 0.0010015 m. The stress field within the two bodies is constant, and whose only nonzero component is Σ_{yy} = 1 001 N/m^2 (recall: $\boldsymbol{\Sigma}$ is the second Piola–Kirchhoff stress tensor). These values agree perfectly with the analytical solution from nonlinear elasticity (here, the numerical results are compared with the nonlinear elastic solution, since they are obtained using a nonlinear computational code).

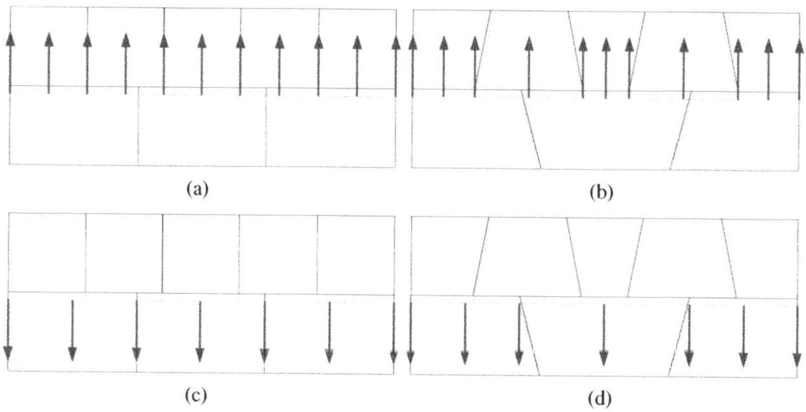

Figure 10.2. *Deformed meshes and contact stresses in the patch test: (a and b) the contactor is the upper body; (c and d): the contactor is the lower body. The displacement has been scaled by a factor of 100. For a color version of this figure, see www.iste.co.uk/levan/contact.zip*

10.2. Hertzian contact problem

Motivation: In this section, we shall consider the Hertzian contact problem. This example has the advantage of being one of the few contact problems to admit an analytical solution, thus allowing us to check the precision of our numerical solution method.

A Hertzian contact problem involves contact between two bodies with simple geometries, and it exists in several variants: cylinder/cylinder, cylinder/plane, sphere/sphere, paraboloid/plane, etc. Here, we consider the cylinder/plane variant in plane strain defined as follows. An elastic cylinder of radius $R = 10$ m lying on its side upon a rigid half-space is subjected to a compressive force F parallel to the axis of symmetry of the problem, as shown in Figure 10.3(a). The cylinder is made of an elastic Saint-Venant Kirchhoff material with $E = 100$ N/m^2 and $\nu = 0.3$. The calculation is performed with two force levels: $F = 1$ N/m and $F = 2$ N/m, and two coefficients of friction between the cylinder and foundation: $\mu = 0$ (without friction) and $\mu = 0.3$.

Given the symmetry of the problem, we only need to study one half of the structure. The half-cylinder is chosen to be the contactor, and the foundation to be the target. Two half-cylinder meshes are used:

– one being a coarse mesh comprised of 198 six-node triangular elements (T6) and 455 nodes;

– the other being a fine mesh comprised of 345 T6 elements and 773 nodes, as shown in Figure 10.3(a).

The rigid foundation is meshed using a single Q8 element, with all nodes being fixed. The meshes are clearly nonmatching.

Numerical integration is performed using seven Gaussian quadrature points for the triangular elements, and seven Gauss points for the line elements of the contact surface. The penalty parameters are $\epsilon_N = \epsilon_T = 100$ N/m^3. The relative convergence tolerance for each residual $\mathbf{R}$ and $\mathbf{R}_\Lambda$ from the equations [9.4] of the static contact problem are fixed to be 10^{-6}.

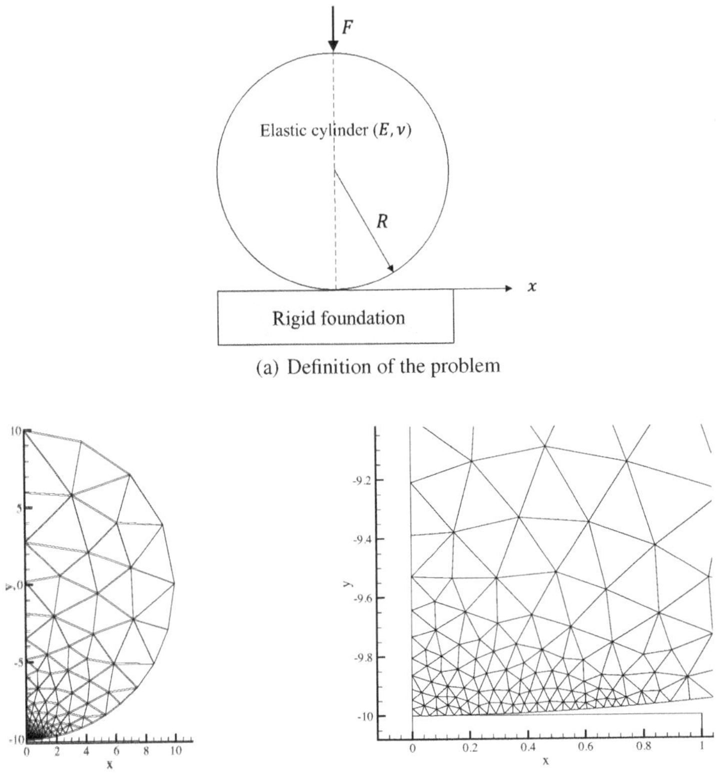

(a) Definition of the problem

(b) The fine mesh

(c) A magnified view of the contact area

Figure 10.3. *2D Hertzian contact problem*

10.2.1. *Frictionless contact case*

• Analytical solution.

The analytical solution in the frictionless contact case with small elastic strains is a classical one. The contact pressure distribution on the contact surface is given by [67, 71]:

$$p_c(x) = \frac{2F}{\pi b^2}\sqrt{b^2 - x^2} \quad [10.1]$$

where b denotes the contact half-width, whose expression is

$$b = 2\sqrt{\frac{FR(1-\nu^2)}{E\pi}} \quad [10.2]$$

• Numerical contact stresses.

Figure 10.4 displays the form of the contact stress obtained numerically. Figure 10.5 shows that the numerical values agree strongly with their theoretical counterparts: for the fine mesh and $F = 1$ N/m (respectively, $F = 2$ N/m), the maximum pressure at $x = 0$ is within 2% (respectively,

2.8%) of the analytical solution. It is normal for the numerical results to diverge further away from the analytical solution as the force F becomes greater: the contact stresses obtained are indeed from the first nominal Piola-Kirchhoff tensor (see relation [7.25]), whereas the analytical pressure in [10.1] comes from the linear elasticity theory. Therefore, a slight difference between these stresses will always exist, even when the strains are small. A more in-depth calculation shows that the numerical values of the Cauchy stresses are twice as close to their theoretical values. This being said, for the entirety of this book, we have decided to present the results only in terms of Piola-Kirchhoff stresses.

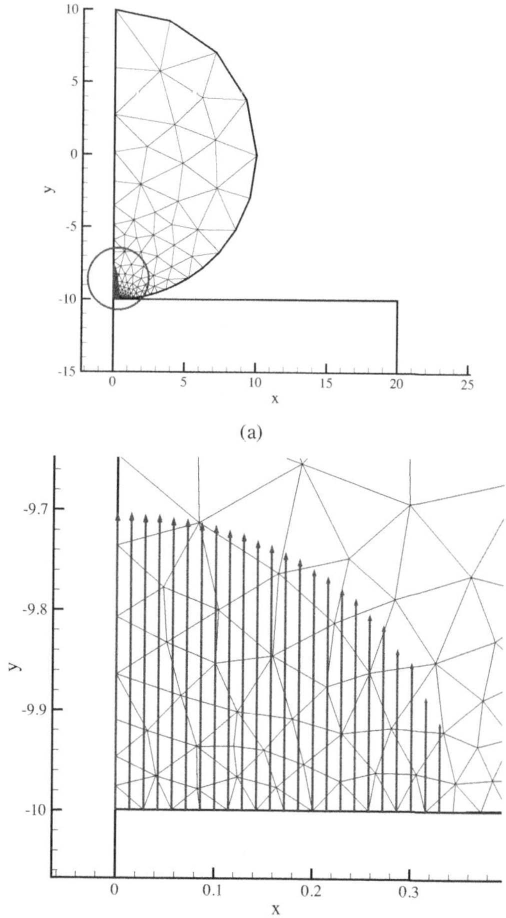

(b) Magnification of the contact zone

Figure 10.4. *Pressure distribution in the frictionless Hertzian contact problem with a coarse mesh. For a color version of this figure, see www.iste.co.uk/levan/contact.zip*

In addition, we notice that, regardless of the mesh considered and the force applied, generally there is an element of the cylinder which straddles between the contact and noncontact zones.

For such an element, the contact stress is nonzero where there is contact and is zero elsewhere, which is what the quadratic interpolation of λ (see relation [8.16]) cannot represent. It turns out that the numerical results are not affected by this difficulty.

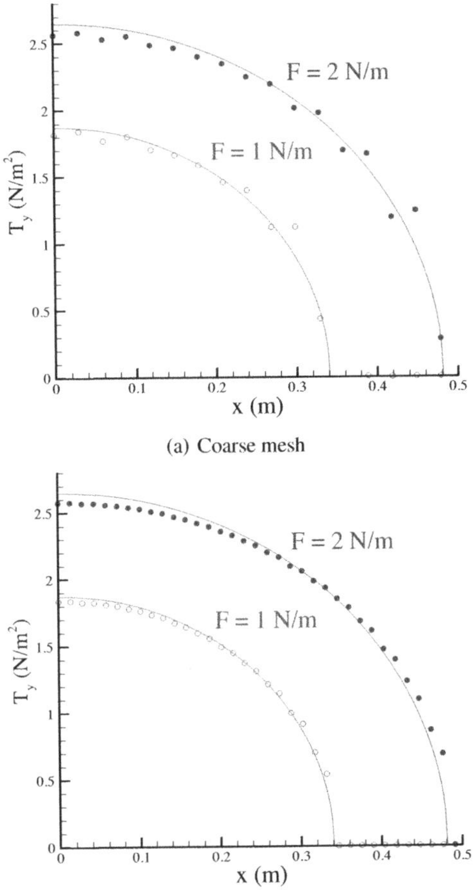

(b) Fine mesh

Figure 10.5. *Contact stress in the frictionless Hertzian contact problem. ∘ or •: numerical results; continuous line: theoretical results. For a color version of this figure, see www.iste.co.uk/levan/contact.zip*

• The resultant of the contact stresses.

Let us calculate the reaction from the rigid plane onto the cylinder by integrating the contact stress vectors over the contact surface $S_c^{(1)}$. The results presented in Table 10.1 show that the reaction balances the applied force $-F\mathbf{y}$ as predicted.

• Maximum shear in the cylinder.

Now, let us focus on the maximum shear in the cylinder, which is defined by $\tau_{max} = \frac{1}{2}(\sigma_{III} - \sigma_I)$, where $\sigma_I \leq \sigma_{II} \leq \sigma_{III}$ are the principal stresses. From the analytical solution, the maximum value of τ_{max}, for $F = 1$ N/m, is attained at the point $(x, y) = (0, -10 + 0.78b) = (0, -9.7345)$, which is located slightly within the interior of the cylinder (and not actually on the contact surface itself as we could be led to believe) and is equal to $0.3\frac{2F}{\pi b} = 0.5611$ N/m^2.

Step	F	Resultant in the y direction	Resultant in the x direction
1	$5.E - 01$	$0.500004E + 00$	$-0.201121E - 19$
2	$1.E + 00$	$0.100000E + 01$	$0.885984E - 18$

Table 10.1. *The resultant of the contact stresses for the Hertzian contact problem*

The contours of τ_{max} obtained numerically are represented in Figure 10.6. Note that the maximum value of τ_{max} is found inside the region $0 < x < 0.02$, $-9.8 < y < -9.6$ and equals 0.5895 N/m^2, which is a value within about 5% of the theoretical value.

10.2.2. *Case of frictional contact with* $\mu = 0.3$

- Analytical solution.

The analytical solution with friction was obtained by Spence [127, 67], who assumed that the influence of friction on the normal contact stress is negligible, so that the equations which govern the normal and tangential stresses may be decoupled. The contact half-width b is given by the same formula [10.2] as that in the frictionless case, and it equals $b = 0.3404$ m if $F = 1$ N/m. The contact region comprises a stick region $|x| \leq c$ and a slip region $c \leq x \leq b$. The extent of the slip region c can be obtained by solving the equation below, yielding c/b as a function of both the coefficient of friction μ and Poisson's ratio ν:

$$\frac{K(c/b)}{K'(c/b)} = \frac{tan^{-1}\mu}{\frac{1}{2}\frac{1+\beta}{1-\beta}} \qquad [10.3]$$

where $\beta \equiv \dfrac{1-2\nu}{2(1-\nu)}$, $K(k)$ denotes the complete elliptic integral of the first kind, $K(k) \equiv \displaystyle\int_0^{\pi/2} \frac{d\phi}{\sqrt{1-k^2 sin^2\phi}}$, $0 \leq k \leq 1$ and $K'(k) = K(\sqrt{1-k^2})$. With $F = 1$ N/m, $\mu = 0.3$ and $\nu = 0.3$, we find that $c/b = 0.7006$, hence $c = 0.2385$ m.

The normal contact stress is given by the same formula [10.1] as that in the frictionless case, and the tangential contact stress is calculated via

$$q(x) = \begin{cases} \dfrac{2\mu F x}{\pi b^2}\left[\dfrac{1}{K(c/b)}\displaystyle\int_{x/b}^{c/b} \frac{K(\frac{c}{b}, sin^{-1}\frac{t}{c/b})}{t^2\sqrt{1-t^2}}dt + \frac{b}{c}\sqrt{1-\left(\frac{c}{b}\right)^2}\right] & , |x| \leq c \\ \mu\, p(x) & , c \leq |x| \leq b \end{cases} \qquad [10.4]$$

- Numerical contact stresses.

Numerical results are uniquely presented with the fine mesh. The total force $F = 1$ N/m is applied over 50 equal steps. The iterative Newton–Raphson scheme converges after three iterations on average, with a maximum of seven iterations for the first step. The contact stress vector is displayed in Figure 10.7(a), which shows a gradual transition from stick to slip along the line of contact. Figure 10.7(b) shows that the normal and tangential contact stresses, as well as the position of the boundary that separates the stick and slip regions, are in agreement with Spence's theoretical solution.

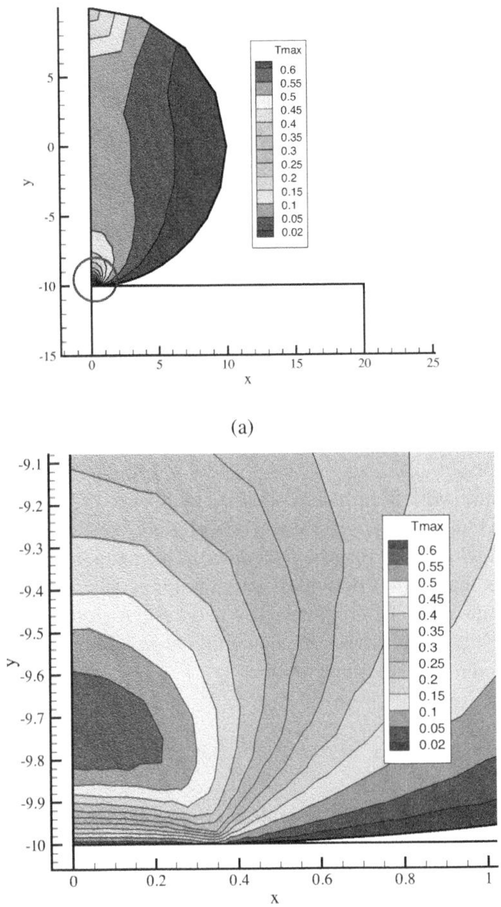

(a)

(b) Blow-up of the contact region

Figure 10.6. *Contours of the maximum shear. For a color version of this figure, see www.iste.co.uk/levan/contact.zip*

• Verification of Coulomb's law.

Before verifying that the finite element calculation satisfies Coulomb's criterion $||T_T|| \leq \mu T_N$ at every contact point or not, let us calculate the ratio $\dfrac{T_x}{\mu T_y}$ for each node of the contact surface in the two cases, $F = 0.5$ and 1 N/m. Observe that the Coulomb inequality is satisfied everywhere (see Figure 10.8). At the nodes where there is slip, the equality $|T_x| = \mu T_y$ is satisfied within 0.1%.

• Influence of the number of load steps on the contact stresses.

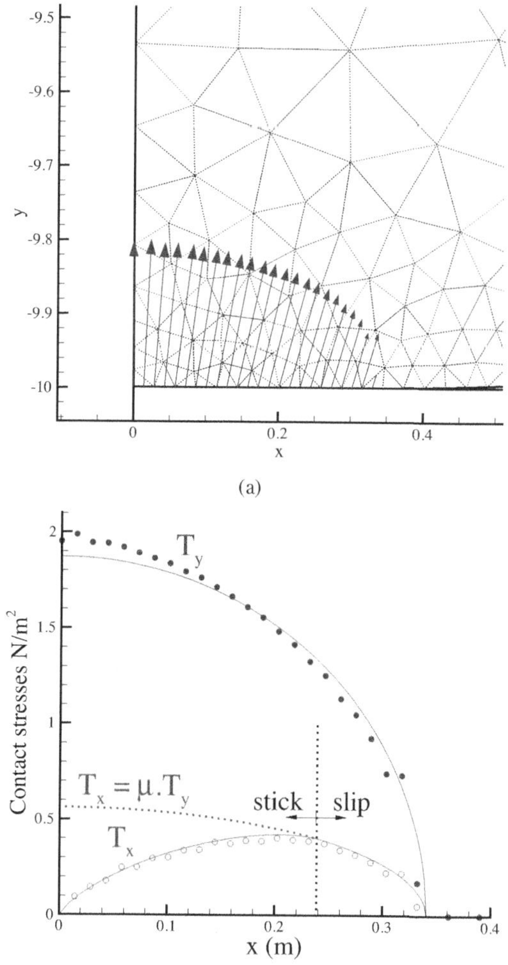

Figure 10.7. *Contact stress of the Hertzian contact problem with friction (F = 1 N/m, μ = 0.3). (a) General shape, (b) ○ or •: numerical results, continuous line: theoretical results. For a color version of this figure, see www.iste.co.uk/levan/contact.zip*

In order to show the influence that the number of steps used has on the numerical results, let us perform two supplementary calculations, one in which the force F = 1 N/m is applied in a single step instead of 50, and in the other the same force is applied in 10 steps. As can be seen in Figure 10.9, the contact width, the contact pressure, as well as the tangential contact stress in the slip region, are obtained again with good precision using either 1 or 10 load steps. On the other hand, the tangential contact stress in the stick region turns out to be rather sensitive to the step number used and requires a sufficiently larger number of steps to calculate it sufficiently well. This is in agreement with Pietrzak's observation [111].

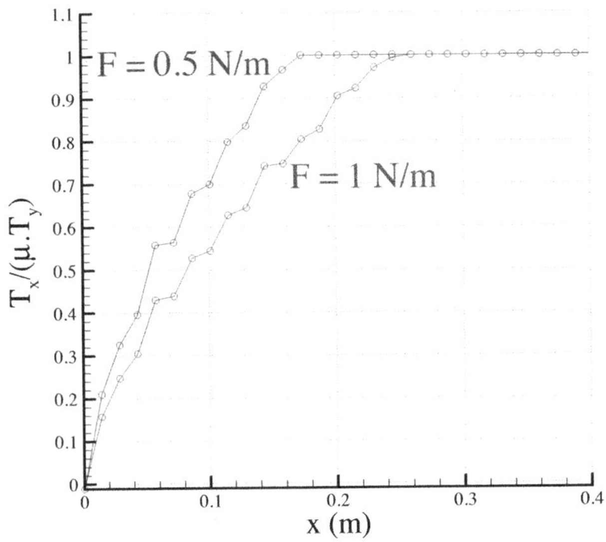

Figure 10.8. *Verification of Coulomb's law. For a color version of this figure, see www.iste.co.uk/levan/contact.zip*

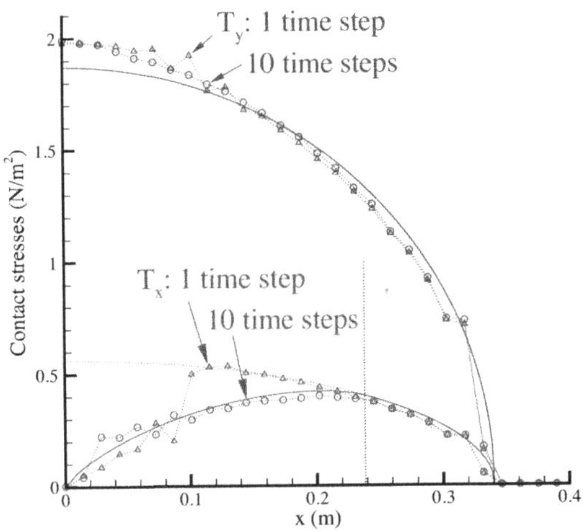

Figure 10.9. *Hertzian contact problem with friction ($F = 1\ N/m$, $\mu = 0.3$). The influence of the number of load steps on the contact stresses. △ or ○: numerical results, continuous line: theoretical solution. For a color version of this figure, see www.iste.co.uk/levan/contact.zip*

10.3. Rolling disk

Motivation: one of the essential ingredients in a contact calculation involving friction is the slip velocity $\mathbf{V}_T$ (or the slip $\mathbf{V}_T\Delta t$), which is estimated using the discrete expression [8.56]: $\mathbf{V}_T\Delta t = \mathbf{y} - \phi^{(?)}(\mathbf{Y}_{n-1}, t)$. We are going to test the validity of this expression in the following example, which involves both large displacements and large rotations.

Consider an elastic disk with center C, of radius $R = 1$ m, and compressed between two parallel rigid plates (see Figure 10.10). The disk is made from a hyperelastic Saint-Venant Kirchhoff material with $E = 2 \times 10^{11}$ N/m^2 and $\nu = 0.3$. Contact between the bodies takes place with friction. The lower plate is stationary, whereas the upper plate is mobile.

The problem is modeled in plane strain and the calculated is performed in two steps:

1. In the first step, we lower the upper plate by $R/100$ along $\mathbf{y}$, which is a sufficiently small displacement to establish contact between the disk and the plates, without deforming the disk too much. This contact allows us to eliminate any rigid-body motions of the disk along $\mathbf{y}$ from here onwards.

2. In the second step, we maintain the previously prescribed vertical displacement, and further prescribe a displacement U along $\mathbf{x}$ on the upper plate, so that the disk rolls.

We will use the coefficient of friction $\mu = 0.1$ between the disk and plates, which is sufficient to guarantee contact without slip.

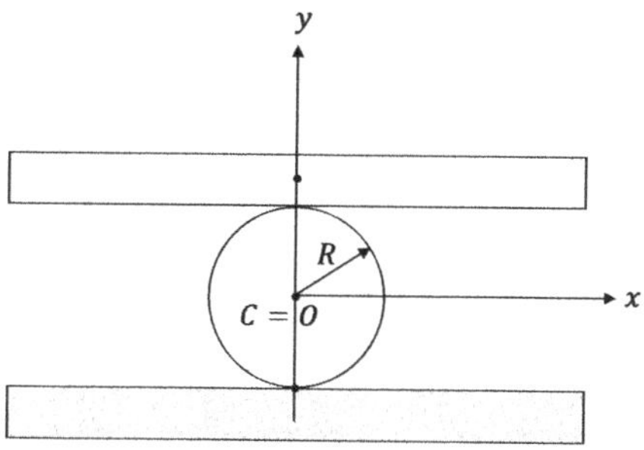

(a) Initial configuration

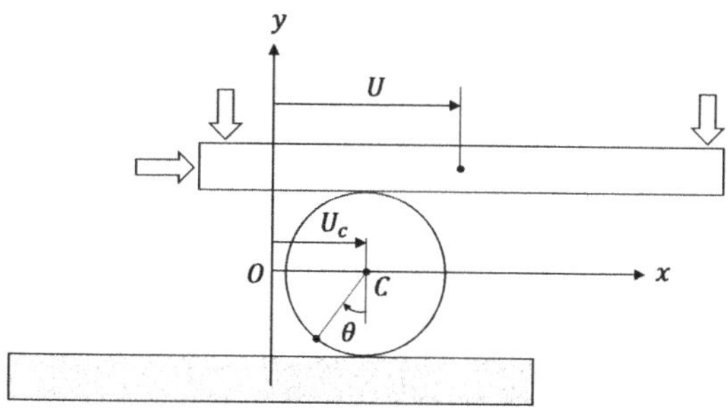

(b) Current configuration. Double arrow = prescribed displacement

Figure 10.10. *Rolling disk*

• Analytical solution.

As the disk is slightly deformed, we can calculate its displacement and rotation with the help of rigid-body mechanics. Assume that, in the initial configuration, the disk center C is located at the origin O. We define the angle of rotation θ to be equal to the angle between the vector $-\mathbf{y}$ and a given radial segment of the disk, measured anti-clockwise. The chosen radial segment is such that the angle θ is zero in the initial configuration.

By expressing the condition for nonslip contact at the upper and lower contact points, we find the horizontal displacement U_c of the disk center and the rotation angle θ of the disk as a function of the horizontal displacement U of the upper plate:

$$U_c = \frac{U}{2} \qquad \theta = \frac{U}{2R}$$

The disk center advances by half as much as the upper plate does. Therefore, for example, when the upper plate is displaced horizontally by πR, the disk center is displaced by $\pi R/2$, and the disk itself turns by a quarter of a complete rotation.

• Numerical results.

The disk is chosen to be the contactor and the two plates to be the targets. The disk is meshed using 148 T6 elements and 325 nodes; each rigid plate using a single Q8 element. Numerical integration is performed using 7 Gaussian quadrature points in the B3 elements on the contact surface. The penalty parameters used are $\epsilon_N = \epsilon_T = 10^{10}$ N/m^3. The calculation is performed in 30 steps, amounting to a total displacement of $U = 3.1416$ m $(= \pi)$ for the upper plate. Convergence is attained within three to eight iterations. Figure 10.11 shows the disk position in the initial state and after 10, 20 and 30 steps. To help visualize the disk rotation more clearly, the position of the radial segment that defines the angle θ has been highlighted.

U (m)	U_c (m)	Theoretical U_c (m)	Error (%)	θ (°)	Theoretical θ (°)	Error (%)
1.0472	0.52349	0.52360	-0.02	30.150	30	0.5
2.0944	1.0471	1.0471	-0.01	60.283	60	0.5
3.1416	1.5707	1.5708	-0.01	90.404	90	0.4

Table 10.2. *Displacement U_c of the disk center and rotation angle θ of the disk as a function of displacement U of the upper plate*

Figure 10.12 and Table 10.2 show that the numerical results are in strong agreement with the theoretical ones.

Figure 10.13 is an adaption from Figure 8.7 to the problem studied here, with the same notation as in section 8.11. The contactor surface $S_c^{(1)}$ is the disk boundary, and the target surface $S_c^{(2)}$ consists of the lower face of the upper plate and the upper face of the lower plate. Figure 10.13(b) (respectively, Figure 10.13(a)) shows the configuration at the current iteration of the current numerical step n (respectively, at the previous step $n-1$).

Let us consider a particle of the disk which is in contact with the upper plate at the current iteration of the current step n. Its position is the point $\mathbf{x}$ in Figure 10.13(b) and its projection onto the target surface is the point $\mathbf{y}$, which coincides with it.

At the previous step $n-1$, the same particle was located at the point $\mathbf{x}_{n-1}$ in Figure 10.13(a), and its projection onto the target surface was the point $\mathbf{y}_{n-1}$. The particle of the upper plane which was located at $\mathbf{y}_{n-1}$ at the previous step occupies the position $\phi^{(2)}(\mathbf{Y}_{n-1}, t)$ in the current configuration. As mentioned in section 8.11, in general, the two points $\mathbf{y}$ and $\phi^{(2)}(\mathbf{Y}_{n-1}, t)$ are distinct.

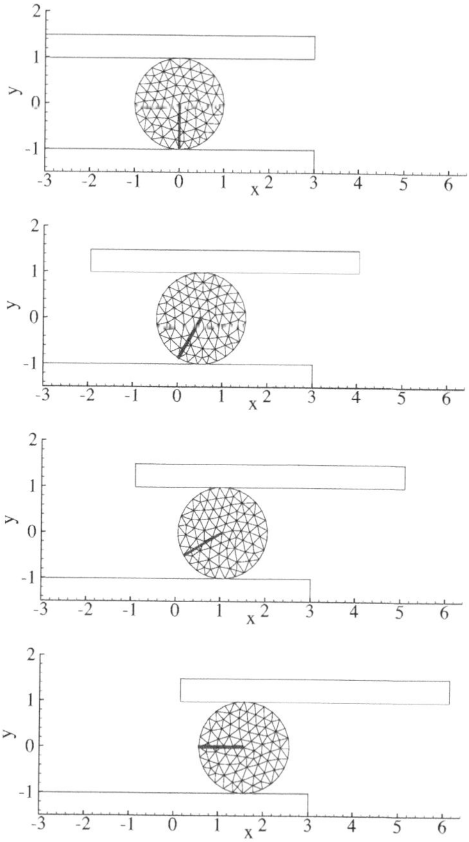

Figure 10.11. *Initial position and positions after 10, 20 and 30 steps. Each plate is meshed using one Q8 element. For a color version of this figure, see www.iste.co.uk/levan/contact.zip*

The slip $\mathbf{V}_T\Delta t$ – at point $\mathbf{x}$ and at the current iteration of the current step – is calculated by the expression [8.56]: $\mathbf{V}_T\Delta t = \mathbf{y} - \phi^{(2)}(\mathbf{Y}_{n-1}, t)$. Imposing contact without slip at $\mathbf{x}$ at each numerical step amounts to canceling out $\mathbf{V}_T\Delta t$ when step n converges, that is to say when the equality $\mathbf{y} = \phi^{(2)}(\mathbf{Y}_{n-1}, t)$ is realized.

We shall redo the calculations using a finer mesh for the target surface: we will keep the same mesh for the disk, but this time each rigid plate will be meshed into 12 Q8 elements rather than just a single one. Calculating the projection points $\mathbf{y}$ and $\mathbf{Y}_{n-1}$ is more complicated now than with the first mesh, since these points may belong to two different elements of the target surface now.

Figure 10.14 shows the disk position in the initial state and also after 10, 20 and 30 steps. Despite the differences between the meshes, the displacements yielded are identical up to the sixth significant figure. The set of results obtained shows that the discrete expression [8.56] can therefore be validated.

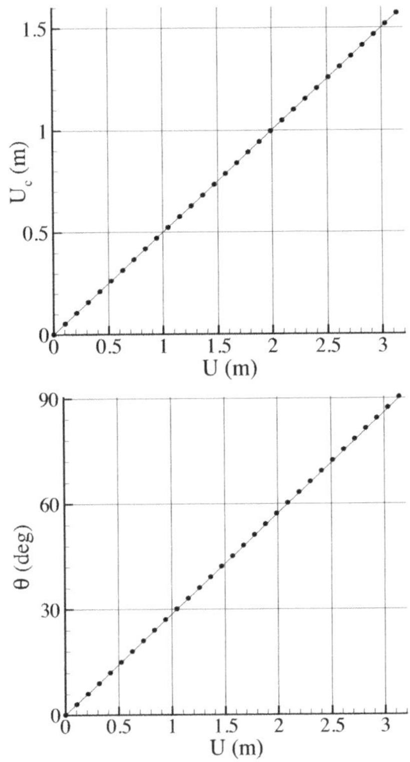

Figure 10.12. *Displacement U_c of the disk center and rotation angle θ of the disk as a function of the upper plate displacement U.*
•: numerical values; —: theoretical solution

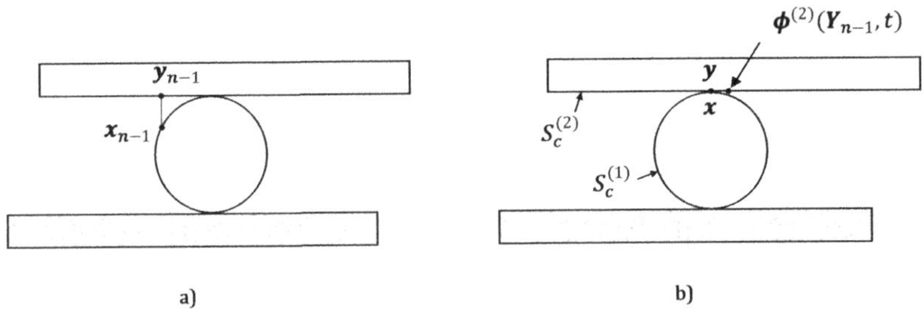

Figure 10.13. *Illustration of the slip $\boldsymbol{V}_T \Delta t = \boldsymbol{y} - \phi^{(2)}(\boldsymbol{Y}_{n-1}, t)$. (a) Configuration at the previous step $n-1$; (b) configuration at the current iteration of the current step n*

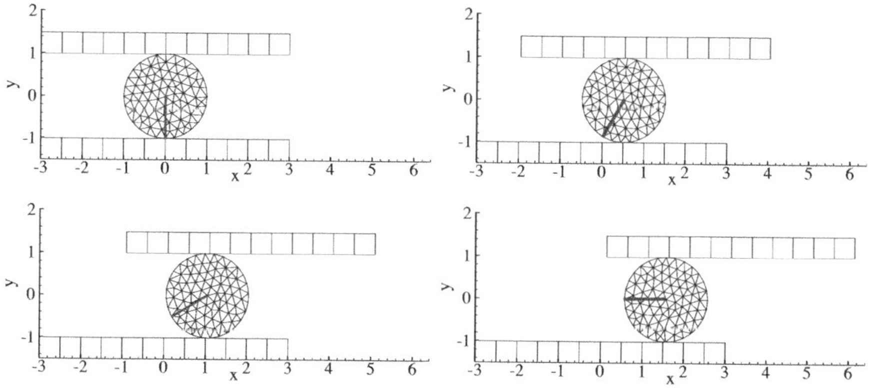

Figure 10.14. *Initial position and the position after 10, 20 and 30 steps. Calculation using a finer mesh for the plate. For a color version of this figure, see www.iste.co.uk/levan/contact.zip*

10.4. Contact between two beams

In this section, we study the problem of two straight adjoining beams, one of which is subjected to a dead or follower force. The deformation of the loaded beam entails that of the other beam because of the contact between them.

Motivation: this example demonstrates that the numerical formulation has the capacity to deal with contact involving large slips.

The two beams have the same geometry, with length $L = 1$ m along **x** and with a rectangular cross-section – height 0.05 m along **y**, and thickness 0.1 m along **z** (see Figure 10.15). They are made from the same isotropic hyperelastic Saint-Venant Kirchhoff material with $E = 2 \times 10^6$ N/m^2 and $\nu = 0.3$. The two beams are clamped at the end $X = 0$ in such a way that, in the reference configuration, the lower face of the upper beam (chosen to be the contactor) and the upper face of the lower beam (the target) occupy the same position.

Each beam is meshed using 15 twenty-node hexahedral elements (H20), and the 15 elements on the contact surface are eight-node quadrilaterals (Q8). Numerical integration is performed using $3 \times 3 \times 3$ Gaussian quadrature points for the H20 and 7×7 for the Q8 on the contact surface.

In the following, we will distinguish between two types of the load applied onto the contactor: the first is a dead force, and the second a follower force. For each load, we will study the frictionless contact as well as the contact with a coefficient of friction $\mu = 0.3$. In either case, the penalty parameters will be taken to be $\epsilon_N = \epsilon_T = 10^5$ N/m^3.

10.4.1. *Dead load*

Let us apply a dead surface force, directed along $-\mathbf{y}$ and uniformly distributed over the cross-section at $X = 1$ m of the upper beam. The resultant load will be denoted by $-F\mathbf{y}$. The applied force, zero at the initial instant, is increased progressively afterwards. The load control chosen is the arc-length method, where the arc-length increment is increased regularly by a factor of 1.1. The convergence tolerance is 10^{-6} for the residual force vector $\{\mathbf{R}\}$, 10^{-6} for the residual arc length and 10^{-6} for the residual contact law vector $\{\mathbf{R}_\Lambda\}$.

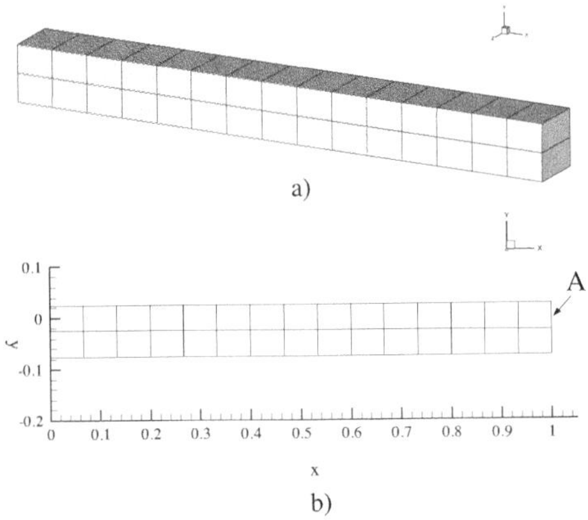

Figure 10.15. *Reference configuration of two beams in contact. (a) Mesh viewed in 3D, (b) mesh viewed in the* **xy** *plane*

Without friction, the calculation is achieved in 29 steps, with the resultant force at the final step equal to $F = 62.61$ N, and the calculation converges after seven iterations on average. Figure 10.16 shows the deformed shapes and the contact stresses exerted on the upper beam, in the **xy** plane view (the 3D view would be more complete yet less readable). With friction $\mu = 0.3$, the calculation is performed in 42 steps, with the resultant force at the final step equal to $F = 86.15$ N, and the calculation converges after 10 iterations on average. Figure 10.17 shows the deformed shapes and their corresponding contact stresses.

Contact immediately occurs between the two beams as soon as the force F is applied and it lasts over the entire lower surface of the upper beam (contactor) until the final step. During the first steps, contact stresses are high close to the extreme right cross-section, where force F is applied; away from this section, they are weak and quasi-invisible in Figures 10.16 and 10.17. Once the deformations start to become more significant, the contact stresses are largely predominant in the vicinity of the clamped end.

- Verification of the inequalities $T_N \geq 0$ and $\|\mathbf{T}_T\| \leq \mu T_N$.

In the frictionless case, the numerical results obtained show that the inequality $T_N \geq 0$ in [5.3] is satisfied globally. At certain steps and at a few nodes of the contact surface, a few negative normal contact stresses T_N may exist, which are the abnormal stresses mentioned in section 8.9. Fortunately, there are not too many of these nodes and the stresses $T_N < 0$ in question are of the order of 1% or 0.1% of the maximal stress $T_N > 0$ that is found in the considered step.

In the frictional case, Coulomb's inequality $\|\mathbf{T}_T\| \leq \mu T_N$ in [5.9] is satisfied at every step and at every node of the contact surface. The abnormal stresses $T_N < 0$, while they do exist, are negligible just like in the frictionless case, except at the final step when the force F is maximal (Figure 10.17(f)). Figure 10.18 shows a blown-up view near the clamped end in Figure 10.17(f), where we can see that a majority of the normal stresses are positive, whereas some are visibly negative, albeit only a few and also relatively small.

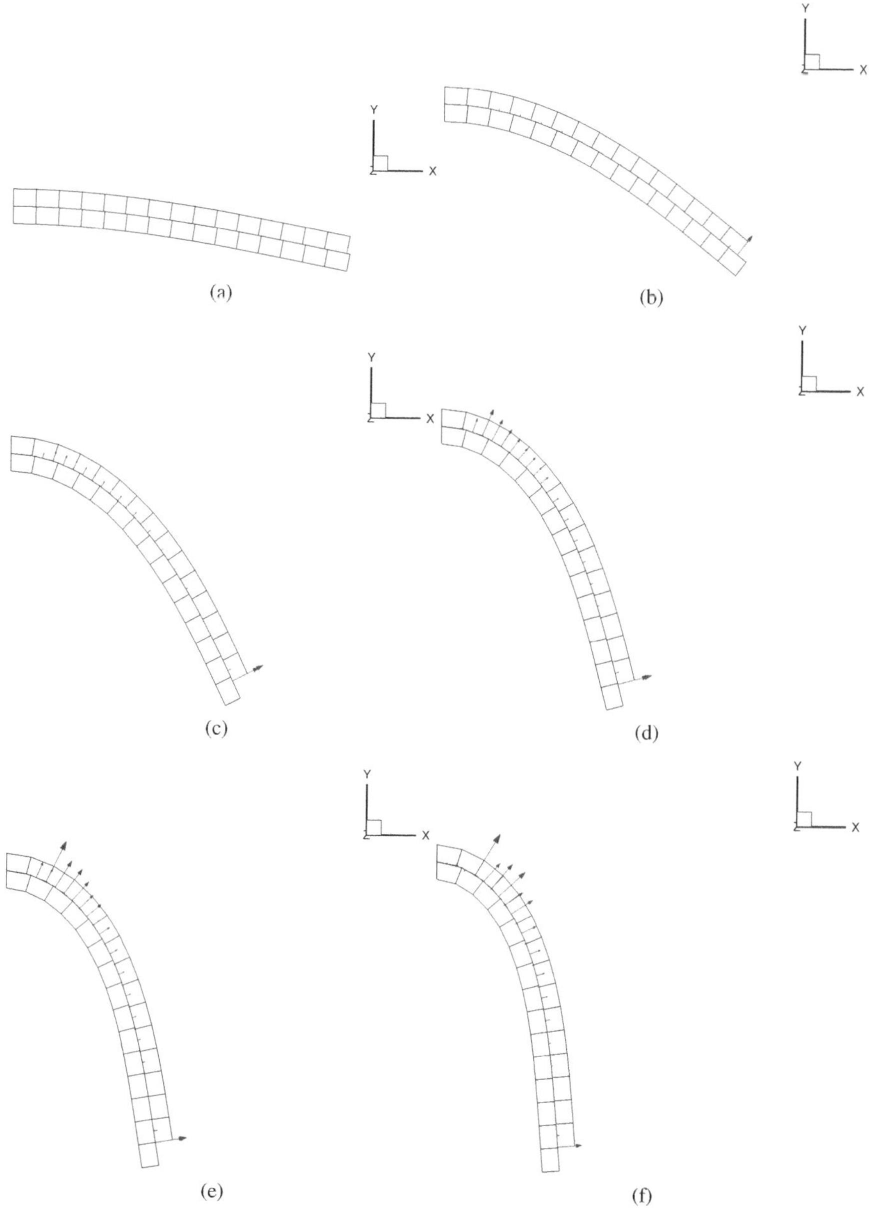

Figure 10.16. *Contact between two beams. Dead force and without friction. (a)* $F = 1.625$ *N, (b)* $F = 6.914$ *N, (c)* $F = 17.17$ *N, (d)* $F = 29.21$ *N, (e)* $F = 41.10$ *N, (f)* $F = 62.61$ *N. For a color version of this figure, see www.iste.co.uk/levan/contact.zip*

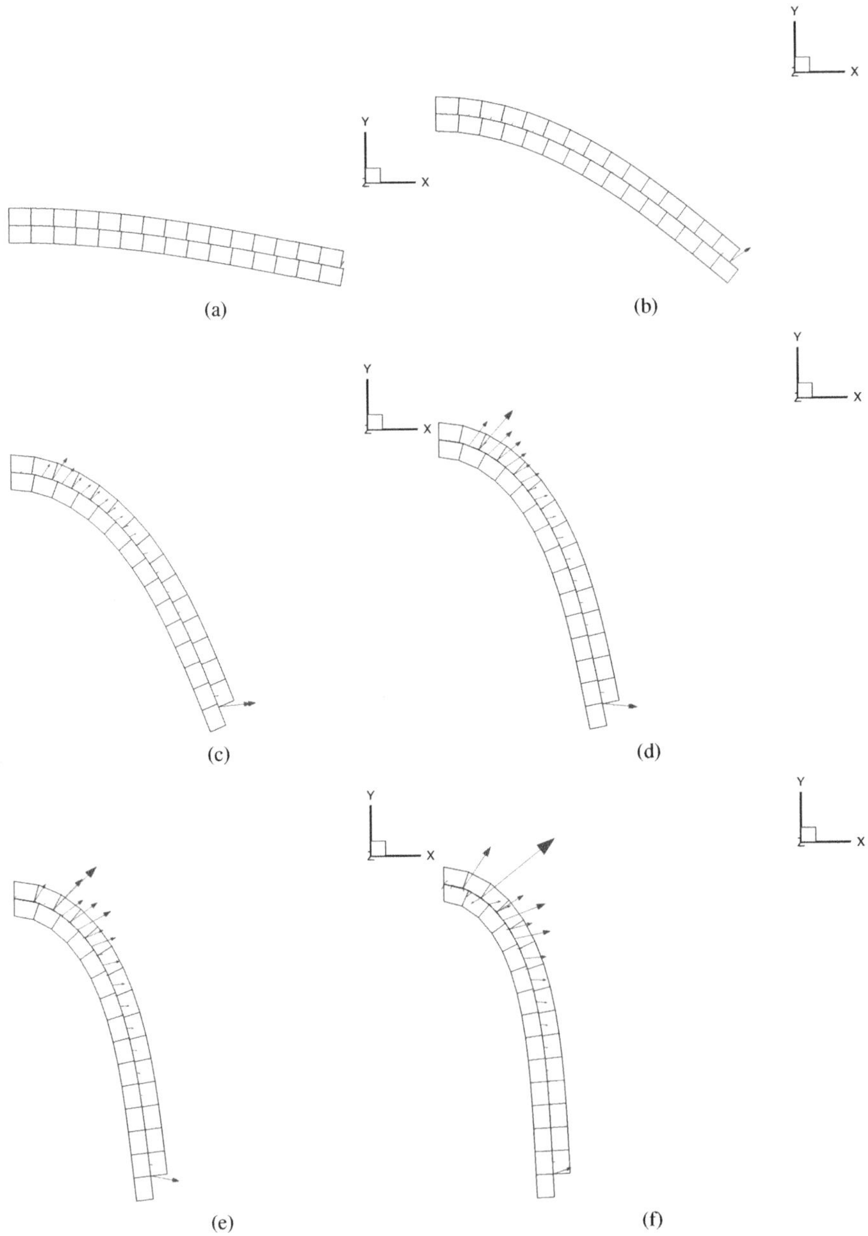

Figure 10.17. *Contact between two beams. Dead load and with friction* $\mu = 0.3$. *(a)* $F = 1.728$ *N, (b)* $F = 7.267$ *N, (c)* $F = 21.05$ *N, (d)* $F = 37.19$ *N, (e)* $F = 54.09$ *N, (f)* $F = 86.15$ *N. For a color version of this figure, see www.iste.co.uk/levan/contact.zip*

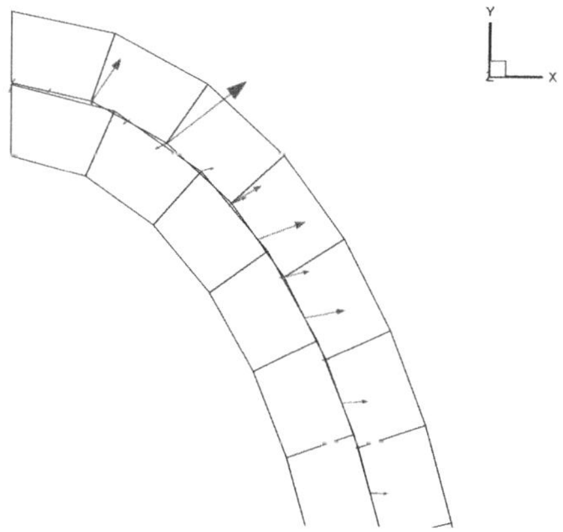

Figure 10.18. *Blown-up view near the clamped end in Figure 10.17f. For a color version of this figure, see www.iste.co.uk/levan/contact.zip*

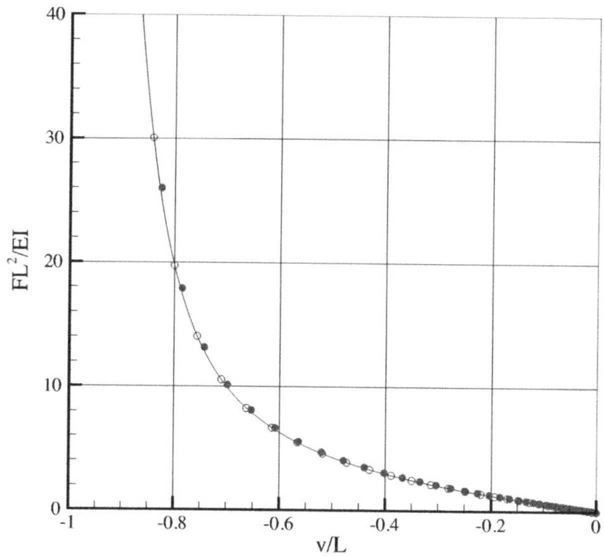

Figure 10.19. *Contact between two beams. Deflection at point* A *versus force* F*.* ○*: without friction,* •*: with friction* $\mu = 0.3$*. For a color version of this figure, see www.iste.co.uk/levan/contact.zip*

Near the clamped end, the curvature of the deformed beam is important and the fairly coarse mesh does not allow for the contact surfaces to fit together well. Within this region, there may be a detachment between the contact surfaces, and the abnormal stresses $T_N < 0$ may come from elements that *straddle* between the contact and noncontact regions. It is necessary to refine the mesh close to the clamped end in order to have more precise results.

• In the presence of friction, the contact status – slip or stick – is too complicated to describe. Basically, at each step, we find that there is slip at most of nodes and at some steps, we find that there is stick at some of the nodes.

• Figure 10.19 shows the deflection at point A (the center of the cross-section $X = 1$ m of the upper beam in Figure 10.15) as a function of force F. Due to slip in the frictional case, the curve with friction and the curve without coincide. If friction had prevented slip from occurring at a few particular steps, then a greater force would be required to attain the same deflection compared to the frictionless case, and the two curves would then differ.

10.4.2. *Follower load*

Now, let us replace the dead force considered above with a uniformly distributed pressure on the last Q8 element located on the upper face of the upper beam (contactor), next to the cross-section at $X = 1$ m. The resultant of the pressure load will be denoted by F. Pressure is a follower load as it constantly remains perpendicular to the surface over which it is applied. It will change direction as the beam is deformed, contrary to the dead load which maintained a fixed direction throughout. The pressure is zero at the initial time instant and progressively increases thereafter. The load control method and the convergence tolerances are the same as those in the dead load case.

Without friction, the calculation is achieved in 34 steps, with the resultant force at the final step equal to $F = 73.94$ N, and the calculation converges in nine iterations on average. Figure 10.20 shows the deformed shapes and the corresponding contact stresses. We observe that, at the beginning, contact takes place all over the lower surface of the upper beam (contactor) but that, at a later stage of the deformation, the absence of friction leads to a loss of contact over a significant region of the initial contact surface, with this loss of contact being visible in Figures 10.20(e) and 10.20(f).

With friction $\mu = 0.3$, the calculation is performed in 47 steps, the resultant force at the last step is equal to $F = 78.01$ N, and the calculation converges in 10 iterations on average. Since friction is sufficiently strong, contact persists over a large region of the lower surface of the upper beam (see Figure 10.21). There is loss of contact only in a few barely perceptible places. As in the case with the dead load, the contact status in the presence of friction – slip or stick – is quite complicated to describe. Here, at each step, we see that there is slip at most of the nodes, except for the last few steps when there is more stick than slip.

Figure 10.22 shows the deflection at point A of the upper beam as a function of F. With a follower load, observe that there is a limit point under displacement control.

10.5. Contact of two pressurized membranes

Motivation: to test the numerical formulation through the example of two pressurized membranes in contact, an example which comprises many forms of nonlinearity: contact of course, but also nonlinear geometry due to the large strains and large displacements of the membranes, material nonlinearity due to the neo-Hookean hyperelasticity constitutive law, and nonlinearity due to the pressure load, which is a follower load. This example also provides an occasion to test the performance with membrane finite elements in contact.

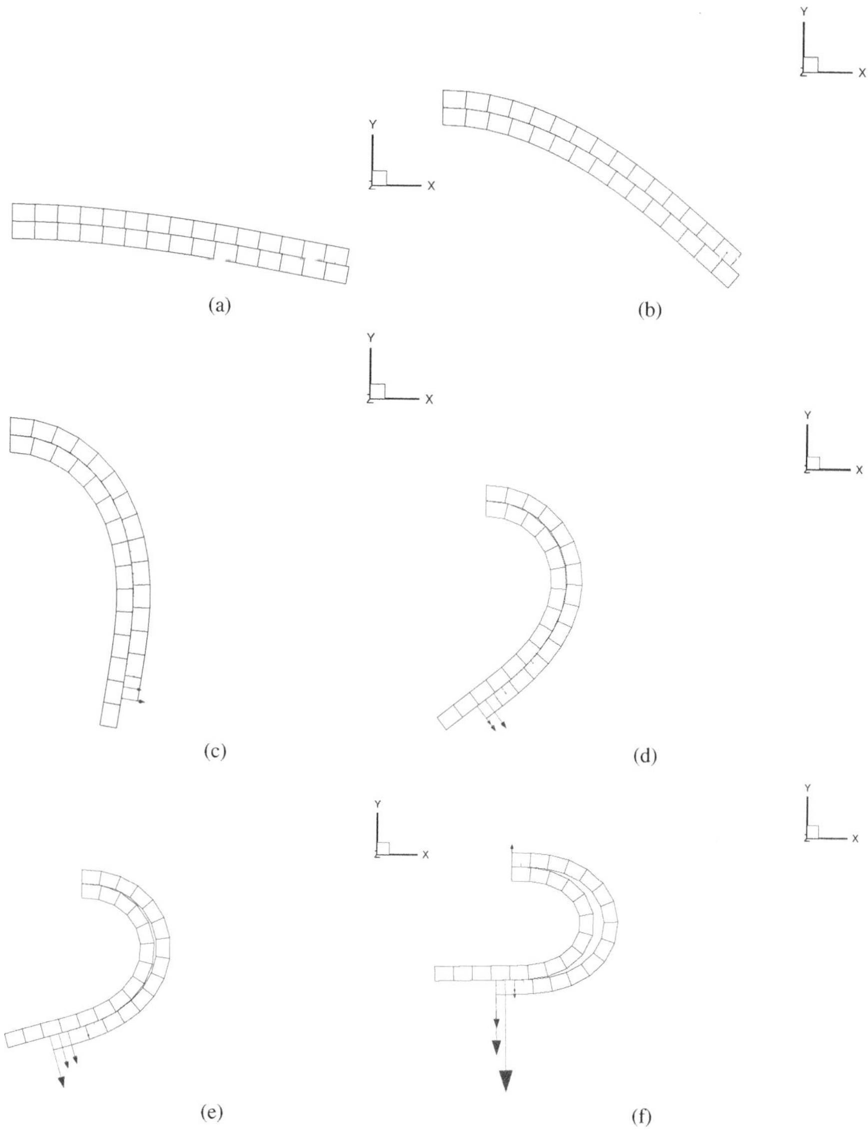

Figure 10.20. *Contact between two beams. Follower load and without friction. (a) $F = 1.607\ N$, (b) $F = 6.757\ N$, (c) $F = 18.88\ N$, (d) $F = 33.35\ N$, (e) $F = 46.44\ N$ and (f) $F = 73.94\ N$. For a color version of this figure, see www.iste.co.uk/levan/contact.zip*

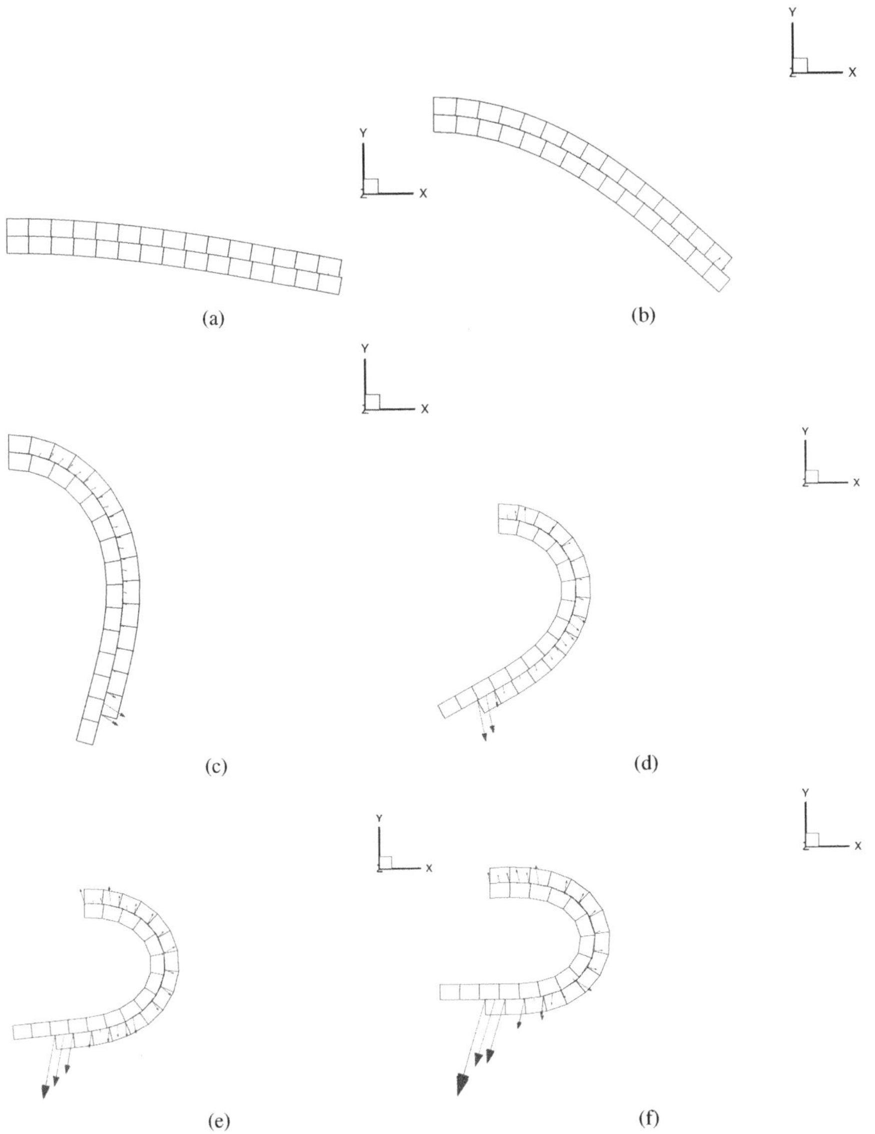

Figure 10.21. *Contact between two beams. Follower load and with friction* $\mu = 0.3$. *(a)* $F = 1.716$ *N, (b)* $F = 6.991$ *N, (c)* $F = 21.07$ *N, (d)* $F = 39.14$ *N, (e)* $F = 58.63$ *N and (f)* $F = 78.01$ *N. For a color version of this figure, see www.iste.co.uk/levan/contact.zip*

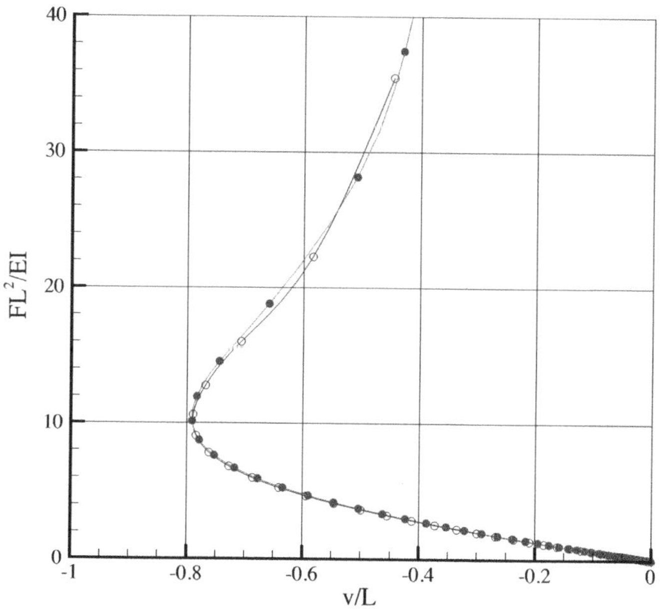

Figure 10.22. *Contact between two beams. Follower load. Deflection at point A versus force F. ∘: Without friction, •: with friction $\mu = 0.3$. For a color version of this figure, see www.iste.co.uk/levan/contact.zip*

The problem is modeled in plane strain in the **xy** plane. In the reference configuration, the two membranes have a thickness of 1 mm, their mid-lines are two line segments with a length of 1 m. The membranes are separated from one another by a distance of $\Delta y = 1$ m in the **y** direction and staggered by Δx along **x** (see Figure 10.23). We will study two cases: $\Delta x = 0$ (*centered membranes*) and then $\Delta x = 1$ m (*staggered membranes*). The boundaries of the membranes are fixed.

The membranes are made of an isotropic hyperelastic neo-Hookean material whose constitutive law is given by [4.7], with $E = 2.10^6$ N/m^2 and $\nu = 0.3$. The two membranes are subjected to the same pressure p which inflates them gradually, causing them to come into contact with one another. The calculations are performed without and with friction; in the latter case, we take the coefficient of friction to be $\mu = 0.5$, which is large enough to simulate frictional contact between the rubber membranes.

Each membrane is meshed using 20 three-node line elements of membrane type (i.e. without bending stiffness), in plane strain with respect to the **xy** plane, and in plane stress with respect to the normal direction to the membrane surface. The upper membrane is chosen to be the contactor, and the lower membrane as the target. Here, the mesh of the contact surface is quite simply the membrane mesh itself. One particularity of membrane structures is that the stiffness matrix is singular at the very start of the computation when the membranes are unstretched. To avoid this singularity, let us artificially apply a weak prestress in the first numerical step, and then remove this prestress immediately at the beginning of the second step.

Numerical integration is performed using three Gaussian quardrature points per line element to calculate the internal force vector, and seven points to calculate the contact force vector. The penalty parameters are $\epsilon_N = \epsilon_T = 2 \times 10^4$ N/m^3. The pressure p is controlled using the arc-length method, where the arc-length increment is regularly increased by a factor of 1.05. The convergence tolerance is 10^{-6} for the residual force vector $\{\mathbf{R}\}$, 10^{-6} for the residual arc-length and 10^{-6} for the residual contact law vector $\{\mathbf{R}_\Lambda\}$.

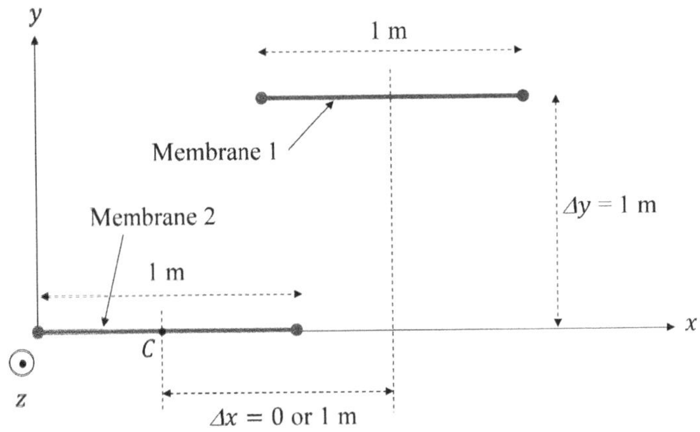

Figure 10.23. *Reference configuration of the two membranes. For a color version of this figure, see www.iste.co.uk/levan/contact.zip*

10.5.1. *Centered membranes*

First, let us consider the case in which the two membranes are centered ($\Delta x = 0$).

In the frictionless case, the calculation is done in 29 steps, where the maximum value of the applied pressure is $p = 7,105$ N/m^2. Figure 10.24(a) shows the deformed shapes and the contact stresses applied on the upper membrane when the pressure attains its maximum value.

Without friction, the normal contact stress T_N is practically uniform over the contact region, as illustrated in Figure 10.24(a). In the absence of an analytical solution, let us perform a quick verification for the normal contact stress value. To do so, consider the relation [3.5], $t_N dS^{(1)} = T_N dS_0^{(1)}$, which relates the reference normal contact stress T_N to the spatial one t_N, and then let us see that this relation is indeed satisfied at the central node of the contact region.

- Consider the pressure $p = 7,105$ N/m^2. The normal contact stress equals the same value: $t_N = 7,105$ N/m^2.

- For this pressure, the numerical computation gives $T_N = 17,561$ N/m^2 at the central node of the contact region.

- Approximate the reference area $dS_0^{(1)}$ by $dS_0^{(1)} = l_0^{(1)} \times 1$ m, where $l_0^{(1)}$ is the length in the reference state of a finite element of $S_0^{(1)}$ adjoining the node in question, and where 1 m is the thickness in the **z** direction (recall that we are in plane strain). Here, we choose one of the two finite elements containing the central node, whose reference length is $l_0^{(1)} = 1\text{ m}/20 = 0.05$ m.

- The current area $dS^{(1)}$ is approximated in the same manner: $dS^{(1)} = l^{(1)} \times 1$ m, where $l^{(1)}$ is the current length of the same finite element. From the numerical results, when the pressure p equals 7, 105 N/m^2, the length $l^{(1)}$ equals $l^{(1)} = 0.1233$ m.

With these values, we find that the relation [3.5] is satisfied within 0.2%.

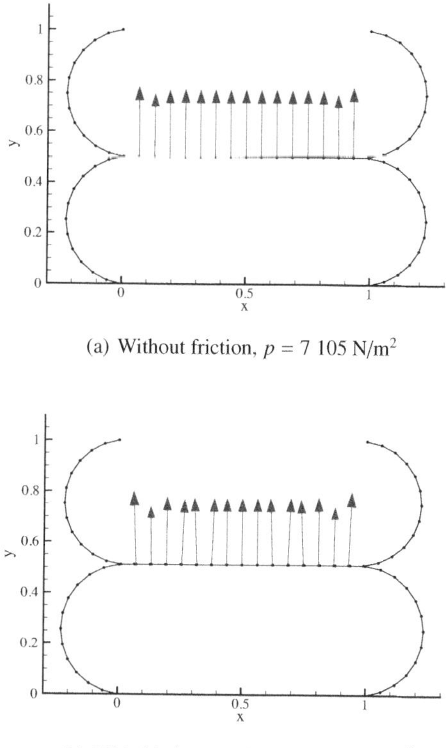

(a) Without friction, $p = 7\ 105$ N/m^2

(b) With friction $\mu = 0.5$, $p = 7\ 104$ N/m^2

Figure 10.24. *Contact between two centered pressurized membranes. For a color version of this figure, see www.iste.co.uk/levan/contact.zip*

In the frictional case, the computation is done in 29 steps, where the maximum pressure value is $p = 7,104$ N/m^2. Figure 10.24(b) shows the deformed shapes and the contact stresses applied on the upper membrane when the pressure attains its maximum value. With friction, stick is present at each node in contact and for each step.

The following remark holds either with or without friction. According to the values chosen for the penalty parameters ϵ_N, ϵ_T, and that for the arc-length increment factor, we observe that, with a high enough pressure, one can unexpectedly pass onto a bifurcated solution branch where the two membranes deform in an asymmetric manner, whereas one should remain on a symmetrical solution (since the problem is symmetrical and no bifurcation treatment is activated). This loss of symmetric can be explained by the fact that, in contact theory, the contactor and target bodies are treated in an asymmetric manner, as has been remarked in section 6.4.

10.5.2. *Membranes staggered along* x

Now let us consider the case where the two membranes are staggered in the **x** direction by $\Delta x = 1$ m.

In the frictionless case, the calculation is done in 35 steps, the maximum pressure value is $p = 4,020$ N/m^2, and the calculation converges after four iterations on average. Figure 10.25 shows the deformed shapes and the contact stresses exerted on the upper membrane. Two particles have been identified in particular, belonging to either of the two membranes, respectively. At the beginning of contact ($p = 2,048$ N/m^2), their positions A_1 and B_1 almost coincide (see Figure 10.25(a)). When the pressure equals $p = 4,020$ N/m^2, their positions A_2 and B_2 are far apart as shown in Figure 10.25(b). The distance A_2B_2 is an approximation of the slip $\mathbf{V}_T\Delta t$ (see definition [8.53]) calculated for the particle identified by A_1 and between the two time instants corresponding to the two pressures just mentioned.

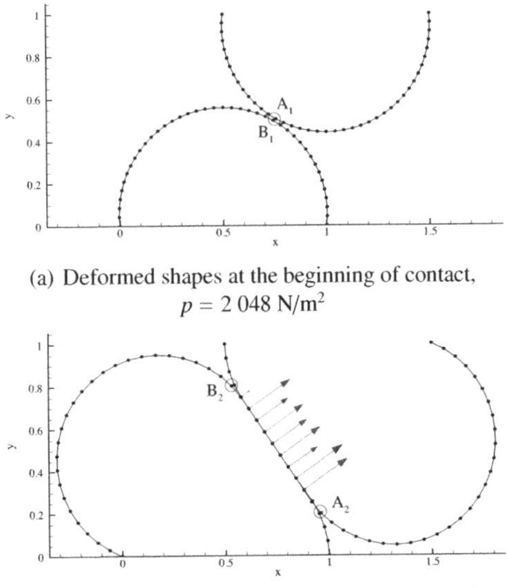

(a) Deformed shapes at the beginning of contact, $p = 2\,048$ N/m^2

(b) Deformed shapes at pressure $p = 4\,020$ N/m^2

Figure 10.25. *Contact between two pressurized membranes shifted by* 0.5 *m. Frictionless case. For a color version of this figure, see www.iste.co.uk/levan/contact.zip*

In the frictional case with $\mu = 0.5$, the calculation is done in 33 steps, the maximum pressure value is $p = 5,541$ N/m^2, and the calculation converges after four iterations on average. Figure 10.26 shows the deformed shapes and the contact stresses exerted on the upper membrane. All contact stresses are drawn using the same scale in Figures 10.24–10.26.

The final deformed shape with friction has the same form as that without friction, yet the particle displacements are completely different. To see this, consider the two particles mentioned above again with positions A_1 and B_1 that almost coincide at the beginning of contact (pressure $p = 2,048$ N/m^2). This time, due to large friction, we note that virtually these two particles do not move away from one another, up to the pressure $p = 5,541$ N/m^2 when the deformation is significant (see Figure 10.26(b)). In other words, the slip $\mathbf{V}_T\Delta t$ of particle A_1 is practically zero.

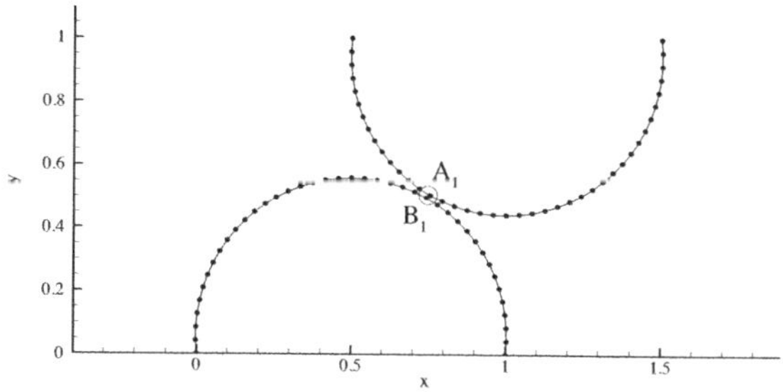

(a) Beginning of contact, $p = 2\ 048\ \text{N/m}^2$

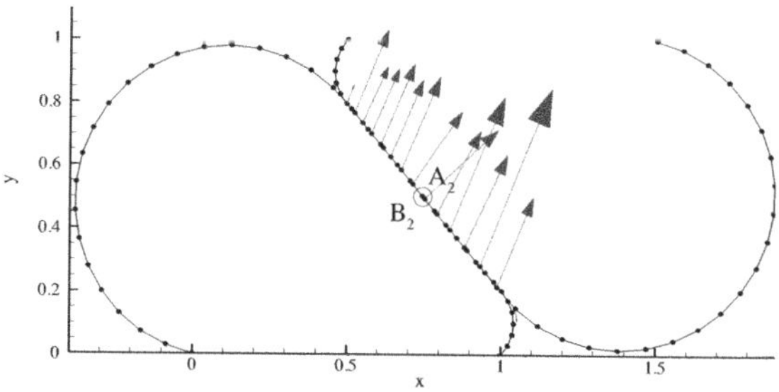

(b) Deformed shapes at pressure $p = 5\ 541\ \text{N/m}^2$

Figure 10.26. *Contact between two pressurized membranes shifted by 0.5 m. Case with friction $\mu = 0.5$. For a color version of this figure, see www.iste.co.uk/levan/contact.zip*

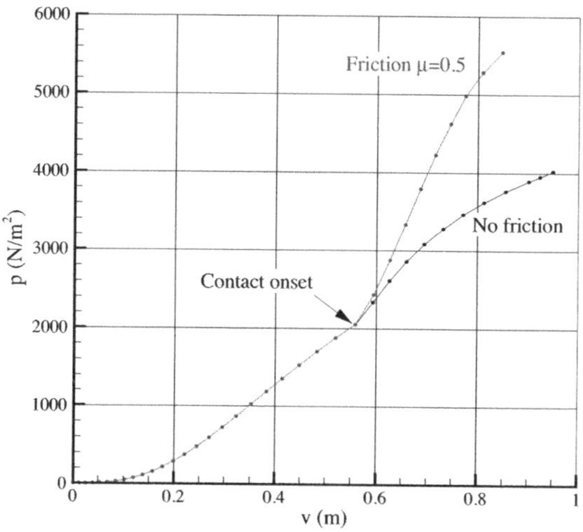

Figure 10.27. *Contact between two pressurized membranes. The deflection at point C versus pressure p. For a color version of this figure, see www.iste.co.uk/levan/contact.zip*

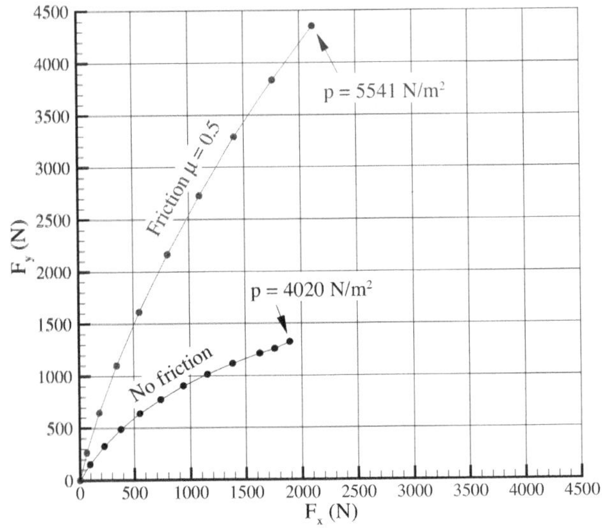

Figure 10.28. *Contact between two pressurized membranes. The F_x and F_y components of the resultant contact force* **F***. For a color version of this figure, see www.iste.co.uk/levan/contact.zip*

Numerical results show that there is stick within a neighborhood of particle A_1 and that there may be slip elsewhere.

Figure 10.27 shows the vertical displacement of the midpoint of the lower membrane (point C in Figure 10.23) as a function of pressure p. The deflection in the presence of friction is smaller than that without friction, as expected.

Figure 10.28 represents the F_x and F_y components of the resultant force **F** of the contact stresses applied to the upper membrane, along the **x**- and **y**-axes, respectively. The final pressure applied in the frictional case ($p = 5,541$ N/m^2) is roughly equal to 1.4 times that in the frictionless case ($p = 4,020$ N/m^2). These two pressures lead to deformations that are of the same order (see Figures 10.25(b) and (b)), yet the contact force **F** produced by the first pressure is more than twice that produced by the second pressure.

10.6. Extrusion of an elastoplastic cylinder

Motivation: we have seen that the weak formulation [7.21] includes the contact laws automatically, and the numerical treatment of contact in Chapter 9 is the same regardless of the constitutive law for the bodies. In particular, this signifies that we do not need to perform any local integration of the friction laws at the level of the contact surface. In this section, the example that we are going to treat now has the aim of testing the numerical formulation in the presence of an elastoplastic constitutive law. We will consider the problem of extrusion of an elastoplastic cylinder into a rigid frustoconical die.

We take the same geometric data as in [71, 77] (see Figure 10.29). In the reference configuration, the cylindrical piece has radius 5.08 cm and length 25.4 cm. The extrusion angle imposed by the die is 5°, and the height of the die is 50.8 cm.

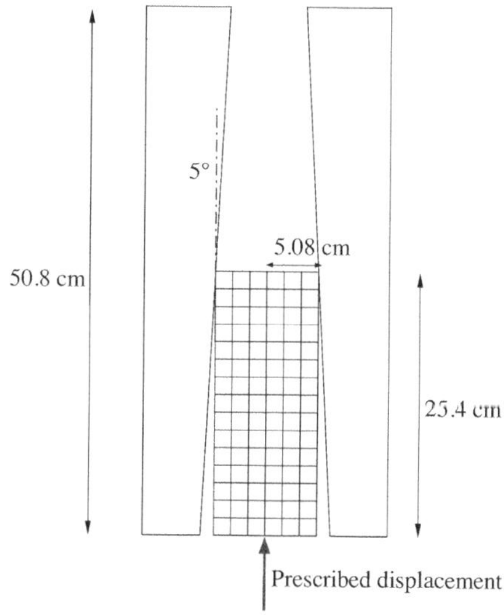

Figure 10.29. *Initial configuration of the extrusion problem. For a color version of this figure, see www.iste.co.uk/levan/contact.zip*

The cylinder is made from a Mooney-type elastoplastic material with linear isotropic hardening, whose constitutive law is described in section 4.2. We take the same material data used in [111], namely, Young's modulus $E = 100,000$ MPa, Poisson's ratio $\nu = 0.3$, initial yield stress $\Sigma_0 = 300$ MPa and the hardening parameters appearing in relation [4.15] are $c_0 = c_1 = c_3 = 0$, $c_2 = 3,560$ MPa.

The cylinder is subjected to a prescribed axial displacement U on its lower base that forces it to progress between the rigid wall of the die. The calculations are performed both with and without friction $\mu = 0.1$.

The problem is modeled axisymetrically. The cylinder (considered to be the contactor) is meshed using 3×15 eight-node quadrilateral elements (Q8), containing 172 nodes in total. The 15 elements of the contact surface are three-node line elements. The rigid die (considered to be the target) is meshed using just a single Q8 element. Numerical integration is performed using 3×3 Gaussian quadrature points for Q8 elements, and seven points for the line elements of the contact surface. The penalty parameters are $\epsilon_N = \epsilon_T = 10^9$ N/m^3.

The computation is performed by incrementing the prescribed axial displacement U using a constant step size of 5 mm (respectively, 2.5 mm) in the frictionless (respectively, frictional) case. The convergence tolerances are 10^{-6} for both the residual vector force $\{\mathbf{R}\}$ and the residual contact law vector $\{\mathbf{R}_\Lambda\}$.

Without friction, the calculation is performed in 40 steps, the axial displacement U attained in the final step is 20 cm, and the calculation converges after five iterations on average. With friction, the calculation is performed in 85 steps, the axial displacement attained in the final step is 21.25 cm and the calculation converges after four iterations on average. Figure 10.30 (respectively, Figure 10.31) shows the deformed shapes corresponding to $U = 6$ cm, $U = 12$ cm and $U = 20$ cm without friction (respectively, corresponding to $U = 6.25$ cm, $U = 12.5$ cm, and $U = 21.25$ cm with friction). The same figures also displays the contours of the equivalent von

Mises stress Σ_{eq} defined by

$$\Sigma_{eq} \equiv \sqrt{\frac{3}{2}\mathbf{S}:\mathbf{S}} \qquad \text{where } \mathbf{S} \equiv \boldsymbol{\Sigma} - \frac{1}{3}(tr\boldsymbol{\Sigma})\mathbf{I} \qquad [10.5]$$

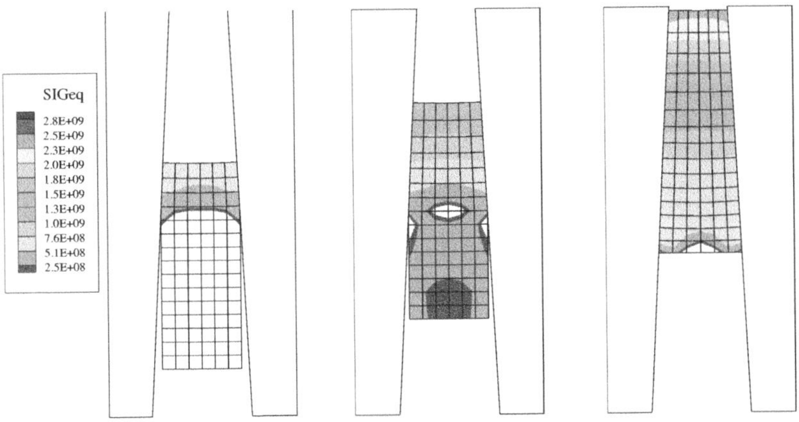

Figure 10.30. *Deformed shapes of the extruded cylinder and the equivalent stresses. Frictionless case. For a color version of this figure, see www.iste.co.uk/levan/contact.zip*

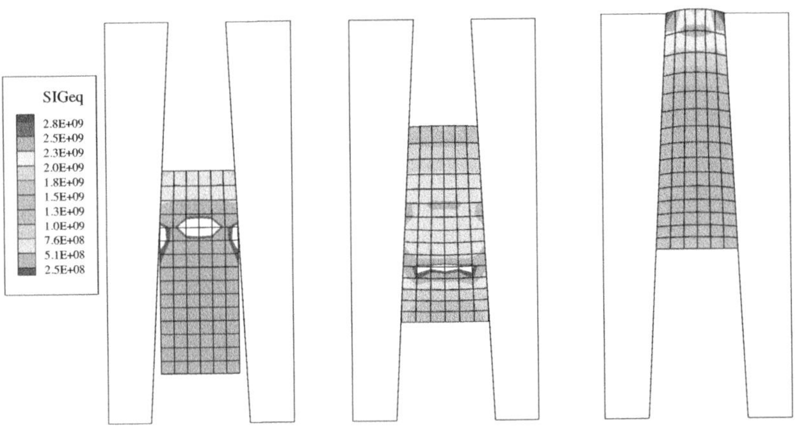

Figure 10.31. *Deformation of the extruded cylinder and the equivalent stresses. Frictional case, $\mu = 0.1$. For a color version of this figure, see www.iste.co.uk/levan/contact.zip*

The tensor **S** is the deviatoric part of the second Piola–Kirchhoff stress tensor $\boldsymbol{\Sigma}$. For convenience of view, we choose not to color the entire structure, this is why we have decided not to represent values of Σ_{eq} that are less than 2×10^8 N/m^2.

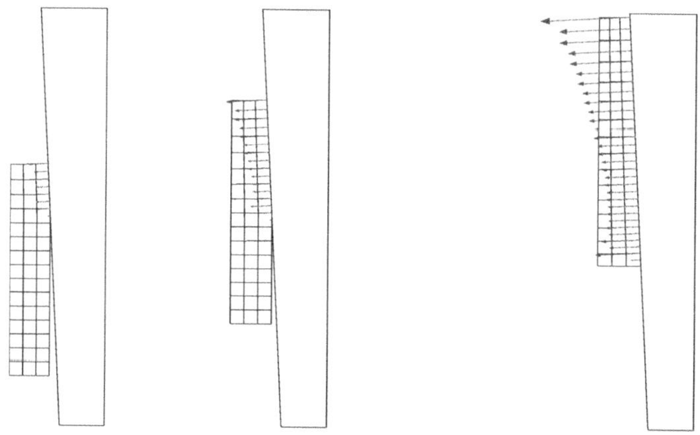

Figure 10.32. *Contact stresses exerted on the cylinder. Frictionless case. For a color version of this figure, see www.iste.co.uk/levan/contact.zip*

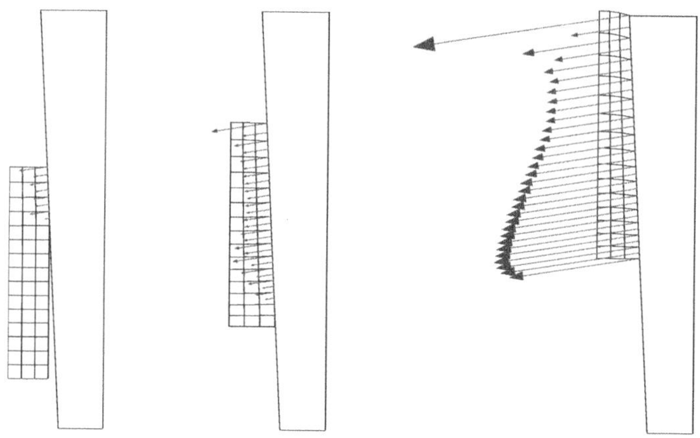

Figure 10.33. *Contact stresses exerted on the cylinder. Frictional case,* $\mu = 0.1$*. For a color version of this figure, see www.iste.co.uk/levan/contact.zip*

Figures 10.32 and 10.33 show that the contact stress vector exerted on the cylinder in both cases without and with friction, respectively. These vectors are represented using the same length scale in the both figures.

As the cylinder advances, the equivalent stress Σ_{eq} is concentrated at the head of the cylinder, which is where the contact stresses are greatest. Toward the lower part of the cylinder, which is still yet to come into contact with the die, the stress state resembles that of a uniaxial compression and the derivative tensor $\mathbf{S}$ can be large. On the other hand, in the part of the cylinder in contact with the die, the compressive hoop stresses due to contact add to the axial stresses, so that at certain points the stress state is almost spherical and the deviatoric tensor $\mathbf{S}$ is almost zero. This explains the appearance of the uncolored regions in Figures 10.30 and 10.31, which signifies that the equivalent stress Σ_{eq} in these regions is weak. We thus come to the same remark made by Kikuchi and Oden ([71], p. 396), who treated the same example with a similar elastoplastic constitutive law. According to the findings of the authors, when a boundary point is brought into contact with the die, occasionally the elements around the point unload temporarily.

When friction is present, numerical results show that there is slip at each node of the contact surface and for every displacement value U, and that Coulomb's equality $\|\mathbf{T}_T\| = \mu T_N$ is satisfied up to the sixth significant figure.

Furthermore, friction induces a more significant force resisting the cylinder's advance. The contact stresses are larger than in frictionless case, and this is particularly visible in Figures 10.32(c) and 10.33(c), which show the cylinder at the end of its advance.

Figure 10.34 displays the thrust force F versus the displacement U of the lower base of the cylinder, in both cases without and with friction. This thrust force becomes larger and larger as the cylinder advances further and as the contact surface between the cylinder and die increases. For the same advance of the cylinder, the thrust force is always more significant with friction than without friction. When U equals 20 cm, F is about six times larger when there is friction. Compared with Laursen's results in [77], the thrust force with friction takes on the same form, but it is normal that the numerical values are notably different, since we have not considered the same material as Laursen.

Figure 10.35 shows the volume variation of the cylinder over the course of its extrusion. It can be seen that the volume decreases progressively as the cylinder advances further, and that it decreases more rapidly in the presence of friction. The absolute value of the relative variation reaches is maximum at the last step, equaling -2.5% without friction and -6.2% with friction.

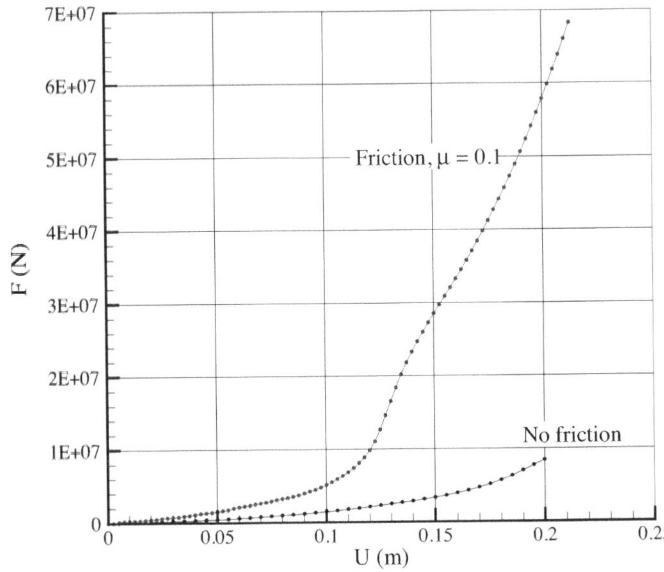

Figure 10.34. *Thrust force versus the cylinder displacement. For a color version of this figure, see www.iste.co.uk/levan/contact.zip*

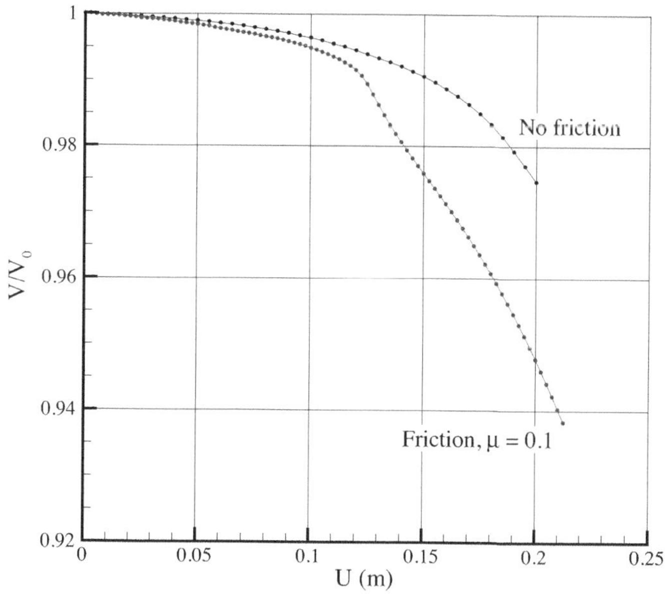

Figure 10.35. *Volume variation of the extruded cylinder. For a color version of this figure, see www.iste.co.uk/levan/contact.zip*

10.7. Interference fit problem

In this section, we will study a cylinder inserted into a truncated-cone cavity of a deformable die block, when the cylinder upper base is subjected to an axial displacement U, taken positive in the opposite direction of $\mathbf{z}$-axis (see Figure 10.36). The problem considered here is known to be more difficult to solve than the extrusion problem from the previous section, since now both contacting bodies are deformable and undergo significant plastic strains.

The geometric data here is the same as in [77]. In the reference configuration, the cylinder has radius 4 cm and height 30 cm. The exterior surface of the die block is cylindrical, with radius 30 cm and height 50 cm. The internal surface of the block is the union of a truncated-cone section (with larger radius 4 cm, smaller radius 3.5 cm and length 15 cm) and a cylindrical section of radius 3.5 cm. The wall angle of the truncated-cone section is therefore 1.9°.

The two bodies are made of a Mooney-type elastoplastic material with linear isotropic hardening, whose constitutive law was described in section 4.2. We will use the same material data as in [111]: namely, Young's modulus $E = 100,000$ MPa, Poisson's ratio $\nu = 0.3$, initial yield stress $\Sigma_0 = 300$ MPa and the work hardening parameters appearing in relation [4.15] are $c_0 = c_1 = c_3 = 0$ and $c_2 = 3,560$ MPa.

The cylinder is considered to be the contactor, and the die block is considered to be the target. Calculations are done both without or with friction ($\mu = 0.2$), and the problem is modeled axisymetrically. The cylinder and the die block are discretized, respectively, using 3×10 and 6×15 eight-node quadrilateral elements (Q8), so that we end up with 430 nodes in total for the mesh. The 10 elements on the contact surface of the cylindrical piece are three-node line elements. The external surface and bottom surface of the die block are simply supported.

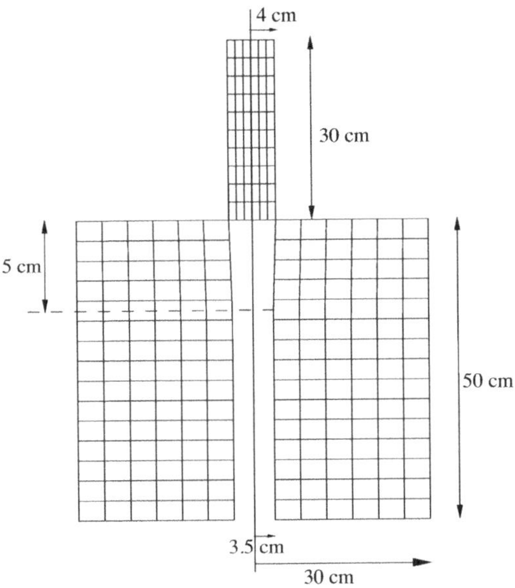

Figure 10.36. *Initial configuration of the interference fit problem*

Numerical integration is performed using 3 × 3 Gaussian quadrature points for the Q8 elements, and seven points for the line elements on the contact surface. The penalty parameters are $\epsilon_N = 10^{11}$ N/m^3, $\epsilon_T = 5 \times 10^{10}$ N/m^3.

The convergence tolerances are 10^{-6} for the residual force vector $\{\mathbf{R}\}$, the residual arc-length and the residual contact law vector $\{\mathbf{R}_\Lambda\}$. Without friction, the prescribed axial displacement U is increased from one step to the next by 1 cm, the computation is performed in 30 steps and converges after five iterations on average. With friction, the displacement U is controlled by the arc-length method, where the arc-length increment is increased between each step by a factor of 1.05. The computation requires 266 steps and is more laborious: over the course of the solution, we are sometimes required to resume the current step with only half the size of the arc-length increment.

Figure 10.37 shows the deformed shapes at the final step, without and with friction. Figure 10.38 displays the contours of the equivalent plastic strain P defined by [4.14]. To make the plots easier to view, only the values of P that are greater than 0.1% have been represented.

As expected, the cylinder advances further in the frictionless case than in the frictional case. Friction causes a larger deformation of the cylinder and a slight bulge at the free surface of the die block around the hole (see Figure 10.37(b)).

In the frictionless case, the equivalent plastic strain P is maximal (22%) toward the head of the cylinder. When there is friction, its maximum is toward the bottom end of the cylinder, and is twice as large (46%). In general, friction causes larger plastic strains throughout the cylinder.

Within the die block, the yielded region is much larger when friction is present. The equivalent plastic strain is smaller than inside the cylinder and remains concentrated close to the contact interface.

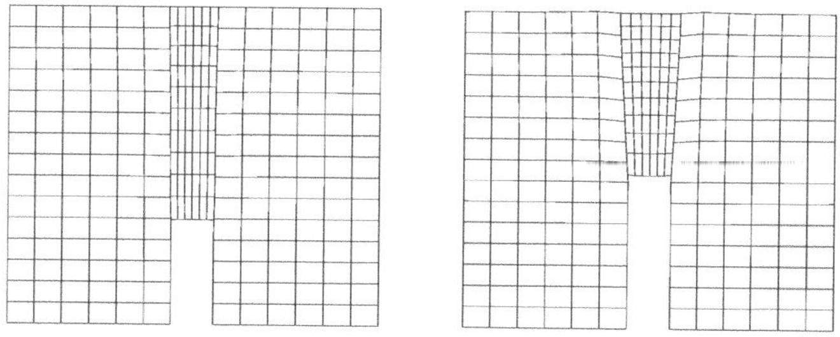

(a) Without friction (b) With friction $\mu = 0.2$

Figure 10.37. *Deformed shapes in the interference fit problem*

Figure 10.39 shows that the contact stresses are significantly higher when there is friction compared to when there is no friction.

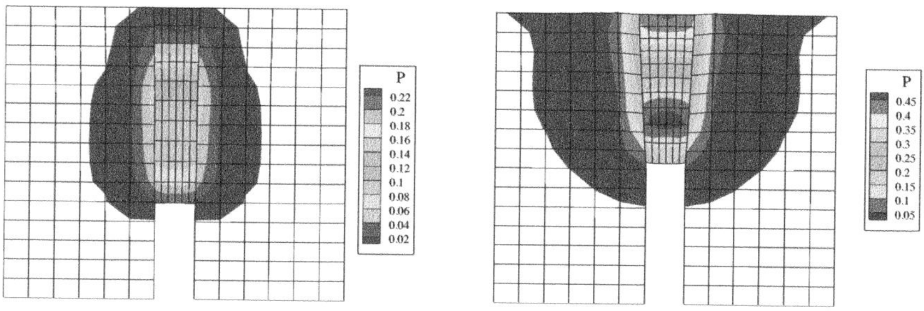

(a) Without friction (b) With friction $\mu = 0.2$

Figure 10.38. *Equivalent plastic strain in the interference fit problem. For a color version of this figure, see www.iste.co.uk/levan/contact.zip*

REMARK.– Figure 10.40 corresponds to another computation in which both the cylinder and die block are made of a neo-Hookean *hyperelastic* material, and no friction is assumed. You may be surprised to see in Figure 10.40(a) that the contact stress vector at the head of the cylinder is not parallel to the other vectors.

In fact, since the friction is zero, the contact stress vector at a point of the contactor surface $S_c^{(1)}$ should be orthogonal to the target surface $S_c^{(2)}$ (see the decomposition [3.3] of the contact stress vector). Figure 10.40(b), which gives a blown-up view of the head of the cylinder, shows that the target surface in this region is *not* planar. As the contact stress vector at the point in question is orthogonal to the target surface, it is therefore not parallel to any other vector. ■

Figure 10.41 shows the resultant thrust force F as a function of displacement U of the upper base of the cylinder, in both cases without and with friction. For the same advance of the cylinder, the thrust force is much greater in the presence of friction. When the cylinder is completely inside the die block, F is more than 20 times greater with friction than without it. Without friction, the force F becomes practically constant after the prescribed displacement equals half the initial length of the cylinder ($U \approx 15$ cm). These findings are the same as in [111] and [77], although the material considered in the latter reference is different.

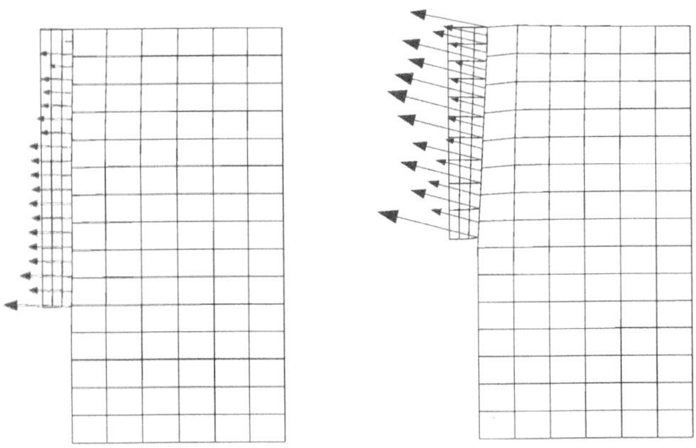

(a) Without friction (b) With friction $\mu = 0.2$

Figure 10.39. *Contact stresses exerted on the cylinder in the interference fit problem: (a) without friction; (b) with friction $\mu = 0.2$. For a color version of this figure, see www.iste.co.uk/levan/contact.zip*

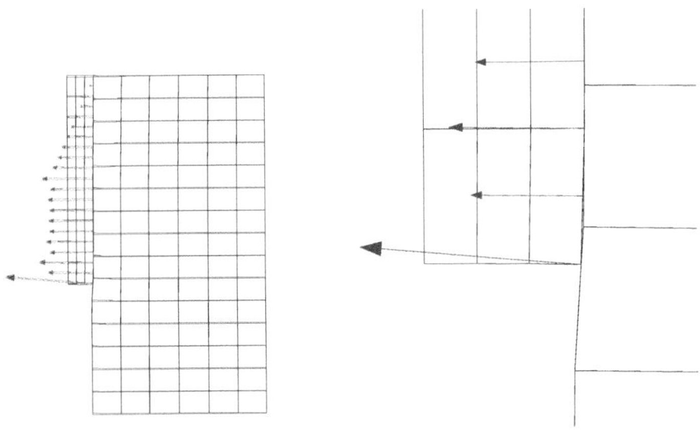

(a) (b) Blow-up of the lower part of the cylinder

Figure 10.40. *Contact stresses in the frictionless case and with a more rigid material. For a color version of this figure, see www.iste.co.uk/levan/contact.zip*

10.7.1. *Abnormal contact stresses*

In the above solution, we have declared that the contactor surface $S_c^{(1)}$ is equal to the entire external surface of the cylinder, and that the target surface $S_c^{(2)}$ is equal to the entire internal wall of the die block (here, these surfaces are represented by two lines). Without friction, at certain steps we may observe nonzero contact stresses outside of the contact region, which are the abnormal contact stresses mentioned in section 8.9. All of these stresses are found at points of the contactor surface that are close to the upper surface of the die block, where an element of $S_c^{(1)}$ is in partial contact. However, these stresses are negligible since, if existing, they are between 10^{-4} and 10^{-2} times the size of the largest contact vector of the step.

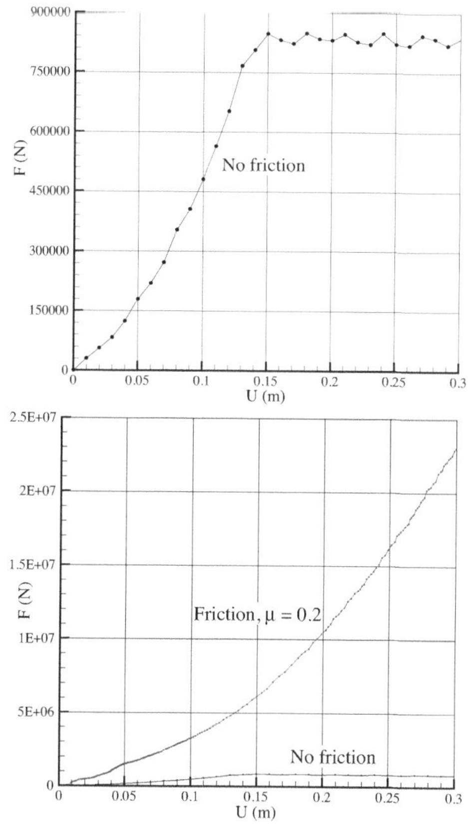

Figure 10.41. *Thrust force versus top displacement for the interference fit problem. For a color version of this figure, see www.iste.co.uk/levan/contact.zip*

These abnormal stresses become more visible in the case with friction, and again they are found at points close to the upper surface of the die block. Figure 10.42 shows the contact stress vectors for a particular computational step, with an blown-up view of the upper part of the cylinder. Nonzero contact stresses are found in the region of $S_c^{(1)}$ which is outside of the die block, at two nodes which lies above the upper surface of the die block. The magnitudes of the abnormal stresses are 6% the largest contact stress vector (located toward the cylinder base).

In Figure 10.42(b), we can also see a negative normal stress λ_N at a node of the contactor surface, which is close to the upper surface of the die block. It should be noted that we have drawn the elements with straight sides, when in fact they are quadratic elements (here, eight-node quadrilaterals), and that these sides may become curved after deformation. Thus, despite the appearance, the contact surfaces may not match around the node in question, and there may be small losses of contact at this node, which explains the existence of the negative normal stress.

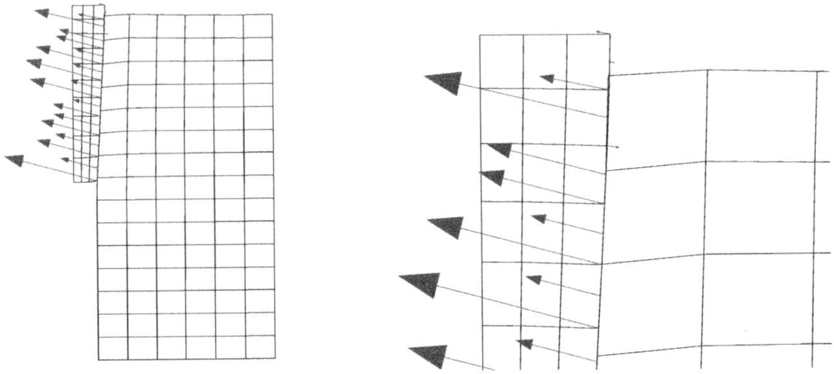

(a) Fit interference with friction $\mu = 0.2$ (b) Blow-up of the upper part of the cylinder

Figure 10.42. *Abnormal contact stresses. For a color version of this figure, see www.iste.co.uk/levan/contact.zip*

10.7.2. *Influence of the mesh*

When refining the mesh, the numerical results improve in a general way. From the contact point of view, as was mentioned in section 8.9, increasing the number of nodes on the contactor surface enriches the space of virtual multipliers λ^*, and the contact laws are better satisfied. The contact surfaces correspond more closely to each other, and the contact stresses become more precise. Furthermore, it is found that the abnormal stress phenomenon occurs less.

Let us compare the results arising from two different meshes: the first is that in Figure 10.36 (or 10.42), and the second contains 2×2 more elements – namely, 6×20 elements – for the cylinder (the die block mesh remains the same).

– Without friction, the abnormal stresses, which were already tiny when using the first mesh, almost disappear when using the finer mesh.

– With friction, there always are some abnormal stresses when using the finer mesh, as was the case when using the first mesh. Their magnitudes are almost the same, but now they appear over a less extensive area than before. Indeed, they are perceptible only on the first element above the upper surface of the die block, which is an element that is twice as small than when using the first mesh.

10.8. Conclusion

The different numerical examples presented in this chapter have allowed us to validate the weak contact formulation proposed in Chapter 7, as well as the computational method described in Chapter 9, by going through many standard examples involving quasi-static contact.

The contact patch test made it possible to verify that contact stresses are transmitted correctly between arbitrary nonmatching contact interfaces.

Convergence with respect to the mesh – that is to say, the fact that the numerical results converge closer toward the theoretical solution as the mesh is further refined – has been confirmed through the Hertzian contact example.

With regard to the contact laws, we have verified (i) that the non-penetrability condition is satisfied to a very high precision (in each example), (ii) and that Coulomb's law is also satisfied accurately at every point (see the Hertzian and the extruded cylinder examples).

Stability of the formulation has been confirmed since, for each example considered, no locking phenomenon was observed nor any spurious contact pressure mode.

The rolling disk was an example of contact where large slips occur, and which admits a simple analytical solution with both large displacements and large rotations. The numerical results obtained are in very good agreement with the analytical solution and they allow us to validate the discrete expression [8.56] for the slip $\mathbf{V}_T \Delta t$, which is an essential ingredient when performing contact computations.

The examples involving the contact of two beams, the extrusion of a cylinder and the interference fit, have shown that the method presented can deal with a combination of large displacements, large strains and large slips.

The contact example involving two membranes shows that the formulation works well with small structures. In the case of two centered membranes in frictionless contact, the numerical contact pressure found agrees very well with the theoretical one.

The formulation works in the same manner regardless of the constitutive laws of the bodies considered. The aforementioned examples make use of hyperelastic laws, whereas the interference fit example makes use of an elastoplastic law.

The collection of results obtained in this chapter allow for the contact formulation in quasistatics to be validated. Only in quasistatics for the moment, since we have ignored any inertial effects in the computational scheme. The computational method and corresponding numerical tests for dynamic contact will be treated in the subsequent chapters.

11

Solution of the Dynamic Contact Problem

In Chapter 9, we saw how to solve the quasi-static contact problem. In this chapter, we are going to focus on dynamics where inertial effects are no longer negligible, and will show how to solve the system of equations [8.47] that govern the dynamic contact problem.

When it comes to its computational solution, dynamic contact/impact is a complex subject and has been intensively studied in the literature. The first section of this chapter will provide an overview of the difficulties that arise, on the one hand, from the dynamic framework and, on the other hand, from the contact, as well as the variety of approaches proposed in either small or large deformations to overcome these difficulties.

In this chapter, our aim is much more limited, as we do not seek to develop a new and more sophisticated time integration scheme. We will merely make use of the Newmark standard algorithm – which is straightforward to implement – in order to discretize (with respect to time) the system of differential equations that govern the dynamic contact problem. We will thus transform the system [8.47] into an algebraic and nonlinear system, which, in turn, can be solved using a Newton–Raphson type iterative scheme.

Through the numerical examples in Chapter 12, we will see that the Newmark time stepping algorithm is capable of producing reasonable numerical results in dynamic contact.

11.1. A brief review of the computational methods in dynamic contact

Here, we are going to provide a brief and non-exhaustive review of the existing work on dynamic contact, by distinguishing – in the order of increasing difficulty – the dynamic contact problem with small perturbations from that with large deformations.

In dynamic contact, one intrinsic difficulty is the discontinuity in velocities. In rigid-body mechanics, the creation of a kinematic constraint (i.e. the shock) gives rise to a velocity discontinuity but breaking a constraint does not produce any velocity discontinuity. The situation is more complicated in the mechanics of deformable bodies. For instance, the impact problem of two identical elastic bars shows that there is a velocity discontinuity after contact *and* after the separation (or the rebound).

The velocity of the particles on the contact surface is discontinuous when there is a sudden change in the contact status (contact or separation). Consequently, the acceleration becomes a Dirac measure during these instants. In terms of numerical computation, this means that the acceleration has very large values at these instants and the numerical velocity and displacement fields undergo unexpected oscillations; this can lead to divergence of the computational scheme and explains the necessity for special treatments to take into account any discontinuity of the velocity field at the moment of impact. Note that it is unhelpful to use any higher order scheme, since it will not necessarily give more precise results.

Another difficulty arises from friction and the use of explicit methods: in the presence of friction, convergence is difficult to obtain with an explicit method in which the tangential contact stresses are computed from the slip velocity at the preceding step (see Jean [66]).

• For dynamic contact problems *with small perturbations*, the Newmark algorithm and its variants are largely used, along with some adjustments to deal with the aforementioned velocity discontinuities.

In the small perturbations framework, Hughes et al. [64] used the Newmark algorithm to solve the *frictionless* dynamic contact/impact problem. These authors apply Newmark's method to each instant apart from two special ones, which they treat separately: the instant of impact and the instant of release (or the instant when rebound occurs). This treatment of the impact consists of updating the velocity, acceleration, and the contact force immediately after the impact; and also immediately after the rebound by using results from the wave propagation theory. This approach requires using the diagonalized mass matrix, which can be obtained from the special numerical integration methods presented in Hughes [63]. By following the idea of Hughes et al., Chen and Tsai [25] proposed another updating formula for the velocity and the acceleration, applicable to the *frictional* contact/impact problem. The advantage of this approach is that this time we use the consistent mass matrix obtained from the finite element calculation, rather than the diagonalized mass matrix which does not really carry any physical meaning. There exist several other analogous works, in which special treatments are done at the impact and rebound instants. The common thread throughout these methods is that the impact and rebound instants must be determined, and these quantities are not always straightforward to accurately calculate.

In other works, the Newmark algorithm is applied at every instant. To treat *frictional* dynamic contact with small perturbations, Chaudhary and Bathe [23] proposed to use Newmark's method with the special values for the parameters $\alpha = 1$ and $\beta = 1/2$. They showed that, in the absence of friction, the total energy and the momentums are conserved when the time step satisfies $\Delta t \leq 0.2/\omega_{min}$, where ω_{min} is the smallest natural frequency of the structure. Taylor and Papadopoulos [133] follow a different approach. First, they showed that the Newmark algorithm is likely to produce oscillations in the relative velocity between two contact surfaces. Then, they used the Newmark algorithm again for the time discretization, but, in order to circumvent the velocity oscillations, they imposed, in addition to the relation $g = 0$ for the normal contact law, two extra relations on the time derivatives of the proximity g, namely $\dot{g} = 0, \ddot{g} = 0$. Two Lagrange multipliers λ_V and λ_A, called "generalized forces", are then introduced to prescribe the constraints $\dot{g} = 0$ and $\ddot{g} = 0$. Khenous et al. [70] also used the Newmark algorithm with $\alpha = \beta = 1/2$ for time integration. He suggested that the discretized mass on the contact surface is responsible for either the energy creation or the energy lost during impact, knowing that Newmark's method with $\alpha = \beta = 1/2$ is meant to be conservative. He therefore proposed to replace this mass with another mass matrix, whose components M_{ij} are set to zero for the nodes located on the contact surfaces $S_{oc}^{(1)}$ and $S_{oc}^{(2)}$.

We can exploit other time integrators that are analogous to Newmark's one. Czekanski and Meguid [31] solved the dynamic contact/impact problem with the help of the generalized-α method of Chung and Hulbert [26], combined with mathematical programming techniques. To obtain the optimal values of the parameters, α_m and α_f, these authors solved the generic problem of two dissimilar stiffness-free masses in contact, and then calculated the values of α_m, α_f so that the energy and the momentums are conserved. The same values were then reused in other, more complicated problems. These authors admitted that this procedure generated a small loss of energy during the impact. Sometimes, it is necessary to numerically dissipate the excessive higher modes (spurious numerical oscillations) in implicit dynamics. In this case, the authors did not use the values for the parameters found previously, but rather different values determined by a more complicated criterion.

• Dynamic contact *with finite deformations* combines the difficulties which arise from two types of problem: (i) difficulties in nonlinear dynamics without contact; (ii) difficulties in dynamic contact with small perturbations, which we have just seen above. In particular, the difficulty arising from friction, which already exists with small deformations, just becomes more penalizing with finite deformations. These difficulties indicate that special treatments are required to take account of the discontinuity in the velocity field at the moment of impact, as is the case for dynamic contact with small perturbations.

Temporal integration schemes that deal with finite deformations are hard to achieve. Newmark's methods can provide acceptable results, but other methods are necessary, should we want greater precision. Among the proposed methods, Moreau's scheme [97] is an alternative one which is sufficiently straightforward yet robust. It works as a first-order scheme and is particularly adapted to dynamic contact problems, since it automatically incorporates any velocity discontinuity in the formulation, and it does not involve the acceleration variable which becomes a Dirac measure during impact. Good results have been obtained in some previous studies, such as Vola et al. [134], Jean [66], and Feng et al. [38] by applying this algorithm.

• *Energy methods* constitute another approach to solving the dynamic contact problem. When the consistency and stability of a time integration scheme cannot be proven – which is particularly true with finite deformations – we turn instead to using energy methods, where now it is a question of the energy and/or momentums conservation. These approaches are specific to mechanics, since there is no equivalent in any other field of physics. Energy conservation is related to the stability.

The mechanical energy of the structure, which is a scalar quantity, is often used to evaluate the quality of a numerical method. It is known that mechanical energy is conserved in the case of a so-called conservative system: if the material is elastic and the external load onto the structure is derivable from a potential, then the mechanical energy of the structure remains constant during motion. A good computational scheme should satisfy this property, that is, it should generate a constant mechanical energy when applied to a conservative system. On the other hand, the momentums are vector quantities – more complicated than energy – and, just like energy, we know that they are conserved under certain conditions. Hence, they can also be used to assess the quality of a numerical method. The idea, therefore, is to design a numerical scheme – which is said to be *conservative* – which conserves the energy and/or the momentums or, failing that, a scheme such that these quantities remain *bounded* over the course of time (in the absence of any external loads, a decreasing energy makes sense physically, whereas an increasing energy would be very surprising).

In the framework of dynamic contact with finite deformations, many authors have proposed algorithms which conserve the total energy of the system with persistent contact (in the absence of friction), or which return, after the rebound, the original energy level prior to the impact. Here, we have used the expression "persistent contact" to emphasize that contact does not take place at a single and isolated time instant, but rather it endures for a specific time period ($g = 0$ and also $\dot{g} = 0$). Laursen and Chawla [78] generalized the work of Simo and Tarnow [122] to the *frictionless* dynamic contact/impact problem. They introduced the notion of algorithmic proximity $\tilde{g}_{n-\frac{1}{2}}$ and the corresponding algorithmic contact force. Their algorithm allows for the energy and momentums to be conserved during any persistent contact. Afterward, the *frictional* dynamic contact/impact problem was solved in a similar way by the same authors [24], who introduced the algorithmic slip $\tilde{\mathbf{V}}_{T,n-1/2}\Delta t$. Energy is conserved in the stick phase, whereas it diminishes in the slip one where, of course, energy loss corresponds to the physical energy dissipation throughout the slip. As we observe in the numerical results that there is always some interpenetration during contact, Love and Laursen [90] subsequently generalized the idea in [79] to the *frictional* contact problem by replacing the equation $E_n - E_{n-1} = 0$ with $E_n - E_{n-1} = -\Delta t \mathcal{D}^c_{n-1/2}$, where $\mathcal{D}^c_{n-1/2}$ is the dissipated energy due to friction. Amero and Petocz [3] chose a slightly different approach to that proposed in [78]. Rather than seeking to conserve the energy during persistent contact, instead they conceived an algorithm such that the energy lost during persistent contact is entirely regained after the rebound, in such a way that the energy returns to its original level prior to impact. Following another route, Betsch generalized the discrete null space method that he proposed for discrete mechanical systems [14] to the *frictionless* contact problem [15].

Energy methods seek either to conserve or bound the energy, and all of them are based upon the energy balance to formulate an algorithm which either conserves or dissipates energy. They do not take into account the velocity discontinuity during the impact, nor the fact that acceleration becomes a Dirac measure at this instant. These methods provide very satisfying results inasmuch as the energy is bounded by construction and the algorithm is stable. However, these formulations are often rather complicated, and are more difficult to implement than standard time integration schemes.

- The choice of computational method for dynamic contact problems is an important question, with clear repercussions on the quality of the results. The previous brief review shows that dynamic contact problems with large deformations remain a difficult subject that has not yet been solved completely to-date.

In this chapter, we choose to carry out the time discretization with the help of the Newmark algorithm. We will see that this algorithm – which is straightforward to implement – is sufficient to obtain results that are quite acceptable in the first approach. The objective here is to show that the weak formulation [7.21] can also be applied to dynamic contact problems, rather than to test the performance of a time stepping scheme, nor is it to obtain the best scheme possible. Moreau's scheme can also be implemented in parallel with Newmark's one, provided that certain reasonable adjustments are taken inside the computational code.

11.2. Solution of the dynamic contact problem via Newmark's algorithm

In the next section, we discretize, with respect to time, the equation of motion $[8.47]_1$ with the help of Newmark's algorithm. In doing so, we transform the nonlinear differential system [8.47] of the dynamic contact problem into a nonlinear algebraic system, which, in turn, can be solved using a Newton–Raphson type iterative scheme. In his exposition on the time integration of second-order differential systems, Curnier [29] provided an elegant synthesis of Newmark's family which unifies the implicit/explicit versions, as well as the solutions for the displacement, velocity, and acceleration via a simple algorithm. We are going to adapt Curnier's formulation to the dynamic contact problem.

The system [8.47] of equations for the dynamic contact problem is to be solved with the initial conditions:

at the initial instant $t_0 = 0$, $\quad \mathbf{U}_0 \equiv \mathbf{U}(t_0) = \mathbf{0} \quad \mathbf{V}_0 \equiv \dot{\mathbf{U}}(t_0)$ is prescribed $\quad$ and $\quad \mathbf{\Lambda}(t_0) = \mathbf{0}$

The initial values for the multipliers $\boldsymbol{\Lambda}$ are zero, since we assume that each body studied is not in contact with another at the initial time instant.

• If we were to solve the problem analytically, we would search for a mapping $t \mapsto (\mathbf{U}(t), \boldsymbol{\Lambda}(t))$ which satisfies [8.47] at every time instant $t \in [0, t_{max}]$. Here, since we aim to solve the problem numerically, such a mapping is not realizable. The numerical solution of the system [8.47] should be obtained *incrementally* (i.e. obtained step-by-step). We will seek therefore to determine *the values* of $(\mathbf{U}, \mathbf{V}, \mathbf{A}, \boldsymbol{\Lambda})$ (where $\mathbf{V} \equiv \dot{\mathbf{U}}$ is the velocity and $\mathbf{A} \equiv \ddot{\mathbf{U}}$ the acceleration) for multiple time instants $t_1, \dots, t_n$ in succession:

$$(\mathbf{U}(t_1), \mathbf{V}(t_1), \mathbf{A}(t_1), \boldsymbol{\Lambda}(t_1)), \dots, (\mathbf{U}(t_n), \mathbf{V}(t_n), \mathbf{A}(t_n), \boldsymbol{\Lambda}(t_n)), \dots$$

Thus, the computational solution is performed after several steps. The value of the solution at increment $n-1$ is then used to calculate that for the increment n. The overall calculation takes place as described in the following loop:

Incremental loop $n \geq 1$

- Calculate $t_n = t_{n-1} + h$, where $\boxed{h = t_n - t_{n-1}}$ is *the time step*, which is a given constant > 0.
- Knowing:

  (i) the solutions for the previous instant t_{n-1}: displacement $\mathbf{U}(t_{n-1})$, velocity $\mathbf{V}(t_{n-1})$, acceleration $\mathbf{A}(t_{n-1})$ and multiplier $\boldsymbol{\Lambda}(t_{n-1})$ (plus any additional plastic quantities);

  (ii) the external load applied on the mechanical system at the instant t_n or, more precisely, the values of the dead loads and the values of the (follower) pressures.

  Calculate the solution for the instant t_n: displacement $\mathbf{U}(t_n)$, velocity $\mathbf{V}(t_n)$, acceleration $\mathbf{A}(t_n)$ and multiplier $\boldsymbol{\Lambda}(t_n)$ (plus any additional plastic quantities).

  This calculation requires a nonlinear differential system to be solved and, to do this, we will use Newmark's time integration algorithm, combined with the iterative Newton–Raphson scheme.

End of the incremental loop

Therefore, we will construct numerically – beginning with known initial values $\mathbf{U}_0, \mathbf{V}_0$ – a sequence of values for the solution $(\mathbf{U}, \mathbf{V}, \mathbf{A}, \boldsymbol{\Lambda})$ at successive instants (see Figure 11.1):

t_0	$t_1 = t_0 + h$	$t_2 = t_0 + 2h$	$\cdots$	$t_{n-1} = t_0 + (n-1)h$	$t_n = t_0 + nh$	$\cdots$
$\mathbf{U}(t_0)$	$\mathbf{U}(t_1)$	$\mathbf{U}(t_2)$	$\cdots$	$\mathbf{U}(t_{n-1})$	$\mathbf{U}(t_n)$	$\cdots$
$\mathbf{V}(t_0)$	$\mathbf{V}(t_1)$	$\mathbf{V}(t_2)$	$\cdots$	$\mathbf{V}(t_{n-1})$	$\mathbf{V}(t_n)$	$\cdots$
$\mathbf{A}(t_0)$	$\mathbf{A}(t_1)$	$\mathbf{A}(t_2)$	$\cdots$	$\mathbf{A}(t_{n-1})$	$\mathbf{A}(t_n)$	$\cdots$
$\boldsymbol{\Lambda}(t_0) = \mathbf{0}$	$\boldsymbol{\Lambda}(t_1)$	$\boldsymbol{\Lambda}(t_2)$	$\cdots$	$\boldsymbol{\Lambda}(t_{n-1})$	$\boldsymbol{\Lambda}(t_n)$	$\cdots$

Note that the velocity $\mathbf{V}$ does not appear in the equation system [8.47] of the dynamic contact problem, but it will appear in the kinematic equations later on, hence it is indeed part of the formulation of the problem, in the same way that $\mathbf{U}, \mathbf{A}$, and $\boldsymbol{\Lambda}$ are.

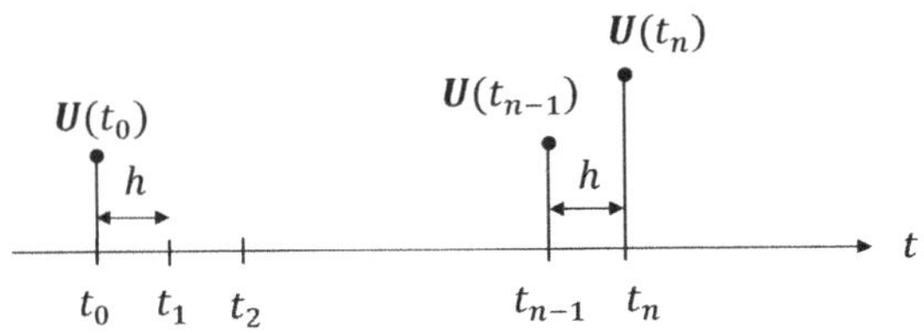

Figure 11.1. *Solution **U** at distinct discrete time instants*

NOTATION.– For the sake of brevity, the values at a given time instant t_n, $n = 0, 1, 2, \ldots$ will be denoted using the index n:

$$\forall n = 0, 1, 2, \ldots, \quad \boxed{\mathbf{U}_n \equiv \mathbf{U}(t_n) \quad \mathbf{V}_n \equiv \dot{\mathbf{U}}(t_n) \quad \mathbf{A}_n \equiv \ddot{\mathbf{U}}(t_n) \quad \mathbf{\Lambda}_n \equiv \mathbf{\Lambda}(t_n)}$$

The same applies to any other possible plastic quantity involved in the problem.

11.2.1. *Initializing the incremental loop: the vectors U_0, V_0, A_0 and Λ_0*

For each step $n \geq 1$, we have assumed that the values of $\mathbf{U}_{n-1}, \mathbf{V}_{n-1}, \mathbf{A}_{n-1}$ and $\mathbf{\Lambda}_{n-1}$ are known.

1. If $n > 1$, that is, $n-1 > 0$, then $(\mathbf{U}_{n-1}, \mathbf{V}_{n-1}, \mathbf{A}_{n-1}, \mathbf{\Lambda}_{n-1})$ is simply the value of the quadruple $(\mathbf{U}, \mathbf{V}, \mathbf{A}, \mathbf{\Lambda})$ obtained at the previous step $n - 1 > 0$.

2. If $n = 1$, then $(\mathbf{U}_{n-1}, \mathbf{V}_{n-1}, \mathbf{A}_{n-1}, \mathbf{\Lambda}_{n-1}) = (\mathbf{U}_0, \mathbf{V}_0, \mathbf{A}_0, \mathbf{\Lambda}_0)$ is the initial value of $(\mathbf{U}, \mathbf{V}, \mathbf{A}, \mathbf{\Lambda})$, necessary for initializing the incremental loop.

   The initial displacement $\mathbf{U}_0$ is zero, the initial velocity $\mathbf{V}_0$ is prescribed and the initial multiplier $\mathbf{\Lambda}_0$ is zero. As for the initial acceleration $\mathbf{A}_0$, it is computed from $\mathbf{U}_0$ and $\mathbf{V}_0$ by writing the equation of motion $[8.47]_1$ at t_0:

   $$\mathbf{M}.\ddot{\mathbf{U}}_0 + \mathbf{\Psi}(\mathbf{U}_0) = \mathbf{\Phi}(\mathbf{U}_0, t_0) + \mathbf{\Phi}_{\text{contact}}(\mathbf{U}_0, \mathbf{\Lambda}_0)$$

   As we have assumed that the bodies being studied are not in contact at the initial time instant, we get $\mathbf{\Phi}_{\text{contact}}(\mathbf{U}_0, \mathbf{\Lambda}_0) = \mathbf{0}$ (by applying [9.86], which is still valid here in dynamics). Consequently:

   $$\boxed{\mathbf{A}_0 = \mathbf{M}^{-1}.(\mathbf{\Phi}(\mathbf{U}_0, t_0) - \mathbf{\Psi}(\mathbf{U}_0))} \qquad [11.1]$$

11.2.2. *Calculation for a step $n \geq 1$: calculating U_n, V_n, A_n, Λ_n*

Let us consider a given computation step $n \geq 1$ and determine the value $(\mathbf{U}_n, \mathbf{V}_n, \mathbf{A}_n, \mathbf{\Lambda}_n)$ of the solution $(\mathbf{U}, \mathbf{V}, \mathbf{A}, \mathbf{\Lambda})$ at time instant t_n. In demanding that equation [8.47] must be satisfied at the time instant t_n, we obtain

$$\begin{aligned} &\mathbf{M}\mathbf{A}_n + \mathbf{\Psi}(\mathbf{U}_n) = \mathbf{\Phi}(\mathbf{U}_n, t_n) + \mathbf{\Phi}_{\text{contact}}(\mathbf{U}_n, \mathbf{\Lambda}_n) \\ &\mathbf{R}_{\Lambda}(\mathbf{U}_n, \mathbf{\Lambda}_n) = \mathbf{0} \end{aligned} \qquad [11.2]$$

According to Newmark's view, we must consider the previous system as two *algebraic* equations for the four independent unknowns $\mathbf{U}_n, \mathbf{V}_n, \mathbf{A}_n$ and $\mathbf{\Lambda}_n$. Rather than two functions $t \mapsto \mathbf{U}(t)$ and $t \mapsto \mathbf{\Lambda}(t)$, we look instead for the four unknowns $\mathbf{U}_n, \mathbf{V}_n, \mathbf{A}_n, \mathbf{\Lambda}_n$ at each time instant t_n.

• Since the system [11.2] provides only two equations for four unknowns, Newmark's idea was to adjoin two additional equations of finite-difference type onto the preceding ones, which are called *kinematic equations*, and which provide an approximation for the displacement $\mathbf{U}_n$ and for the velocity $\mathbf{V}_n$ at the instant t_n:

$$\begin{aligned} \mathbf{U}_n &= \mathbf{U}_{n-1} + (1-\alpha)h\mathbf{V}_{n-1} + \alpha h\mathbf{V}_n + (\frac{1}{2} - \alpha)h^2\mathbf{A}_{n-1} \\ \mathbf{V}_n &= \mathbf{V}_{n-1} + (1-\beta)h\mathbf{A}_{n-1} + \beta h\mathbf{A}_n \end{aligned} \qquad [11.3]$$

where α and β are certain constant parameters chosen beforehand.

The relations [11.2] and [11.3] constitute a system of four *algebraic* equations with the four unknowns $\mathbf{U}_n$, $\mathbf{V}_n$, $\mathbf{A}_n$ and $\mathbf{\Lambda}_n$.

The two kinematic equations [11.3] are linear but, as the system [11.2] is nonlinear, the set of equations is nonlinear globally and we are going to solve it by means of the iterative Newton–Raphson scheme.

A simple idea is to conserve just a single unknown (called the principal kinematic unknown) from the three kinematic unknowns $\mathbf{U}_n, \mathbf{V}_n, \mathbf{A}_n$, before eliminating the other two (called the secondary kinematic unknowns):

- Exploit the two kinematic equations [11.3], which are linear, in order to express the two secondary kinematic unknowns as functions of the principal kinematic unknown.
- With the help of these expressions, recast the system [11.3] for the contact problem as a nonlinear one involving the principal kinematic unknown and the multiplier, which we can solve using the Newton–Raphson scheme.

However, this procedure leads to rather complicated expressions. Furthermore, these expressions will differ depending on which principal kinematic unknown was chosen. Consequently, the expression for the tangent matrix and the update formulae for each iteration will differ, depending on which principal kinematic unknown was chosen. There is no unified formula, depending on whether we take $\mathbf{U}_n$, $\mathbf{V}_n$, or $\mathbf{A}_n$ to be the principal kinematic unknown.

We will present a slightly different procedure now, which will lead to very simple expressions and make it possible to unify the three types of solution (in $\mathbf{U}_n$, $\mathbf{V}_n$, or $\mathbf{A}_n$) within a single formalism in an elegant way, and which also gives rise to a simple final algorithm.

• Let us solve the system [11.2] and [11.3] using the iterative Newton–Raphson scheme. Suppose that we are at the iteration $i \geq 1$ of step n and that, from the previous iteration $i-1$, we obtained an approximation $(\mathbf{U}_n^{i-1}, \mathbf{V}_n^{i-1}, \mathbf{A}_n^{i-1}, \mathbf{\Lambda}_n^{i-1})$ for the desired solution *such that* $(\mathbf{U}_n^{i-1}, \mathbf{V}_n^{i-1}, \mathbf{A}_n^{i-1})$ *satisfies the kinematic equations* [11.3]. We do not demand that $(\mathbf{U}_n^{i-1}, \mathbf{V}_n^{i-1}, \mathbf{A}_n^{i-1}, \mathbf{\Lambda}_n^{i-1})$ satisfies the system [11.2] because, if this was the case, then the quadruple would be a solution of the step n and we could just continue onto the next step.

We look for a new approximation $(\mathbf{U}_n^i, \mathbf{V}_n^i, \mathbf{A}_n^i, \mathbf{\Lambda}_n^i)$ of the solution that satisfies the following four equations [11.2]–[11.3]:

$$
\begin{aligned}
&\mathbf{M}.\mathbf{A}_n^i + \mathbf{\Psi}(\mathbf{U}_n^i) = \mathbf{\Phi}(\mathbf{U}_n^i, t_n) + \mathbf{\Phi}_{\text{contact}}(\mathbf{U}_n^i, \mathbf{\Lambda}_n^i)\\
&\mathbf{R}_\Lambda(\mathbf{U}_n^i, \mathbf{\Lambda}_n^i) = \mathbf{0}\\
&\mathbf{U}_n^i = \mathbf{U}_{n-1} + (1-\alpha)h\mathbf{V}_{n-1} + \alpha h\mathbf{V}_n^i + (\frac{1}{2} - \alpha)h^2\mathbf{A}_{n-1}\\
&\mathbf{V}_n^i = \mathbf{V}_{n-1} + (1-\beta)h\mathbf{A}_{n-1} + \beta h\mathbf{A}_n^i
\end{aligned}
\qquad [11.4]
$$

Let us introduce the quantities $\delta\mathbf{U}, \delta\mathbf{V}, \delta\mathbf{A}$ and $\delta\mathbf{\Lambda}$ defined as follows:

$$
\forall i \geq 1, \boxed{\begin{aligned}
\delta\mathbf{U} &= \mathbf{U}_n^i - \mathbf{U}_n^{i-1}\\
\delta\mathbf{V} &= \mathbf{V}_n^i - \mathbf{V}_n^{i-1}\\
\delta\mathbf{A} &= \mathbf{A}_n^i - \mathbf{A}_n^{i-1}\\
\delta\mathbf{\Lambda} &= \mathbf{\Lambda}_n^i - \mathbf{\Lambda}_n^{i-1}
\end{aligned}}
\qquad [11.5]
$$

Knowing $(\mathbf{U}_n^i, \mathbf{V}_n^i, \mathbf{A}_n^i, \mathbf{\Lambda}_n^i)$ is equivalent to knowing $(\delta\mathbf{U}, \delta\mathbf{V}, \delta\mathbf{A}, \delta\mathbf{\Lambda})$. The notation [11.5] allows for [11.4] to be transformed into an equivalent system with the unknowns $\delta\mathbf{U}, \delta\mathbf{V}, \delta\mathbf{A}$ and $\delta\mathbf{\Lambda}$:

$$
\begin{aligned}
&\mathbf{M}.(\mathbf{A}_n^{i-1} + \delta\mathbf{A}) + \mathbf{\Psi}(\mathbf{U}_n^{i-1} + \delta\mathbf{U}) = \mathbf{\Phi}(\mathbf{U}_n^{i-1} + \delta\mathbf{U}, t_n) + \mathbf{\Phi}_{\text{contact}}(\mathbf{U}_n^{i-1} + \delta\mathbf{U}, \mathbf{\Lambda}_n^{i-1} + \delta\mathbf{\Lambda})\\
&\mathbf{R}_\Lambda(\mathbf{U}_n^{i-1} + \delta\mathbf{U}, \mathbf{\Lambda}_n^{i-1} + \delta\mathbf{\Lambda}) = \mathbf{0}\\
&\mathbf{U}_n^{i-1} + \delta\mathbf{U} = \mathbf{U}_{n-1} + (1-\alpha)h\mathbf{V}_{n-1} + \alpha h(\mathbf{V}_n^{i-1} + \delta\mathbf{V}) + (\frac{1}{2} - \alpha)h^2\mathbf{A}_{n-1}\\
&\mathbf{V}_n^{i-1} + \delta\mathbf{V} = \mathbf{V}_{n-1} + (1-\beta)h\mathbf{A}_{n-1} + \beta h(\mathbf{A}_n^{i-1} + \delta\mathbf{A})
\end{aligned}
\qquad [11.6]
$$

According to the Newton–Raphson scheme, equations [11.6] are expanded into a Taylor series up to first order about the solution $(\mathbf{U}_n^{i-1}, \mathbf{V}_n^{i-1}, \mathbf{A}_n^{i-1}, \mathbf{\Lambda}_n^{i-1})$ that was obtained from the previous iteration. The force vectors and the residual contact law vector in $[11.6]_{1-2}$ give

$$
\begin{aligned}
\mathbf{\Psi}(\mathbf{U}_n^{i-1} + \delta\mathbf{U}) - \mathbf{\Phi}(\mathbf{U}_n^{i-1} + \delta\mathbf{U}, t_n) &= \mathbf{\Psi}(\mathbf{U}_n^{i-1}) - \mathbf{\Phi}(\mathbf{U}_n^{i-1}, t_n) +\\
&\quad \frac{\partial \mathbf{\Psi}}{\partial \mathbf{U}}(\mathbf{U}_n^{i-1}).\delta\mathbf{U} - \frac{\partial \mathbf{\Phi}}{\partial \mathbf{U}}(\mathbf{U}_n^{i-1}, t_n).\delta\mathbf{U} + o(|\delta\mathbf{U}|)\\
\mathbf{\Phi}_{\text{contact}}(\mathbf{U}_n^{i-1} + \delta\mathbf{U}, \mathbf{\Lambda}_n^{i-1} + \delta\mathbf{\Lambda}) &= \mathbf{\Phi}_{\text{contact}}(\mathbf{U}_n^{i-1}, \mathbf{\Lambda}_n^{i-1}) + \frac{\partial \mathbf{\Phi}_{\text{contact}}}{\partial \mathbf{U}}(\mathbf{U}_n^{i-1}, \mathbf{\Lambda}_n^{i-1}).\delta\mathbf{U} +\\
&\quad \frac{\partial \mathbf{\Phi}_{\text{contact}}}{\partial \mathbf{\Lambda}}(\mathbf{U}_n^{i-1}, \mathbf{\Lambda}_n^{i-1}).\delta\mathbf{\Lambda} + o(|\delta\mathbf{U}|, |\delta\mathbf{\Lambda}|)\\
\mathbf{R}_\Lambda(\mathbf{U}_n^{i-1} + \delta\mathbf{U}, \mathbf{\Lambda}_n^{i-1} + \delta\mathbf{\Lambda}) &= \mathbf{R}_\Lambda(\mathbf{U}_n^{i-1}, \mathbf{\Lambda}_n^{i-1}) + \frac{\partial \mathbf{R}_\Lambda}{\partial \mathbf{U}}(\mathbf{U}_n^{i-1}, \mathbf{\Lambda}_n^{i-1}).\delta\mathbf{U} +\\
&\quad \frac{\partial \mathbf{R}_\Lambda}{\partial \mathbf{\Lambda}}(\mathbf{U}_n^{i-1}, \mathbf{\Lambda}_n^{i-1}).\delta\mathbf{\Lambda} + o(|\delta\mathbf{U}|, |\delta\mathbf{\Lambda}|)
\end{aligned}
$$

Or

$$
\begin{aligned}
\mathbf{\Psi}(\mathbf{U}_n^{i-1} + \delta\mathbf{U}) - \mathbf{\Phi}(\mathbf{U}_n^{i-1} + \delta\mathbf{U}, t_n) &= \mathbf{\Psi}(\mathbf{U}_n^{i-1}) - \mathbf{\Phi}(\mathbf{U}_n^{i-1}, t_n) + (\mathbf{K}_{classical})_n^{i-1}.\delta\mathbf{U} + o(|\delta\mathbf{U}|)\\
\mathbf{\Phi}_{\text{contact}}(\mathbf{U}_n^{i-1} + \delta\mathbf{U}, \mathbf{\Lambda}_n^{i-1} + \delta\mathbf{\Lambda}) &= \mathbf{\Phi}_{\text{contact}}(\mathbf{U}_n^{i-1}, \mathbf{\Lambda}_n^{i-1}) - (\mathbf{K}_{UU})_n^{i-1}.\delta\mathbf{U} - (\mathbf{K}_{U\Lambda})_n^{i-1}.\delta\mathbf{\Lambda}\\
&\quad + o(|\delta\mathbf{U}|, |\delta\mathbf{\Lambda}|)\\
\mathbf{R}_\Lambda(\mathbf{U}_n^{i-1} + \delta\mathbf{U}, \mathbf{\Lambda}_n^{i-1} + \delta\mathbf{\Lambda}) &= \mathbf{R}_\Lambda(\mathbf{U}_n^{i-1}, \mathbf{\Lambda}_n^{i-1}) + (\mathbf{K}_{\Lambda U})_n^{i-1}.\delta\mathbf{U} + (\mathbf{K}_{\Lambda\Lambda})_n^{i-1}.\delta\mathbf{\Lambda}\\
&\quad + o(|\delta\mathbf{U}|, |\delta\mathbf{\Lambda}|)
\end{aligned}
\qquad [11.7]
$$

In the previous expression, the tangent matrix $\mathbf{K}_{classical}$ and the contact tangent matrices $\mathbf{K}_{UU}, \mathbf{K}_{U\Lambda}, \mathbf{K}_{\Lambda U}, \mathbf{K}_{\Lambda\Lambda}$ are defined by the same relations [9.10] and [9.11] as in statics.

Moreover, the indices $(\cdots)_n^{i-1}$ enclosing a matrix indicate that it had been calculated at $(\mathbf{U}_n^{i-1}, \mathbf{\Lambda}_n^{i-1}, t_n)$:

NOTATION.–

$$\boxed{(\mathbf{K}_{classical})_n^{i-1} \equiv \mathbf{K}_{classical}(\mathbf{U}_n^{i-1}, t_n)}$$

$$\begin{array}{|l|l|} (\mathbf{K}_{UU})_n^{i-1} \equiv \mathbf{K}_{UU}(\mathbf{U}_n^{i-1}, \mathbf{\Lambda}_n^{i-1}) & (\mathbf{K}_{U\Lambda})_n^{i-1} \equiv \mathbf{K}_{U\Lambda}(\mathbf{U}_n^{i-1}, \mathbf{\Lambda}_n^{i-1}) \\ (\mathbf{K}_{\Lambda U})_n^{i-1} \equiv \mathbf{K}_{\Lambda U}(\mathbf{U}_n^{i-1}, \mathbf{\Lambda}_n^{i-1}) & (\mathbf{K}_{\Lambda\Lambda})_n^{i-1} \equiv \mathbf{K}_{\Lambda\Lambda}(\mathbf{U}_n^{i-1}, \mathbf{\Lambda}_n^{i-1}) \end{array}$$

Given that the term $\mathbf{M}.(\mathbf{A}_n^{i-1} + \delta\mathbf{A})$ in $[11.6]_1$ as well as relations $[11.6]_{3\text{-}4}$ are already linear in $\delta\mathbf{U}$, $\delta\mathbf{V}$ and $\delta\mathbf{A}$, their first-order Taylor expansions produce the same expressions. Regarding the terms in question, the Taylor expansion provides an exact result, rather than just an approximation.

Furthermore, let us prove the following property:

THEOREM.– The two equations $[11.6]_{3\text{-}4}$ simplify to $\delta\mathbf{U} = \alpha h\, \delta\mathbf{V}$ and $\delta\mathbf{V} = \beta h\, \delta\mathbf{A}$.

PROOF. Indeed, since $(\mathbf{U}_n^{i-1}, \mathbf{V}_n^{i-1}, \mathbf{A}_n^{i-1})$ is assumed to satisfy the kinematic equations [11.3], we have

$$\begin{aligned} \mathbf{U}_n^{i-1} &= \mathbf{U}_{n-1} + (1-\alpha)h\mathbf{V}_{n-1} + \alpha h\mathbf{V}_n^{i-1} + (\tfrac{1}{2} - \alpha)h^2\mathbf{A}_{n-1} \\ \mathbf{V}_n^{i-1} &= \mathbf{V}_{n-1} + (1-\beta)h\mathbf{A}_{n-1} + \beta h\mathbf{A}_n^{i-1} \end{aligned} \qquad [11.8]$$

By substituting these relations to $[11.6]_{3\text{-}4}$, we obtain the desired result, QED.

We could have obtained the same result by subtracting the kinematic equations [11.3] written at iteration $i-1$ from those at iteration i, namely, [11.8] and $[11.4]_{3\text{-}4}$. ■

Finally, by taking the previous results into account and by ignoring any first order or higher in [11.7], the system [11.6] becomes four linear equations to be solved at the iteration i for the four unknowns $\delta\mathbf{U}, \delta\mathbf{V}, \delta\mathbf{A}$ and $\delta\mathbf{\Lambda}$:

$$\begin{aligned} &\mathbf{M}.\delta\mathbf{A} + (\mathbf{K}_{classical} + \mathbf{K}_{UU})_n^{i-1}.\delta\mathbf{U} + (\mathbf{K}_{U\Lambda})_n^{i-1}.\delta\mathbf{\Lambda} = -\mathbf{R}_n^{i-1} \\ &(\mathbf{K}_{\Lambda U})_n^{i-1}.\delta\mathbf{U} + (\mathbf{K}_{\Lambda\Lambda})_n^{i-1}.\delta\mathbf{\Lambda} = -\mathbf{R}_{\Lambda,n}^{i-1} \\ &\delta\mathbf{U} = \alpha h\, \delta\mathbf{V} \\ &\delta\mathbf{V} = \beta h\, \delta\mathbf{A} \end{aligned} \qquad [11.9]$$

where *the residuals* $\mathbf{R}_n^{i-1}$ and $\mathbf{R}_{\Lambda,n}^{i-1}$ are defined by

NOTATION.–

$$\boxed{\begin{aligned} \mathbf{R}_n^{i-1} &\equiv \mathbf{M}.\mathbf{A}_n^{i-1} + \mathbf{\Psi}(\mathbf{U}_n^{i-1}) - \mathbf{\Phi}(\mathbf{U}_n^{i-1}, t_n) - \mathbf{\Phi}_{\text{contact}}(\mathbf{U}_n^{i-1}, \mathbf{\Lambda}_n^{i-1}) \\ \mathbf{R}_{\Lambda,n}^{i-1} &\equiv \mathbf{R}_\Lambda(\mathbf{U}_n^{i-1}, \mathbf{\Lambda}_n^{i-1}) \end{aligned}} \qquad [11.10]$$

We are led to the system [11.9], which is linear, and whose last two equations are very simple. By using $[11.9]_{3\text{-}4}$, we can choose any two unknowns out of the three $\delta\mathbf{U}, \delta\mathbf{V}, \delta\mathbf{A}$ to be eliminated, and we easily arrive at just two equations with two unknowns: one out of the three $(\delta\mathbf{U}, \delta\mathbf{V}, \delta\mathbf{A})$ and $\delta\mathbf{\Lambda}$. Thus, there are three possible ways to solve equations [11.9]:

Definition.– Depending on whether we keep $\delta\mathbf{U}, \delta\mathbf{V}$, or $\delta\mathbf{A}$, we say that one solves for the displacement, for the velocity, or for the acceleration.

– *Solution for the displacement*: keep $\delta\mathbf{U}$ and eliminate $\delta\mathbf{V}$ and $\delta\mathbf{A}$ by $\delta\mathbf{V} = \frac{\delta\mathbf{U}}{\alpha h}, \delta\mathbf{A} = \frac{\delta\mathbf{U}}{\alpha\beta h^2}$.

– *Solution for the velocity*: keep $\delta\mathbf{V}$ and eliminate $\delta\mathbf{U}$ and $\delta\mathbf{A}$ by $\delta\mathbf{U} = \alpha h \delta\mathbf{V}, \delta\mathbf{A} = \frac{\delta\mathbf{V}}{\beta h}$.

– *Solution for the acceleration*: keep $\delta\mathbf{A}$ and eliminate $\delta\mathbf{U}$ and $\delta\mathbf{V}$ by $\delta\mathbf{U} = \alpha\beta h^2 \delta\mathbf{A}, \delta\mathbf{V} = \beta h \delta\mathbf{A}$.

To unify the three solution methods, we say that one solves for $\delta\mathbf{W}$, where $\delta\mathbf{W}$ is equal to either $\delta\mathbf{U}$, $\delta\mathbf{V}$ or $\delta\mathbf{A}$, depending on which solution type was chosen:

– *Solution for the displacement*: keep $\delta\mathbf{W} = \delta\mathbf{U}$ and eliminate $\delta\mathbf{V}$ and$\delta\mathbf{A}$ by $\delta\mathbf{V} = \frac{\delta\mathbf{W}}{\alpha h}, \delta\mathbf{A} = \frac{\delta\mathbf{W}}{\alpha\beta h^2}$.

– *Solution for the velocity*: keep $\delta\mathbf{W} = \delta\mathbf{V}$ and eliminate $\delta\mathbf{U}$ and $\delta\mathbf{A}$ by $\delta\mathbf{U} = \alpha h \delta\mathbf{W}, \delta\mathbf{A} = \frac{\delta\mathbf{W}}{\beta h}$.

– *Solution for the acceleration*: keep $\delta\mathbf{W} = \delta\mathbf{A}$ and eliminate $\delta\mathbf{U}$ and $\delta\mathbf{V}$ by $\delta\mathbf{U} = \alpha\beta h^2 \delta\mathbf{W}, \delta\mathbf{V} = \beta h \delta\mathbf{W}$.

Thus, in any case, $\delta\mathbf{U}$, $\delta\mathbf{V}$ and $\delta\mathbf{A}$ are related to $\delta\mathbf{W}$ by

$$\boxed{\begin{array}{l} \delta\mathbf{A} = m\,\delta\mathbf{W} \\ \delta\mathbf{V} = c\,\delta\mathbf{W} \\ \delta\mathbf{U} = k\,\delta\mathbf{W} \end{array}} \qquad [11.11]$$

where m, c, k are coefficients which change, depending on which type of solution was chosen:

Solution	Coefficients m, c, k		
For the displacement	$m = \dfrac{1}{\alpha\beta h^2}$	$c = \dfrac{1}{\alpha h}$	$k = 1$
For the velocity	$m = \dfrac{1}{\beta h}$	$c = 1$	$k = \alpha h$
For the acceleration	$m = 1$	$c = \beta h$	$k = \alpha\beta h^2$

[11.12]

By substituting [11.11] into $[11.9]_{1\text{-}2}$, we obtain two equations with two unknowns $\delta\mathbf{W}$ and $\delta\mathbf{\Lambda}$ to be solved at iteration i:

$$\begin{array}{rcl} [m\mathbf{M} + k(\mathbf{K}_{classical} + \mathbf{K}_{UU})_n^{i-1}] \,.\, \delta\mathbf{W} + (\mathbf{K}_{U\Lambda})_n^{i-1} \,.\, \delta\mathbf{\Lambda} & = & -\mathbf{R}_n^{i-1} \\ k(\mathbf{K}_{\Lambda U})_n^{i-1} \,.\, \delta\mathbf{W} + (\mathbf{K}_{\Lambda\Lambda})_n^{i-1} \,.\, \delta\mathbf{\Lambda} & = & -\mathbf{R}_{\Lambda,n}^{i-1} \end{array}$$

or, equivalently, by writing the matrices without the indices $(\cdots)_n^{i-1}$ for brevity:

$$\boxed{\left[\begin{array}{c|c} \mathbf{S} \equiv m\,\mathbf{M} + k\,(\mathbf{K}_{classical} + \mathbf{K}_{UU}) & \mathbf{K}_{U\Lambda} \\ \hline k\,\mathbf{K}_{\Lambda U} & \mathbf{K}_{\Lambda\Lambda} \end{array}\right] \left\{\begin{array}{c} \delta\mathbf{W} \\ \delta\mathbf{\Lambda} \end{array}\right\} = -\left\{\begin{array}{c} \mathbf{R}_n^{i-1} \\ \mathbf{R}_{\Lambda,n}^{i-1} \end{array}\right\}} \qquad [11.13]$$

For the same reason as in statics (see section 9.3.3), in general, the system [11.13] is not invertible as is, and hence it must be solved blockwise. We are going to eliminate $\delta\mathbf{W}$ in $[11.13]_1$, assuming that the matrix $\mathbf{S}$ is invertible:

$$\boxed{\delta\mathbf{W} = -\mathbf{S}^{-1}.\mathbf{R}_n^{i-1} - \mathbf{S}^{-1}.\mathbf{K}_{U\Lambda}.\delta\mathbf{\Lambda}} \qquad [11.14]$$

By inserting this equation in $[11.13]_2$, we arrive at an equation with a single unknown $\delta\mathbf{\Lambda}$:

$$\boxed{\left(\mathbf{K}_{\Lambda\Lambda} - k\,\mathbf{K}_{\Lambda U}.\mathbf{S}^{-1}.\mathbf{K}_{U\Lambda}\right).\delta\mathbf{\Lambda} = -\mathbf{R}^{i-1}_{\Lambda,n} + k\,\mathbf{K}_{\Lambda U}.\mathbf{S}^{-1}.\mathbf{R}^{i-1}_{n}} \qquad [11.15]$$

The matrix $(\mathbf{K}_{\Lambda\Lambda} - \mathbf{K}_{\Lambda U}.\mathbf{S}^{-1}.\mathbf{K}_{U\Lambda})$, which is the Schur complement of matrix $\mathbf{S}$, is invertible in practice. Inverting the linear system [11.15] yields $\delta\mathbf{\Lambda}$, then $\delta\mathbf{W}$ from [11.14].

Once $\delta\mathbf{W}$ and $\delta\mathbf{\Lambda}$ have been obtained, the displacement, velocity, acceleration and the contact multipliers are updated using [11.5] and [11.11]:

$$\boxed{\begin{aligned} \mathbf{U}^{i}_{n} &= \mathbf{U}^{i-1}_{n} + k\,\delta\mathbf{W} \\ \mathbf{V}^{i}_{n} &= \mathbf{V}^{i-1}_{n} + c\,\delta\mathbf{W} \\ \mathbf{A}^{i}_{n} &= \mathbf{A}^{i-1}_{n} + m\,\delta\mathbf{W} \\ \mathbf{\Lambda}^{i}_{n} &= \mathbf{\Lambda}^{i-1}_{n} + \delta\mathbf{\Lambda} \end{aligned}} \qquad [11.16]$$

We pass onto the next iteration, resuming the system [11.13] where, this time, the matrices and the residuals are calculated with the new approximation $(\mathbf{U}^{i}_{n}, \mathbf{V}^{i}_{n}, \mathbf{A}^{i}_{n}, \mathbf{\Lambda}^{i}_{n})$ in the place of the previous one $(\mathbf{U}^{i-1}_{n}, \mathbf{V}^{i-1}_{n}, \mathbf{A}^{i-1}_{n}, \mathbf{\Lambda}^{i-1}_{n})$. The iterations are thus repeated until the residuals $\mathbf{R}^{i-1}_{n}$ and $\mathbf{R}^{i-1}_{\Lambda,n}$ are less than predetermined tolerances, before then stopping the current step n.

11.2.3. *Initializing the iterative loop: the vectors $U^0_n, V^0_n, A^0_n, \Lambda^0_n$*

At each iteration $i \geq 1$ of increment n, we need to know $(\mathbf{U}^{i-1}_{n}, \mathbf{V}^{i-1}_{n}, \mathbf{A}^{i-1}_{n}, \mathbf{\Lambda}^{i-1}_{n})$.

- If $i > 1$, that is, $i - 1 > 0$, then $(\mathbf{U}^{i-1}_{n}, \mathbf{V}^{i-1}_{n}, \mathbf{A}^{i-1}_{n}, \mathbf{\Lambda}^{i-1}_{n})$ is simply the value of the quadruple $(\mathbf{U}, \mathbf{V}, \mathbf{A}, \mathbf{\Lambda})$ acquired from the preceding iteration $i - 1 > 0$.

- If $i = 1$, then $(\mathbf{U}^{i-1}_{n}, \mathbf{V}^{i-1}_{n}, \mathbf{A}^{i-1}_{n}, \mathbf{\Lambda}^{i-1}_{n}) = (\mathbf{U}^{0}_{n}, \mathbf{V}^{0}_{n}, \mathbf{A}^{0}_{n}, \mathbf{\Lambda}^{0}_{n})$. This value, which serves as a prediction to initialize the iterative loop, is not known yet and must be specified. This is what we are going to do now.

For $\mathbf{\Lambda}^{0}_{n}$, we simply choose $\mathbf{\Lambda}^{0}_{n} = \mathbf{\Lambda}_{n-1}$.

Now, let us examine the vectors $\mathbf{U}^{0}_{n}, \mathbf{V}^{0}_{n}, \mathbf{A}^{0}_{n}$. Recall that we have assumed that $\forall i \geq 1$, $(\mathbf{U}^{i-1}_{n}, \mathbf{V}^{i-1}_{n}, \mathbf{A}^{i-1}_{n})$ *satisfies the kinematic equations* [11.3]. Hence in particular, $(\mathbf{U}^{0}_{n}, \mathbf{V}^{0}_{n}, \mathbf{A}^{0}_{n})$ should satisfy the kinematic equations. Specifically, from [11.8]:

$$\begin{aligned} \mathbf{U}^{0}_{n} &= \mathbf{U}_{n-1} + (1-\alpha)h\mathbf{V}_{n-1} + \alpha h\mathbf{V}^{0}_{n} + (\tfrac{1}{2} - \alpha)h^2\mathbf{A}_{n-1} \\ \mathbf{V}^{0}_{n} &= \mathbf{V}_{n-1} + (1-\beta)h\mathbf{A}_{n-1} + \beta h\mathbf{A}^{0}_{n} \end{aligned} \qquad [11.17]$$

Since we have three quantities that should satisfy two equations, there are infinitely many possibilities. Here is how to proceed:

1. choose $\mathbf{A}^{0}_{n} = \mathbf{A}_{n-1}$;

2. by substituting this into $[11.17]_2$, we obtain $\mathbf{V}^{0}_{n} = \mathbf{V}_{n-1} + h\mathbf{A}_{n-1}$;

3. by substituting these results into $[11.17]_1$, we obtain $\mathbf{U}^{0}_{n} = \mathbf{U}_{n-1} + h\mathbf{V}_{n-1} + \frac{h^2}{2}\mathbf{A}_{n-1}$.

In conclusion, we choose the following predictor relations:

$$\boxed{\begin{array}{lcl} \mathbf{U}_n^0 & = & \mathbf{U}_{n-1} + h\mathbf{V}_{n-1} + \frac{h^2}{2}\mathbf{A}_{n-1} \\ \mathbf{V}_n^0 & = & \mathbf{V}_{n-1} + h\mathbf{A}_{n-1} \\ \mathbf{A}_n^0 & = & \mathbf{A}_{n-1} \\ \hline \mathbf{\Lambda}_n^0 & = & \mathbf{\Lambda}_{n-1} \end{array}} \qquad [11.18]$$

Note that, in dynamics, we do not set $\mathbf{U}_n^0 = \mathbf{U}_{n-1}$ nor $\mathbf{V}_n^0 = \mathbf{V}_{n-1}$! In statics, the prediction [11.18] reduces to $\mathbf{U}_n^0 = \mathbf{U}_{n-1}$ and $\mathbf{\Lambda}_n^0 = \mathbf{\Lambda}_{n-1}$, which is the natural choice [9.18].

REMARK.– Let us return to the hypothesis that $\forall i \geq 1$, $(\mathbf{U}_n^{i-1}, \mathbf{V}_n^{i-1}, \mathbf{A}_n^{i-1})$ *satisfies the kinematic equations* [11.3]. A posteriori, we can see that this hypothesis is satisfied in our formulation. Indeed, rewriting the hypothesis in the equivalent form "$\forall i \geq 0$, $(\mathbf{U}_n^i, \mathbf{V}_n^i, \mathbf{A}_n^i)$ satisfies the kinematic equations [11.3]":

– For $i = 0$, we have just seen that $(\mathbf{U}_n^0, \mathbf{V}_n^0, \mathbf{A}_n^0)$ is forced to satisfy [11.17].

– For $i \geq 1$, the triple $(\mathbf{U}_n^i, \mathbf{V}_n^i, \mathbf{A}_n^i)$ satisfies $[11.4]_{3\text{-}4}$ by construction. Relations $[11.6]_{3\text{-}4}$ and $[11.9]_{3\text{-}4}$ are simply transformations of $[11.4]_{3\text{-}4}$ into different yet equivalent expressions.
■

11.3. Solution algorithm

The solution of the dynamic contact problem is described in Algorithm 11.1. This algorithm can be used directly in the case of prescribed time steps, and it can easily be adapted to the case of an automatic time incrementation after some simple modifications.

The three types of solution (for **U**, for **V**, or for **A**) differ only by the coefficients m, c and k in the matrix $\mathbf{S} = m\mathbf{M} + c\mathbf{C}_n^{i-1} + k\mathbf{K}_n^{i-1}$ and in the corrector relations. The expression for the residual vector $\mathbf{R}_n^{i-1}$ remains the same for the three types.

The three solution types are equivalent, except in two cases:

1. In quasi-static problems, we cannot speak of solving for the velocity nor for the acceleration, since $\mathbf{M} = \mathbf{C} = \mathbf{0}$, $\mathbf{V} = \delta\mathbf{V} = \mathbf{0}$ and $\mathbf{A} = \delta\mathbf{A} = \mathbf{0}$. We must therefore solve for the displacement.

2. In the case $\alpha = 0$ (explicit scheme), the coefficients m and c are not defined in the solution for the displacement. We must therefore use the solution for the velocity or for the acceleration.

Apart from these two cases, the three types of solution are equivalent and thus interchangeable, with the choice of one type over another being simply a question of taste.

Taylor and Papadopoulos [133] showed that, in the presence of contact, Newmark's algorithm may provoke oscillations in the relative velocity between two contact surfaces. In this work, we will not carry out an analytical study of the stability nor of the precision, which is an important yet difficult topic in the framework of contact with large deformations.

Choose the Newmark parameters α, β and the type of solution (solution for the displacement, velocity or acceleration).

Compute the mass matrix $\mathbf{M}$.

Initialize the incremental loop:

Set $\mathbf{U}_0 = \mathbf{0}$, read $\mathbf{V}_0$.

Compute the initial acceleration: $\mathbf{A}_0 = \mathbf{M}^{-1}.(\mathbf{\Phi}(\mathbf{U}_0, t_0) - \mathbf{\Psi}(\mathbf{U}_0))$.

For contact, set $\mathbf{\Lambda}_0 = \mathbf{0}$.

Knowing $\mathbf{U}_0$, and hence the initial positions of $S_c^{(1)}$ and $S_c^{(2)}$, perform a projection calculation to determine the elements $e_0^{(2)}$ and the pre-images $\boldsymbol{\eta}_0$ corresponding to the integration points $\mathbf{X} \in S_{0c}^{(1)}$ [1].

If there is plasticity, initialize the plastic quantities.

Incremental loop $n = 1, 2, \ldots$

Read t_n and determine the time step $h = t_n - t_{n-1}$.

Compute the external force vector $\mathbf{\Phi}$ due to the dead loads at t_n.

$i = 0$.

Initialize the iterative loop with predictor:

$$\begin{cases} \mathbf{U}_n^0 = \mathbf{U}_{n-1} + h\mathbf{V}_{n-1} + \frac{h^2}{2}\mathbf{A}_{n-1} \\ \mathbf{V}_n^0 = \mathbf{V}_{n-1} + h\mathbf{A}_{n-1} \\ \mathbf{A}_n^0 = \mathbf{A}_{n-1} \end{cases} \quad \text{and} \quad \mathbf{\Lambda}_n^0 = \mathbf{\Lambda}_{n-1}$$

Iterative loop

$i = i + 1$

Knowing $\mathbf{U}_n^{i-1}, \mathbf{A}_n^{i-1}$, we compute $\mathbf{L}(\mathbf{A}_n^{i-1}) \equiv \mathbf{M}.\mathbf{A}_n^{i-1}$ and the internal force vector $\mathbf{\Psi}(\mathbf{U}_n^{i-1})$.

If external pressures exist, we include their contribution in $\mathbf{\Phi}$. Then, we obtain $\mathbf{\Phi}(\mathbf{U}_n^{i-1}, t_n)$.

Compute the contact force vector $\mathbf{\Phi}_{\text{contact}}(\mathbf{U}_n^{i-1}, \mathbf{\Lambda}_n^{i-1})$ and the residual contact law vector $\mathbf{R}_{\Lambda,n}^{i-1}$.

This computation uses two results from the preceding step $n-1$: the elements $e_{n-1}^{(2)}$ and the points $\boldsymbol{\eta}_{n-1}$ corresponding to the integration points $\mathbf{X} \in S_{0c}^{(1)}$ [2].

During this computation, a projection calculation is performed and, among other things, we obtain the $e^{(2)}$'s and the $\boldsymbol{\eta}$'s. This information will then be stored at the end of the step.

Compute the residual $\mathbf{R}_n^{i-1} \equiv \mathbf{L}(\mathbf{A}_n^{i-1}) + \mathbf{\Psi}(\mathbf{U}_n^{i-1}) - \mathbf{\Phi}(\mathbf{U}_n^{i-1}, t_n) - \mathbf{\Phi}_{\text{contact}}(\mathbf{U}_n^{i-1}, \mathbf{\Lambda}_n^{i-1})$.

Check the convergence: if $i > 1$ and if $||\mathbf{R}_n^{i-1}||$ and $||\mathbf{R}_{\Lambda,n}^{i-1}|| \leq$ the given tolerances, then exit the iterative loop.

Knowing α, β, h and the type of solution chosen, compute the coefficients m, c, and k according to [11.12].

Compute the tangent matrices:

Calculate $\mathbf{K} \equiv \underbrace{\frac{\partial \mathbf{\Psi}}{\partial \mathbf{U}}(\mathbf{U}_n^{i-1})}_{\text{due to internal stresses}} - \underbrace{\frac{\partial \mathbf{\Phi}}{\partial \mathbf{U}}(\mathbf{U}_n^{i-1}, t_n)}_{\text{due to possible external pressures}} + \underbrace{\mathbf{K}_{UU}(\mathbf{U}_n^{i-1}, \mathbf{\Lambda}_n^{i-1})}_{\text{due to contact}}$.

In addition to $\mathbf{K}_{UU}$, calculate the other contact tangent matrices $\mathbf{K}_{U\Lambda}$, $\mathbf{K}_{\Lambda U}$, and $\mathbf{K}_{\Lambda\Lambda}$ at $(\mathbf{U}_n^{i-1}, \mathbf{\Lambda}_n^{i-1})$.

Compute $\mathbf{S} \equiv m\mathbf{M} + k\mathbf{K}$.

Algorithm 11.1. *Solution algorithm for the dynamic contact problem*

Obtain $\delta\mathbf{\Lambda}$ by solving $\left(\mathbf{K}_{\Lambda\Lambda} - k\,\mathbf{K}_{\Lambda U}.\mathbf{S}^{-1}.\mathbf{K}_{U\Lambda}\right).\delta\mathbf{\Lambda} = -\mathbf{R}^{i-1}_{\Lambda,n} + k\,\mathbf{K}_{\Lambda U}.\mathbf{S}^{-1}.\mathbf{R}^{i-1}_n$.

Deduce $\delta\mathbf{W} = -\mathbf{S}^{-1}.\mathbf{R}^{i-1}_n - \mathbf{S}^{-1}.\mathbf{K}_{U\Lambda}.\delta\mathbf{\Lambda}$.

Correct $\left\{\begin{array}{l} \mathbf{U}^i_n = \mathbf{U}^{i-1}_n + k\,\delta\mathbf{W} \\ \mathbf{V}^i_n = \mathbf{V}^{i-1}_n + c\,\delta\mathbf{W} \\ \mathbf{A}^i_n = \mathbf{A}^{i-1}_n + m\,\delta\mathbf{W} \end{array}\right.$ and $\mathbf{\Lambda}^i_n = \mathbf{\Lambda}^{i-1}_n + \delta\mathbf{\Lambda}$.

End of the iterative loop

Update: $\mathbf{U}_n = \mathbf{U}^{i-1}_n$, $\mathbf{V}_n = \mathbf{V}^{i-1}_n$, $\mathbf{A}_n = \mathbf{A}^{i-1}_n$, $\mathbf{\Lambda}_n = \mathbf{\Lambda}^{i-1}_n$ (also update the plastic quantities).
Store specific results provided by the last projection calculation: the elements $e^{(2)}$ and the antecedents $\boldsymbol{\eta}$ corresponding to the integration points $\mathbf{X} \in S^{(1)}_{0c}$. These quantities will play the role of the $e^{(2)}_{n-1}$ and $\boldsymbol{\eta}_{n-1}$ in the next step.

End of the incremental loop

Algorithm 11.1. (continued)

NOTES ON ALGORITHM 11.1.– *Idem* notes on Algorithm 9.1 in statics.

11.4. Summary

In this chapter, we have shown how to solve the equation system [8.47] for the dynamic contact problem. The principle results are the following:

1. The time discretization of the equation of motion [8.47] is performed with the help of Newmark's algorithm. At each time step n, we are led to a system of four algebraic nonlinear equations [11.2]-[11.3] for the four unknowns $\mathbf{U}_n$, $\mathbf{V}_n$, $\mathbf{A}_n$ and $\mathbf{\Lambda}_n$:

$$\begin{array}{l} \mathbf{M}.\mathbf{A}_n + \mathbf{\Psi}(\mathbf{U}_n) = \mathbf{\Phi}(\mathbf{U}_n, t_n) + \mathbf{\Phi}_{\text{contact}}(\mathbf{U}_n, \mathbf{\Lambda}_n) \\ \mathbf{R}_\Lambda(\mathbf{U}_n, \mathbf{\Lambda}_n) = \mathbf{0} \\ \mathbf{U}_n = \mathbf{U}_{n-1} + (1-\alpha)h\mathbf{V}_{n-1} + \alpha h\mathbf{V}_n + (\frac{1}{2} - \alpha)h^2\mathbf{A}_{n-1} \\ \mathbf{V}_n = \mathbf{V}_{n-1} + (1-\beta)h\mathbf{A}_{n-1} + \beta h\mathbf{A}_n \end{array}$$

   This system has been solved using the iterative Newton–Raphson method.

2. The linear system to solve for each Newton–Raphson iteration is [11.13]:

$$\left[\begin{array}{c|c} \mathbf{S} = m\,\mathbf{M} + k\,(\mathbf{K}_{classical} + \mathbf{K}_{UU}) & \mathbf{K}_{U\Lambda} \\ \hline k\,\mathbf{K}_{\Lambda U} & \mathbf{K}_{\Lambda\Lambda} \end{array}\right] \left\{\begin{array}{c} \delta\mathbf{W} \\ \delta\mathbf{\Lambda} \end{array}\right\} = -\left\{\begin{array}{c} \mathbf{R}^{i-1} \\ \mathbf{R}^{i-1}_\Lambda \end{array}\right\}$$

   All submatrices of the tangent matrix, except for the mass matrix $\mathbf{M}$, are identical to the ones used in statics.

   The computation Algorithm 11.1 unifies the implicit or explicit versions, as well as solutions for the displacement, speed and acceleration.

As said in the introduction to this chapter, time discretization is a difficult subject in nonlinear dynamics in general, and it becomes even more so when contact comes into play. This subject has been addressed in numerous studies in the literature and considerable progress has been made in the past few years in the field of dynamic contact/impact computation; without any doubt, it will continue to mobilize the efforts of researchers with a view to obtaining more efficient and robust formulations. In this work, we have not aimed to conceive a new time integration algorithm that is capable of competing with the sophisticated methods that currently exist. Since it is simple to implement, Newmark's algorithm was chosen to temporally discretize the differential system [8.47].

The solution method presented in this chapter will be applied in Chapter 12 to treat numerical examples involving dynamic contact.

12

Numerical Examples of Dynamic Contact

Continuing on from the quasi-static examples in Chapter 10, in this chapter, we are going to treat *dynamic* examples with the help of the numerical methods described in Chapter 11.

12.1. Impact of two elastic rods

Let us consider the impact problem of two elastic rods, which admits an analytical solution for small strains and which is often used in the literature to test numerical dynamic contact formulations.

The two rods have the same geometry, with length L along the $\mathbf{x}$-axis and with cross-sectional area S (see Figure 12.1). They are made of a Saint-Venant Kirchhoff hyperelastic material with Young's modulus E and density ρ. The Poisson ratio ν is assumed to be zero so that there are no displacements perpendicular to the $\mathbf{x}$-axis, and that only axial displacements can appear within the rods. Thus, the problem reduces to a one-dimensional one where the displacements depend only on the x abscissa of the particles. The cross-sections therefore remain plane and are perpendicular to the $\mathbf{x}$-axis.

The initial gap between the two rods is g_0. Prior to impact, the rod no. 1 (on the left) is traveling in uniform translation along $\mathbf{x}$ with velocity V_0, whereas the rod no. 2 (on the right) remains stationary. We assume that contact takes place without friction.

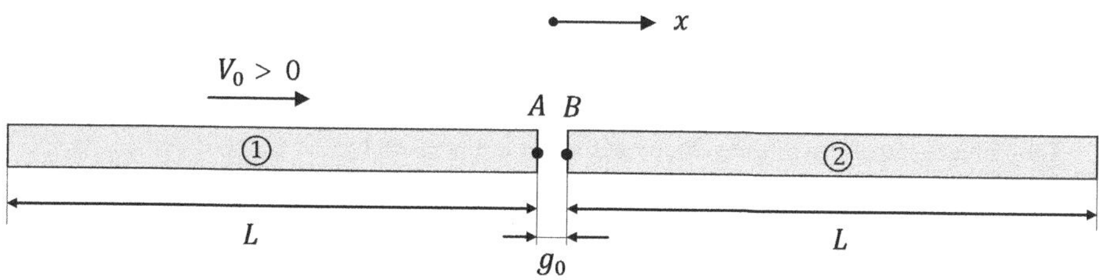

Figure 12.1. *Initial configuration of the impact problem of two rods*

We denote by A the center of the right cross-section of the first rod, and by B the center of the left cross-section of the second rod.

12.1.1. *Analytical solution*

Here, we will summarize the principal analytical results, which are well known in the literature, for the impact problem of two identical elastic rods. We assume small elastic strains and reduce the problem to one with one-dimensional wave propagation in the rods. We will take the initial instant equal to the shock instant; in other words the impact begins at $t = 0$.

Denoting the axial displacement in rod no. 1 by $U^{(1)}(x,t)$ and the force applied to the cross-section at A by $F(t)$, exerted by rod no. 2 onto rod no. 1, the equation of motion for rod no. 1 is written as

$$\rho S\,\ddot{U}^{(1)} - ES\frac{\partial^2 U^{(1)}}{\partial x^2} = F(t)\,\delta(x - x_A) \qquad [12.1]$$

where $\delta(x - x_A)$ denotes the Dirac measure concentrated at the point x_A.

The boundary conditions for rod no. 1 are as follows: zero axial stresses on the left and right cross-sections (by writing the right-hand side of [12.1], we have counted $F(t)$ as a *distributed* force along the rod). The initial conditions for the same rod are as follows: zero displacement relative to the reference position and velocity V_0 at every point.

Likewise, denoting by $U^{(2)}(x,t)$ the axial displacement in the rod no. 2, the equation of motion for this rod, subjected to force $-F(t)$ at the cross-section B, is written as

$$\rho S\,\ddot{U}^{(2)} - ES\frac{\partial^2 U^{(2)}}{\partial x^2} = -F(t)\,\delta(x - x_B)$$

The initial and boundary conditions on rod no. 2 are zero.

Let us calculate the response of the first rod to the force $F(t)$, starting from the time instant $t = 0$, assuming, in the first instance, that the rod is stationary at the initial instant. Let $F(p) = \mathcal{L}\{F(t)\}$ be the Laplace transform of $F(t)$. The Laplace transform of the displacement is given by

$$U^{(1)}(x,p) = F(p)\left[\frac{c}{pES}\frac{\cosh\frac{p}{c}(L - x)}{\sinh\frac{PL}{c}}\right]$$

where $\boxed{c = \sqrt{\frac{E}{\rho}}}$ is the wave propagation speed within each rod. In particular, at point A with abscissa $x = 0$:

$$U^{(1)}(0,p) = F(p)\left[\frac{c}{pES}\cosh\frac{PL}{c}\right] \qquad [12.2]$$

The Laplace transform of the velocity at the point A is given by

$$\mathcal{L}\{\dot{U}^{(1)}(0,t)\} = p\mathcal{L}\{U^{(1)}(0,t)\} - U^{(1)}(0,0^+) = F(p)\frac{c}{ES}\cosh\frac{PL}{c} = F(p)\frac{c}{ES}\frac{1 + e^{-\tau p}}{1 - e^{-\tau p}}$$

having set $\boxed{\tau = \frac{2L}{c}}$. After substituting in $\frac{1}{1 - e^{-\tau p}} = \sum_{n=0}^{\infty}(e^{-\tau p})^n$, we arrive at

$$\mathcal{L}\{\dot{U}^{(1)}(0,t)\} = F(p)\frac{c}{ES}(1 + e^{-\tau p})(1 + e^{-\tau p} + e^{-2\tau p} + \ldots) = F(p)\frac{c}{ES}(1 + 2e^{-\tau p} + 2e^{-2\tau p} + \ldots)$$

that is

$$\mathcal{L}\{\dot{U}^{(1)}(0,t)\} = \frac{c}{ES}\left(\mathcal{L}\{F(t)\} + 2e^{-\tau p}\mathcal{L}\{F(t)\} + 2e^{-2\tau p}\mathcal{L}\{F(t)\} + \ldots\right)$$

The inverse Laplace transform gives

$$\dot{U}^{(1)}(0,t) = \frac{c}{ES}\left(F(t) + 2F(t-\tau) + 2F(t-2\tau) + \ldots\right)$$

Taking into account that the initial velocity of rod no. 1 is V_0, we obtain

$$\dot{U}^{(1)}(0,t) = V_0 + \frac{c}{ES}\left(F(t) + 2F(t-\tau) + 2F(t-2\tau) + \ldots\right) \qquad [12.3]$$

The same reasoning enables us to obtain the velocity of point B in rod no. 2:

$$\dot{U}^{(2)}(0,t) = \frac{c}{ES}\left(-F(t) - 2F(t-\tau) - 2F(t-2\tau) - \ldots\right) \qquad [12.4]$$

Let us recast [12.3]–[12.4] into two equivalent equations by taking the sum and the difference between them. We get

$$\dot{U}^{(1)}(0,t) + \dot{U}^{(2)}(0,t) = V_0 \qquad [12.5]$$

$$\dot{U}^{(1)}(0,t) - \dot{U}^{(2)}(0,t) = V_0 + \frac{2c}{ES}\left(F(t) + 2F(t-\tau) + 2F(t-2\tau) + \ldots\right) \qquad [12.6]$$

The contact law between the two rods is expressed by an equality and an inequality:

$$\begin{array}{ll} 1.\ \text{if } \dot{U}^{(1)}(0,t) - \dot{U}^{(2)}(0,t) \neq 0, & \text{then } F(t) = 0 \text{ (no contact);} \\ 2.\ \text{if } \dot{U}^{(1)}(0,t) - \dot{U}^{(2)}(0,t) = 0, & \text{then } F(t) < 0 \text{ (contact).} \end{array} \qquad [12.7]$$

Relations [12.5]–[12.7] form – in either contact or non-contact cases – three equations with three unknowns $\dot{U}^{(1)}(0,t)$, $\dot{U}^{(2)}(0,t)$ and $F(t)$, supplemented with an inequality that has to be satisfied.

1. In the time interval $0 < t < \tau$.

   Since $t-\tau$, $t-2\tau$, ... are all negative within the interval considered, we have $F(t-\tau) = F(t-2\tau) = \ldots = 0$. Relations [12.5]–[12.7]$_2$ can then be written as

   $$\dot{U}^{(1)}(0,t) + \dot{U}^{(2)}(0,t) = V_0 \qquad \dot{U}^{(1)}(0,t) - \dot{U}^{(2)}(0,t) = V_0 + \frac{2c}{ES}F(t)$$
   $$\dot{U}^{(1)}(0,t) - \dot{U}^{(2)}(0,t) = 0$$

   We obtain therefore

   $$\dot{U}^{(1)}(0,t) = \dot{U}^{(2)}(0,t) = \frac{V_0}{2} \qquad F(t) = -\frac{ES\,V_0}{2c} < 0 \qquad [12.8]$$

   With the inequality $F(t) < 0$ satisfied, the preceding relations form the solution of the problem within the time interval $0 < t < \tau$. The rods come into contact at the points A and B.

2. In the time interval $\tau < t < 2\tau$.

   Suppose that there is contact within this time interval. Since $0 < t-\tau < \tau$, the expression of $F(t-\tau)$ is given by [12.8], namely, $F(t-\tau) = -\frac{ES\,V_0}{2c}$. On the other hand, since $t-2\tau$,

... are all negative within the considered interval, we have $F(t-2\tau) = \ldots = 0$. Relations [12.5]–[12.7]$_2$ can therefore be written as

$$\dot{U}^{(1)}(0,t) + \dot{U}^{(2)}(0,t) = V_0 \qquad \dot{U}^{(1)}(0,t) - \dot{U}^{(2)}(0,t) = V_0 + \frac{2c}{ES}(F(t) - 2\frac{ES\,V_0}{2c})$$
$$\dot{U}^{(1)}(0,t) - \dot{U}^{(2)}(0,t) = 0$$

This implies that $F(t) = \dfrac{ES\,V_0}{2c} > 0$. Since the inequality $F(t) < 0$ is not satisfied, there cannot in fact be any contact. The relations that we must use are [12.5]–[12.7]$_1$:

$$\dot{U}^{(1)}(0,t)+\dot{U}^{(2)}(0,t) = V_0 \qquad \dot{U}^{(1)}(0,t)-\dot{U}^{(2)}(0,t) = V_0+\frac{2c}{ES}(F(t)-2\frac{ES\,V_0}{2c}) \qquad F(t) = 0$$

This time we get

$$\dot{U}^{(1)}(0,t) = 0 \quad \dot{U}^{(2)}(0,t) = V_0 \qquad F(t) = 0 \tag{12.9}$$

As the inequality $\dot{U}^{(1)}(0,t)-\dot{U}^{(2)}(0,t) \neq 0$ is satisfied, the previous relations are the solution to the problem within the time interval $\tau < t < 2\tau$.

The velocity at a current x-abscissa is given by

$$\begin{aligned}\dot{U}^{(1)}(x,t) &= V_0 + \frac{c}{ES}\left(F(t-\tau_1) + F(t-\tau+\tau_1) + F(t-\tau-\tau_1) + F(t-2\tau+\tau_1) + \ldots\right)\\ \dot{U}^{(2)}(x,t) &= \frac{c}{ES}\left(-F(t-\tau_1) - F(t-\tau+\tau_1) - F(t-\tau-\tau_1) - F(t-2\tau+\tau_1) - \ldots\right)\end{aligned} \tag{12.10}$$

where we have set $\tau_1 \equiv \dfrac{x}{c}$. Then we can collect the expressions for $F(t)$ found in [12.8] and [12.9] together as follows:

$$\forall t \in [0, 2\tau], \qquad F(t) = -\frac{ES\,V_0}{2c}\,[H(t) - H(t-\tau)] \tag{12.11}$$

where $H(t)$ is the Heaviside function. By substituting [12.11] into [12.10], we obtain, for any $t \in [\tau, 2\tau]$:

$$\begin{aligned}\dot{U}^{(1)}(x,t) &= V_0\left[1 - \frac{1}{2}(H(t-\tau_1) + H(t-\tau-\tau_1))\right] = 0\\ \dot{U}^{(2)}(x,t) &= \frac{V_0}{2}\,[H(t-\tau_1) + H(t-\tau-\tau_1)] = V_0\end{aligned}$$

3. When $t > 2\tau$.

   Within the interval $]\tau, 2\tau[$ and at every point x, we have just found that $\dot{U}^{(1)}(x,t) = 0$ and $\dot{U}^{(2)}(x,t) = V_0$. Consequently, for $t > 2\tau$, the first rod is stationary, whereas the second rod is moving in uniform translation. There cannot be a second impact afterward and the results found for $\tau < t < 2\tau$ remain valid for $t > 2\tau$.

Now we are going to rewrite the previous solution, assuming that the time instant at which contact starts to occur is t_c, rather than 0.

1. Within the time interval $t_c < t < t_c + \tau$:

$$\boxed{\dot{U}^{(1)}(x_A - g_0, t) = \dot{U}^{(2)}(x_B, t) = \frac{V_0}{2}} \qquad \boxed{F(t) = -\frac{ES\,V_0}{2c} < 0} \qquad [12.12]$$

   There is contact between the two rods. The contact force is negative. The velocities at the contact points are identical, equal to half of V_0.

2. Within the time interval $t > t_c + \tau$:

$$\boxed{\dot{U}^{(1)}(x, t) = 0 \quad \text{and} \quad \dot{U}^{(2)}(x, t) = V_0} \qquad \boxed{F(t) = 0} \qquad [12.13]$$

   There is no longer any contact. The contact force is zero. Rod no. 1 comes to a standstill, whereas rod no. 2 moves in uniform translation with velocity V_0.

   The two rods move away from one another indefinitely, and there is no second impact.

Note that the contact time $\tau = \dfrac{2L}{c}$ is equal to the time taken for the compression wave to travel back and forth along the length of rod no. 1.

12.1.2. *Numerical applications*

Let us calculate the theoretical solution obtained with the same values as in [64] (see Table 12.1).

Length L	Dimension along **y**	Dimension along **z**	Initial gap g_0	Initial velocity V_0 of rod no. 1
10 m	1 m	1 m	– 0.01 m	0.1 m/s

E	ν	ρ
100 N/m^2	0	0.01 kg/m^3

Table 12.1. *Numerical values of the impact problem of two rods*

The cross-sectional area of the rods is therefore $S = 1$ m^2. The wave propagation speed is equal to $c = 100$ m/s. With these chosen numerical values, the previous analytical solution states that:

- contact is established at the time instant $t_c = 0.1$ s;

- contact persists for $\tau = 0.2$ s; the velocity of points A and B is equal to 0.05 m/s, and the contact force equals $F(t) = -0.05$ N;

- at the time instant $t = 0.3$ s, the two rods separate, and rod no. 2 undergoes a uniform translation along **x** with velocity 0.1 m/s, whereas rod no. 1 becomes stationary.

12.1.3. *Numerical solution*

We solve the problem using the Newmark algorithm described in section 11.2.

Each rod is meshed using 20 twenty-node hexahedral elements (H20), such that each contact surface consists of an eight-node quadrilateral (Q8). We choose the rod on the left to be contactor and the one on the right as the target.

Numerical integration is performed using $3 \times 3 \times 3$ Gaussian quadrature points for the internal and external force vectors, and $4 \times 4 \times 4$ points for the mass matrix in the H20 elements. For the elements of the contactor surface, numerical integration is performed using 7×7 Gaussian quadrature points (there are too many in this example). The penalty parameters are $\epsilon_N = \epsilon_T = 10^2$ N/m^3.

The time step is taken to be equal to $\Delta t = 5 \times 10^{-3}$ s. The parameters α and β for the Newmark algorithm are chosen to be equal to those which do not give rise to numerical damping in classical linear dynamics (without contact), namely $\alpha = \beta = \frac{1}{2}$.

The numerical results presented below mainly concern the two particles located at the points A and B, which are the centers of the contact surfaces (see Figure 12.1).

• Motion

Figure 12.2 shows that the axial displacement U at A and B versus time agrees well with the analytical solution, with a relative error, of at most, 5%.

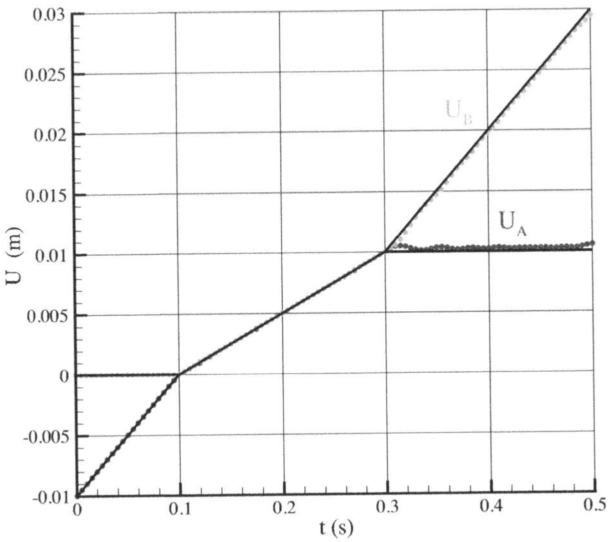

Figure 12.2. *Impact of two rods. Axial displacements at points A and B versus time. •: numerical solution, continuous line ——: theoretical solution. For a color version of this figure, see www.iste.co.uk/levan/contact.zip*

We find that the two rods enter into contact at the instant $t = 0.1$ s, before then separating at the instant $t = 0.3$ s, which agrees with the analytical solution. Figure 12.2 gives the impression that the two rods are attached to one another throughout the contact interval $0.1 \text{ s} \leq t \leq 0.3$ s. In fact, looking closely at the numerical results, there are contact and loss of contact (or release) that alternate a certain number of times during this time interval. This does not agree with the analytical solution.

Figure 12.3 represents the axial velocities V at A and B versus time. During the contact interval, the relative velocity between the two particles A and B strongly oscillate about the theoretical value 0, as contact alternates with release throughout this interval.

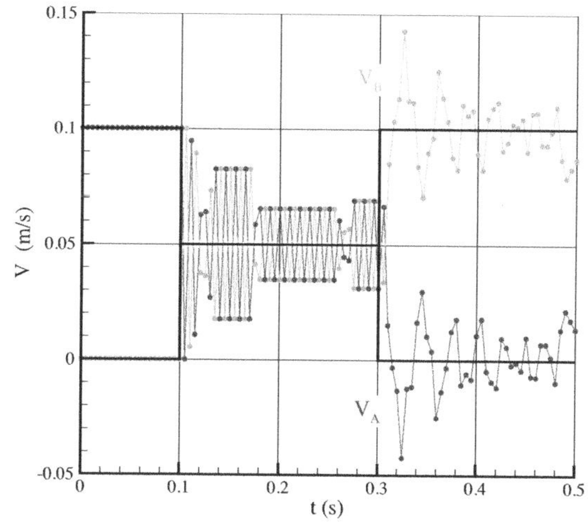

Figure 12.3. *Impact of two rods. Axial velocities at points A and B versus time. •: numerical solution, thick solid line ——: theoretical solution. For a color version of this figure, see www.iste.co.uk/levan/contact.zip*

REMARK.– Even if we had persistent contact during the interval $0.1 \text{ s} \leq t \leq 0.3$ s, the Newmark algorithm would always lead to a discontinuity in the velocities and to one in the accelerations at points A and B (see Taylor and Papadopoulos [133]). Indeed, assuming that the two rods come into contact at time t_{n-1}, then we can write the following relations at this time (the superscripts 1 and 2 refer to rod no. 1 and no. 2, respectively):

– The points A and B touch: $\mathbf{x}^{(1)}_{n-1} = \mathbf{x}^{(2)}_{n-1} \Leftrightarrow \mathbf{X}^{(1)} + \mathbf{U}^{(1)}_{n-1} = \mathbf{X}^{(2)} + \mathbf{U}^{(2)}_{n-1}$.
– The velocity at A is $\mathbf{V}_0$, and the acceleration at $A$is zero: $\mathbf{V}^{(1)}_{n-1} = \mathbf{V}_0, \mathbf{A}^{(1)}_{n-1} = \mathbf{0}$. [12.14]
– The velocity and acceleration at $B$are zero: $\mathbf{V}^{(2)}_{n-1} = \mathbf{A}^{(2)}_{n-1} = \mathbf{0}$.

If the two rods remain in contact at the instant t_n, then we have

$$\mathbf{x}^{(1)}_n = \mathbf{x}^{(2)}_n \quad \Leftrightarrow \quad \mathbf{X}^{(1)} + \mathbf{U}^{(1)}_n = \mathbf{X}^{(2)} + \mathbf{U}^{(2)}_n \qquad [12.15]$$

By expressing $\mathbf{U}^{(1)}_n$ et $\mathbf{U}^{(2)}_n$ in [12.15] via the kinematic equation $[11.3]_1$ and taking [12.14] into account, we obtain the difference between the velocities at A and B at the instant t_n (assuming that $\alpha \neq 0$):

$$\mathbf{V}^{(2)}_n - \mathbf{V}^{(1)}_n = (\frac{1}{\alpha} - 1)\mathbf{V}_0 \neq \mathbf{0} \qquad [12.16]$$

Now, by expressing $\mathbf{V}^{(1)}_n$ and $\mathbf{V}^{(2)}_n$ in [12.16] via the kinematic equation $[11.3]_2$ and again taking [12.14] into account, we obtain the difference between the accelerations at A and B at the instant t_n ($\beta \neq 0$):

$$\mathbf{A}_n^{(2)} - \mathbf{A}_n^{(1)} = \frac{1}{\alpha\beta\Delta t}\mathbf{V}_0 \neq \mathbf{0} \qquad [12.17]$$

Relations [12.16] and [12.17] show the problematic behavior of the Newmark algorithm when it is used in dynamic contact problems. The scheme is not consistent, since the discontinuities in the velocities and accelerations at the points A and B do not tend toward zero, when the time step Δt decreases. ■

Figure 12.3 shows that after the rods separate definitely, the velocity of particle A (respectively, B) oscillates noticeably about the theoretical value 0 (respectively, 0.1 m/s). This phenomenon has already been observed by Laursen et al. [77] and [78] who solved the same problem with the time stepping algorithms of Newmark, Hilber–Hughes–Taylor (HHT), and also using a specific energy method proposed by the authors. These authors found that all methods led to oscillations after the two rods separate. According to their analysis, these oscillations come from spatial discretization and not from the treatment of dynamic contact. These non-physical oscillations can be attenuated by adding in algorithmic damping, but this inevitably leads to a loss of precision.

- Verification of the contact laws

The numerical results show that the inequalities $g \leq 0$ and $T_N \geq 0$ in [5.3] are very well satisfied at nodes on the contact surface, and for all steps. With the time steps where contact takes place, it is found that the proximity g is of the order of 10^{-14} m.

- Contact force

Figure 12.4 represents the resultant contact force $F(t)$ along the $\mathbf{x}$-axis, exerted by the right rod on the left rod. Throughout the contact interval, $t \in [0.1\text{ s}, 0.3\text{ s}]$, the contact force oscillates greatly about the exact value $F = -0.05$ N, which is consistent with the above-mentioned alternation between contact and release.

- Energies

Total energy, defined as the sum of the kinetic and strain energies of the two rods, is indeed conserved as required by the analytical solution, except at a few rare instants that are visible in Figure 12.5, where it undergoes some small increases of 1.4%.

12.2. Disk impacting a rigid plane

The motivation is, after the impact of two rods, in this section, we are going to solve another interesting problem to test the dynamic formulation: the impact of a hyperelastic disk on a fixed rigid plane. Here, the relative motion of the two bodies is more complicated and the velocities considered are sufficiently high to simulate a violent shock.

The disk studied has center $\mathbf{C}$, a radius of $R = 0.1$ m, and a thickness of 0.02 m. It is made of a rubber-like Saint-Venant Kirchhoff material with Young's modulus $E = 10^7$ Pa, Poisson's ratio $\nu = 0.4$ and its reference mass density is $\rho_0 = 1{,}000$ kg/m^3. The plane that the disk strikes is assumed to be rigid.

At the initial time instant, the disk is located above the plane. The midplane of the disk lies in the $\mathbf{Oxz}$ plane, the coordinates of the center $\mathbf{C}$ are $x_C = -0.11$ m, $z_C = 0.11$ m, and the plane is situated at $z = 0$. Prior to impact, the motion of the disk, parallel to the $\mathbf{xz}$ plane,

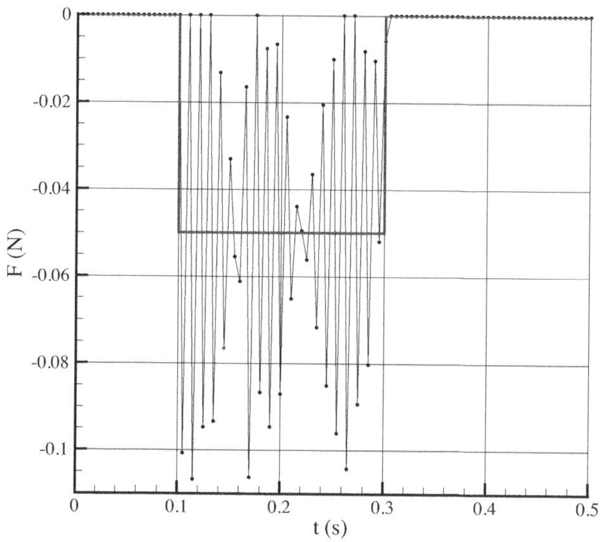

Figure 12.4. *Impact of two rods. Contact force. •: numerical solution, thick solid line ——: theoretical solution. For a color version of this figure, see www.iste.co.uk/levan/contact.zip*

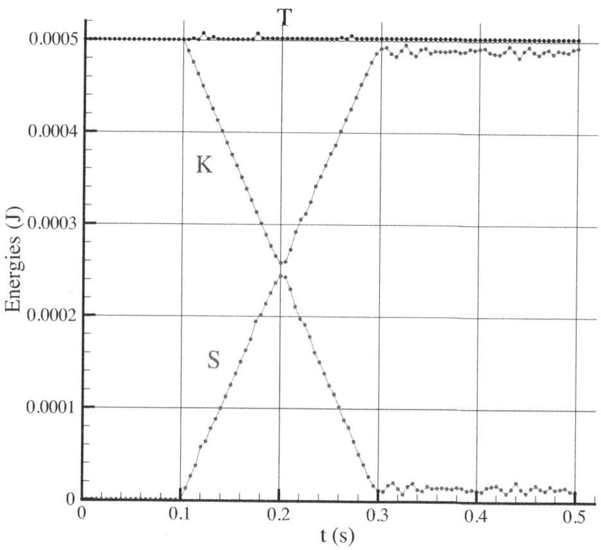

Figure 12.5. *Impact of two rods. Evolution of the energies. K = kinetic energy, S = strain energy, $T = K + S$ = total energy. For a color version of this figure, see www.iste.co.uk/levan/contact.zip*

is composed of a uniform translation with velocity $V = 50$ km/h $= 13.888$ m/s at 45° and a uniform rotation of angular velocity $\dot{\theta}\mathbf{y}$, where $\dot{\theta}$ is equal to one of the three following values: $\dot{\theta} = 0$ (pure translational motion), $\dot{\theta} = -\omega < 0$ (motion with slice effect) and $\dot{\theta} = \omega > 0$ (motion with topspin effect), $\omega = 500$ rev/min (see Figure 12.6).

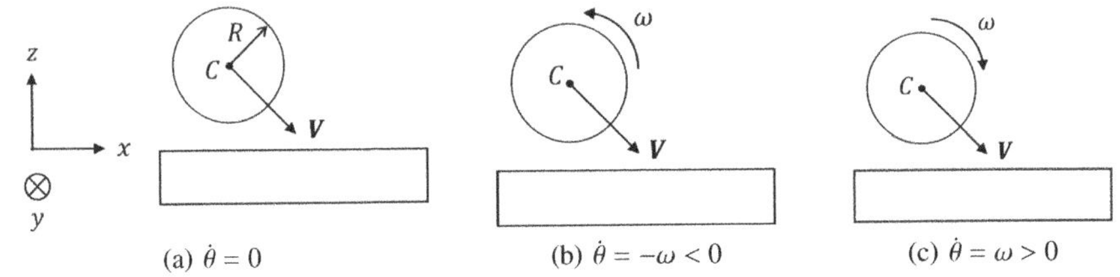

Figure 12.6. *Disk impacting a rigid plane ($\dot{\theta}$ is measured relative to the* **y***-axis)*

We suppose that the contact between the disk and plane takes place either without friction ($\mu = 0$) or with the coefficient of friction $\mu = 0.3$. Gravity is neglected.

The disk is meshed into 24 twenty-node hexahedral elements (H20) and 8 fifteen-node prismatic elements (P15) (see Figure 12.7). The rigid plane is meshed into a single H20 element. The disk is considered to be the contactor and the plane to be the target.

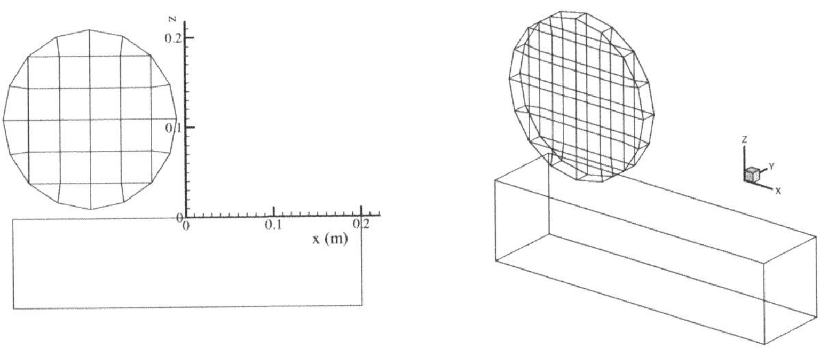

Figure 12.7. *Initial mesh*

Time-stepping is performed using the Newmark algorithm with the parameters $\alpha = \beta = \frac{1}{2}$. In each Q8 element on the contactor surface, we use 7×7 Gaussian quadrature points. The time step is fixed at $\Delta t = 2 \times 10^{-4} s$ and each calculation is performed in 30 steps. The convergence tolerances for the residual force vector $\{\mathbf{R}\}$ and the residual contact law $\{\mathbf{R}_\Lambda\}$ are 10^{-6}.

12.2.1. *Frictionless case*

In the frictionless case, we choose the penalty parameters $\epsilon_N = \epsilon_T = 10^7$ N/m^3. At each step, the iterative solution by the Newton–Raphson scheme converges in six to seven iterations on average.

• Motion

The numerical results show that, without friction, the disk motion is the same for each of the three initial angular velocities ($\dot{\theta} = 0, -\omega$, or $+\omega$). This result is predictable: in the absence of friction, the tangential contact force is always zero, whereas the normal contact force remains the same, regardless of the angular velocity of the disk.

The trajectory of the disk center **C** is shown in Figure 12.8. The contact between the disk and the plane begins when the height z_C is less than 0.1 m; it ends when z_C exceeds roughly the same value. If we consider the trajectory of the center after the rebound as a straight line, then we find that the rebound angle of the disk is very close to the incident angle, namely 45°.

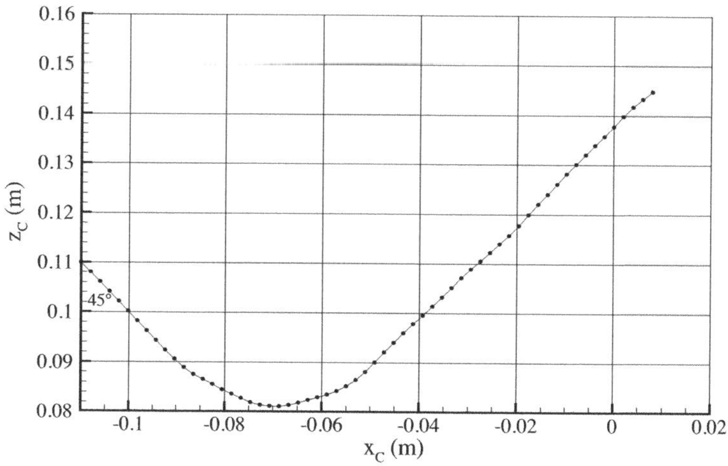

Figure 12.8. *Frictionless impact of a disk onto a rigid plane. Trajectory of the disk center* ***C***

Contact lasts for about 6.2 ms and, during this time, the disk center **C** horizontally shifts over by a distance of 61 mm (as can be seen in Figure 12.8). The maximal vertical displacement of **C** is −29 mm, which means that the lower part of the disk gets strongly compressed.

• Verification of the inequality $g \leq 0$

The numerical results show that the inequality $g \leq 0$ in [5.3] is well satisfied at each node of the contact surface and for each step: at any step when contact takes place, we note that the maximum value of the proximity g over the nodes of the contact area is 0.8 mm, that is, 0.8% of the disk radius.

• Contact stresses

The distribution of contact stresses over space and time is more or less complex. It is shown in Figure 12.9 at the time instant when the disk is compressed at most against the plane. The impact creates stress waves that propagate throughout the disk; thus, in general, the contact stress profile does not resemble that of the static Hertz contact problem seen in section 10.2.

• Verification of the inequality $T_N \geq 0$ in [5.3]

Numerical results show that there exists – at some steps and nodes on the contact surface – negative normal contact stresses T_N, which are the abnormal contact stresses mentioned in section 8.9. The nodes in question are not large in number, and the stresses $T_N < 0$ found are small in comparison to the stresses $T_N > 0$, while still non-negligible.

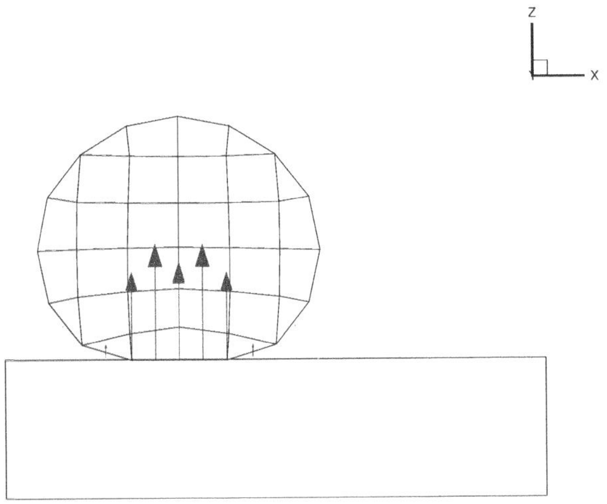

Figure 12.9. *Frictionless impact of a disk onto a rigid plane. Contact stresses at a typical instant. For a color version of this figure, see www.iste.co.uk/levan/contact.zip*

Looking closely at the contact zone, we see that the stresses $T_N < 0$ are right in the middle of this zone and are not due to the presence of the elements in partial contact.

• Energies

The total energy of the disk (kinetic energy + strain energy) must be conserved in frictionless contact, and this must hold regardless of the angular velocity of the disk. Since gravity is neglected, gravitational potential energy does not feature here.

The energies – kinetic, strain and total – are plotted in Figure 12.10. It can be seen that the total energy of the system drops by 0.7% during impact, and then increases by 0.2% after impact.

12.2.2. *Case with friction* $\mu = 0.3$

The calculations with friction are performed with the penalty parameters $\epsilon_N = \epsilon_T = 5 \times 10^7$ N/m^3. At each step, the iterative Newton–Raphson scheme converges in seven to eight iterations on average.

• Motion

The trajectory of the disk center **C** is presented in Figure 12.11. Contact between the disk and the plane begins when the height z_C is less than 0.1 m, and ends when z_C exceeds approximately the same value.

For angular velocities $\dot{\theta} = 0$, $\dot{\theta} = -\omega < 0$ and $\dot{\theta} = \omega > 0$, contact lasts for about 6.4 ms, 6.5 ms and 6.1 ms, respectively. During this time, the center **C** shifts horizontally by 44, 44 and

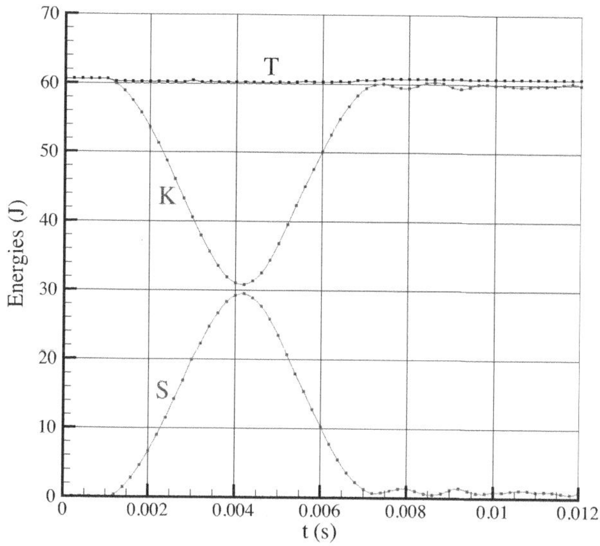

Figure 12.10. *Frictionless impact of a disk onto a rigid plane. Energy evolution. K = kinetic energy, S = strain energy, $T = K + S$ = total energy. For a color version of this figure, see www.iste.co.uk/levan/contact.zip*

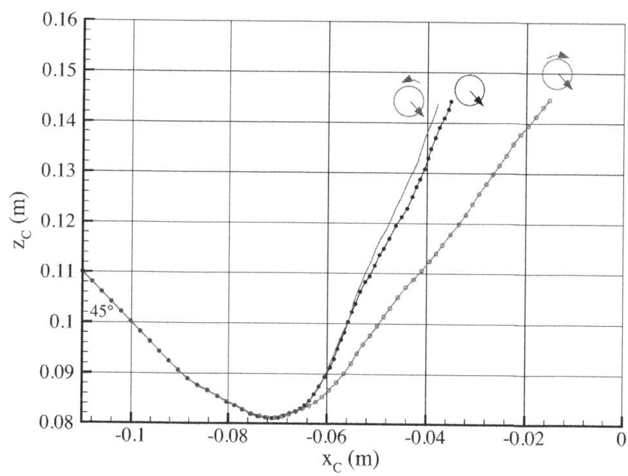

Figure 12.11. *Impact with friction ($\mu = 0.3$) of a disk onto a rigid plane. Trajectory of the disk center for angular velocities $\dot{\theta} = 0$, $\dot{\theta} = -\omega < 0$ and $\dot{\theta} = \omega > 0$. For a color version of this figure, see www.iste.co.uk/levan/contact.zip*

50 mm, respectively, whereas the maximum vertical displacement of **C** is approximately equal to −19 mm in the three cases.

In the presence of friction, the deformed shapes and global motion of the disk after impact differ, according to whether $\dot{\theta} = 0$, $-\omega$, or $+\omega$. In Figure 12.12, we have chosen to present the deformed shapes of the disk in the $\dot{\theta} = -\omega < 0$ case only, where we can observe an interesting effect: friction has the effect of stopping the disk rotation at a particular instant, and then inverting the rotational direction after this instant. The disk therefore rebounds with angular velocity $\dot{\theta} > 0$.

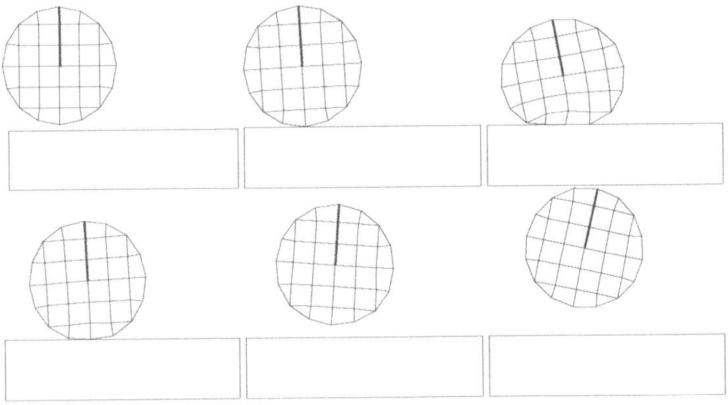

Figure 12.12. *Frictional impact ($\mu = 0.3$) of a disk onto a rigid plane. Deformed shapes over the course of time in the $\dot{\theta} = -\omega < 0$ case. Note the change of rotational direction of the disk after impact. For a color version of this figure, see www.iste.co.uk/levan/contact.zip*

Let us qualitatively explain this inversion of the rotational direction by using some results from rigid-body mechanics. Knowing the translational velocity $\mathbf{V}$ and angular velocity $\dot{\theta}$ of the disk, we can get the relative velocity $\mathbf{V}_I$ at the point $\mathbf{I}$, between the disk and the plane (see Figure 12.13). This velocity $\mathbf{V}_I$ is the standard relative velocity $\mathbf{V}_{standard}$ defined in [2.35]. At the moment of impact, we have $g = 0$ and $\mathbf{V}_{standard} = -\dot{g}\boldsymbol{\nu} + \dot{\eta}^{\alpha}\mathbf{a}_{\alpha}$ coinciding with the relative contact velocity $\mathbf{V}_c$ defined in [2.38]. From definition [2.39], the tangential component of $\mathbf{V}_c = \mathbf{V}_I$ is the slip velocity $\mathbf{V}_T$.

As seen in Figure 12.13, the normal component of velocity $\mathbf{V}_I$ remains the same, regardless of the angular velocity $\dot{\theta}$. On the other hand, the tangential component, that is, the slip velocity $\mathbf{V}_T$, is highest in the $\dot{\theta} = -\omega < 0$ case and smallest in the $\dot{\theta} = \omega > 0$ case.

At the moment when contact is established, the disk is subjected to an impulse due to impact (equal to a very large force integrated over a very short period of time) at the point $\mathbf{I}$. The normal (respectively, tangential) component of the impulse is proportional to the opposite of the normal (respectively, tangential) component of $\mathbf{V}_I$ prior to impact. The normal impulse causes the vertical rebound of the disk, whereas the tangential impulse affects both the horizontal and angular velocities of the disk.

The magnitude of the tangential impulse is therefore maximal in the $\dot{\theta} = -\omega < 0$ case. If the angular velocity ω is sufficiently large, as is the case here, then this tangential impulse may change the rotational direction of the disk.

In the case where $\dot{\theta} = \omega > 0$, this qualitative analysis also allows us to predict that the disk should rotate more quickly after impact.

Finally, if we approximate the trajectory of the disk center $\mathbf{C}$ after the rebound as a straight line, then we find that the rebound angle (from the horizontal) is greater in the $\dot{\theta} = -\omega < 0$ case

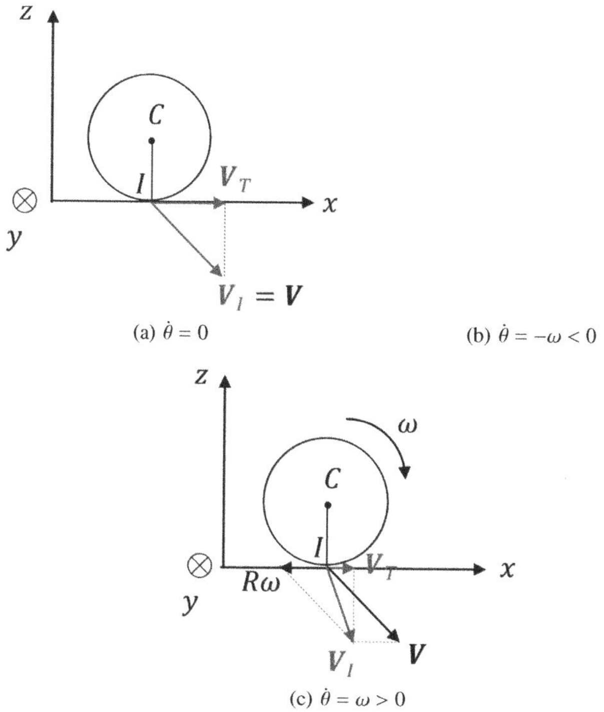

Figure 12.13. *Impact of a disk on a rigid plane. Slip velocity just before impact. For a color version of this figure, see www.iste.co.uk/levan/contact.zip*

than in the $\dot{\theta} = +\omega > 0$ case (see Figure 12.11). This result can be justified by combining together two facts:

(i) On the one hand, the *vertical* velocity of the disk center after the rebound is the same regardless of the angular velocity.

(ii) On the other hand, the larger the tangential component of the impact impulse, the more the *horizontal* velocity is reduced.

• Verification of the inequality $g \leq 0$

The inequality $g \leq 0$ in [5.3] is satisfied at each node of the contact surface and each step: at the steps where contact takes place, it is found that the maximum value of the proximity over the contact nodes is 0.9 mm, that is, 0.9% of the disk radius.

• Contact stresses

The distribution of the contact stresses over both space and the course of time is more or less complex, and is presented in Figure 12.14 for several representative increments.

• Verification of the inequality $T_N \geq 0$ in [5.3]

Whatever the angular velocity ($\dot{\theta} = 0$, $\dot{\theta} = -\omega < 0$, or $\dot{\theta} = \omega > 0$), abnormal stresses $T_N < 0$ can be found and, though small, they are non-negligible and situated in the middle of the contact area – just like in the frictionless case. Besides, there are non-zero contact stresses $\mathbf{T}$ on the elements in partial contact, but these are very small and can be explained, as seen in section 8.9.

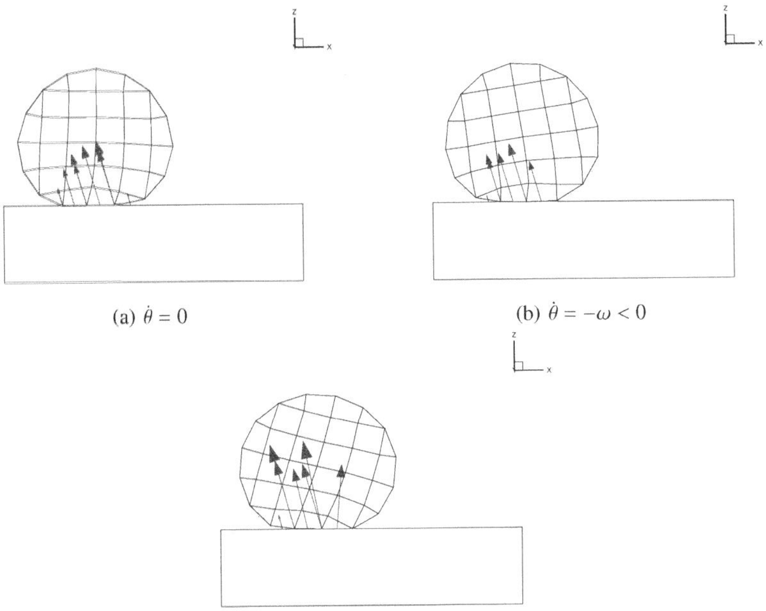

(a) $\dot{\theta} = 0$

(b) $\dot{\theta} = -\omega < 0$

(c) $\dot{\theta} = \omega > 0$

Figure 12.14. *Frictional impact ($\mu = 0.3$) of a disk onto a rigid plane. Contact stresses for a few typical steps. For a color version of this figure, see www.iste.co.uk/levan/contact.zip*

• Energies

Figure 12.15 represents the kinetic, strain and total energies in the case with friction $\mu = 0.3$. Due to friction, the total energy of the system decreases systematically after impact, irrespective of the angular velocity of the disk. Energy loss is the smallest in the $\dot{\theta} > 0$ case and the largest in the $\dot{\theta} < 0$ case.

12.3. Disk falling into a funnel

The motivation is to highlight disk locking due to friction, a phenomenon that did not exist in the preceding example.

In this section, we will consider contact occurring between an elastic disk and a rigid V-shaped funnel (see Figure 12.16). This example is inspired by those treated by Wriggers et al. [142] (a 2D annulus) and Magnain [91] (a 2D disk). Here, we will treat a solid disk in the 3D framework.

The disk has radius 0.1 m and thickness 0.02 m. It is made of a rubber-type Saint-Venant Kirchhoff material with Young's modulus $E = 5 \times 10^7$ Pa, Poisson's ratio $\nu = 0.4$, and its reference density is $\rho_0 = 1{,}000$ kg/m^3. The funnel is composed of two rigid walls that are symmetric relative to the **Oyz** and **Oxz** planes. The trace of the right-hand wall in the **xz** plane is a quadrilateral whose four vertices are $I(0.05, -0.275)$, $J(0.15, -0.275)$, $K(0.15, 0.075)$ and $L(0.12, 0.075)$ (coordinates in m). With these dimensions, the half-opening angle of the funnel is $90° - 78.32° = 11.68°$.

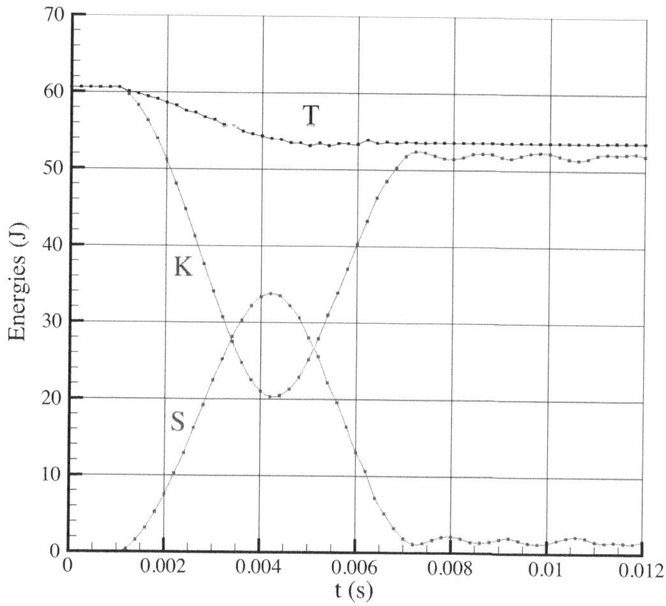

(a) $\dot{\theta} = 0$

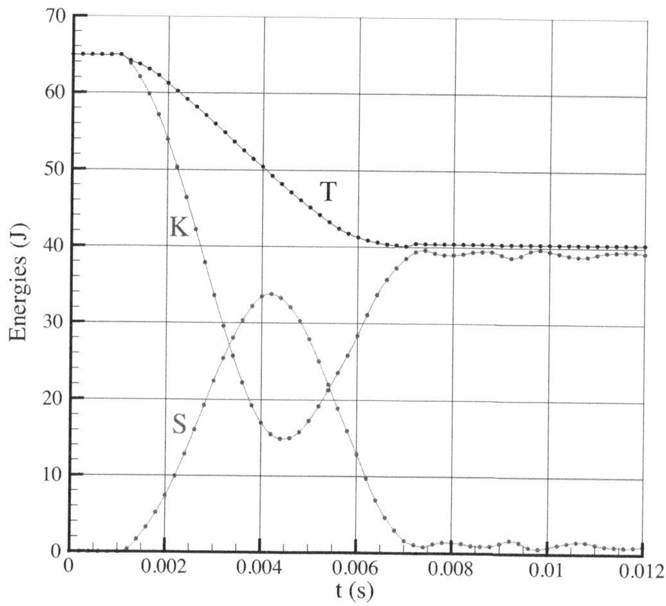

(b) $\dot{\theta} = -\omega < 0$

Figure 12.15. *Frictional impact ($\mu = 0.3$) of a disk onto a rigid plane. Energy evolution. K = kinetic energy, S = strain energy, $T = K + S$ = total energy. For a color version of this figure, see www.iste.co.uk/levan/contact.zip*

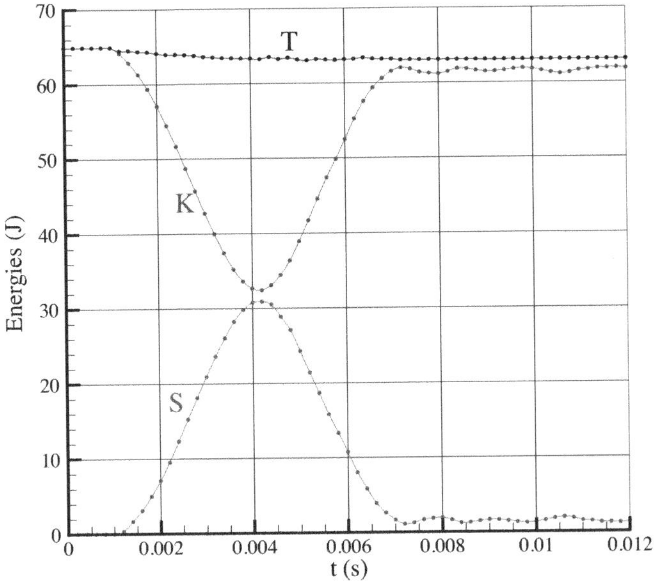

(c) $\dot{\theta} = \omega > 0$

Figure 12.15. *(continued)*

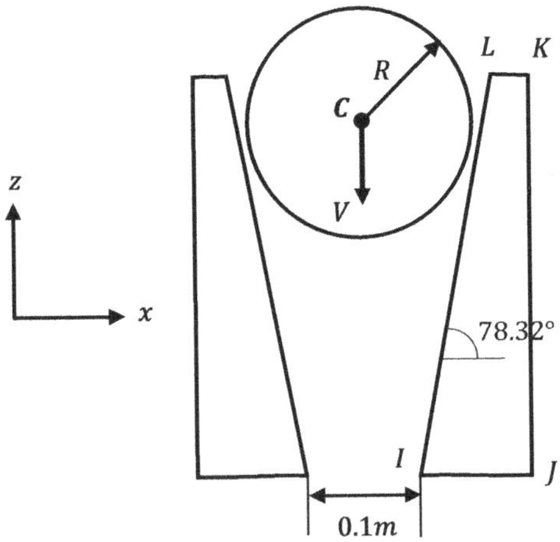

Figure 12.16. *Disk falling into a funnel*

At the initial instant, the disk midplane lies within the **Oxz** plane, and the disk center **C** is located at $(x, y, z) = (0, 0, 0)$. Prior to impact, the disk is propelled into a uniform translational motion in the opposite direction to **z** with velocity $V = 90$ km/h = 25 m/s.

We assume that the contact between the disk and plane takes place either without friction ($\mu = 0$), or with a coefficient of friction $\mu = 0.4$. There is no gravity here.

The entire disk and both the left- and right-hand walls are meshed, although we can just mesh half of the structure, given its symmetry about the central plane. The disk is meshed into 24 twenty-node hexahedral elements (H20) and 8 fifteen-node prismatic elements (P15) (see Figure 12.17). Each wall is meshed using a single H20 element. The disk is considered as the contactor and the two walls as the target.

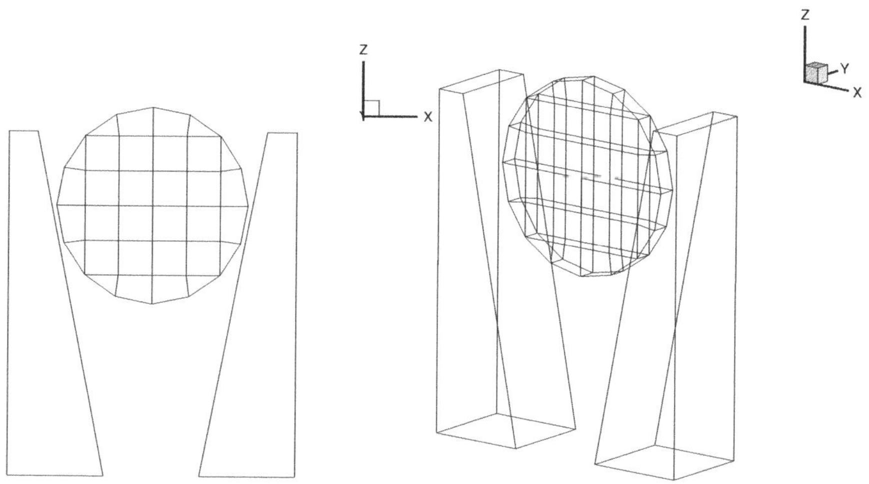

Figure 12.17. *Mesh of the disk falling into a funnel*

Time integration is performed by means of the Newmark algorithm with $\alpha = \beta = \frac{1}{2}$. In each Q8 element on the contactor surface, we take 7×7 Gaussian quadrature points. The time step is $\Delta t = 10^{-4}$ s and each calculation is performed in 130 steps. The convergence tolerances are 10^{-6} for the residual force vector $\{\mathbf{R}\}$ and the residual contact laws $\{\mathbf{R}_\Lambda\}$.

12.3.1. *Frictionless case*

In the frictionless case, we choose the penalty parameters $\epsilon_N = \epsilon_T = 5 \times 10^7$ N/m^3. At each step, the iterative solution using the Newton–Raphson scheme converges in six iterations on average.

- Motion

The disk sinks into the funnel, comes into contact with the two walls for approximately 11 ms, before rebounding back in the **z**-direction.

The position and velocity of the disk center **C** are represented in Figures 12.18 and 12.19 versus time. The disk center reaches its lowest height $z_C = -0.103$ m at $t = 6.2$ ms (this is roughly about halfway through the contact time interval). It stops at this instant before rebounding back, with nearly the same speed at which it arrived but in the opposite direction.

Anticipating Figure 12.21 later on, which depicts the evolution of the kinetic energy, note that it is at the time instant $t = 6$ ms that the kinetic energy of the disk is the lowest, and that the entire disk comes almost to a complete standstill.

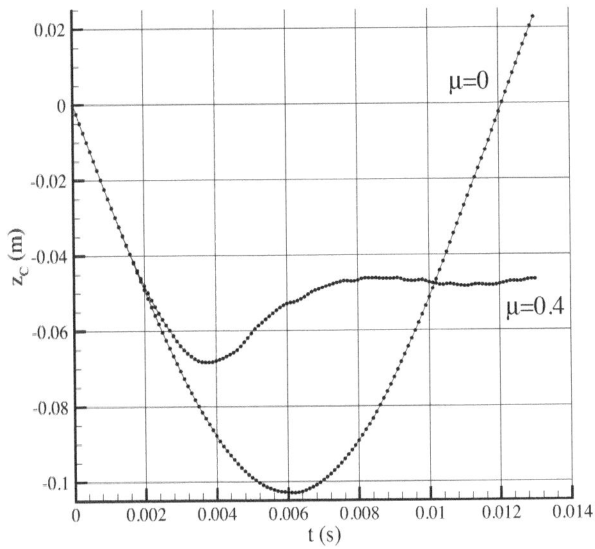

Figure 12.18. *Disk falling into a funnel. Displacement of the disk center versus time*

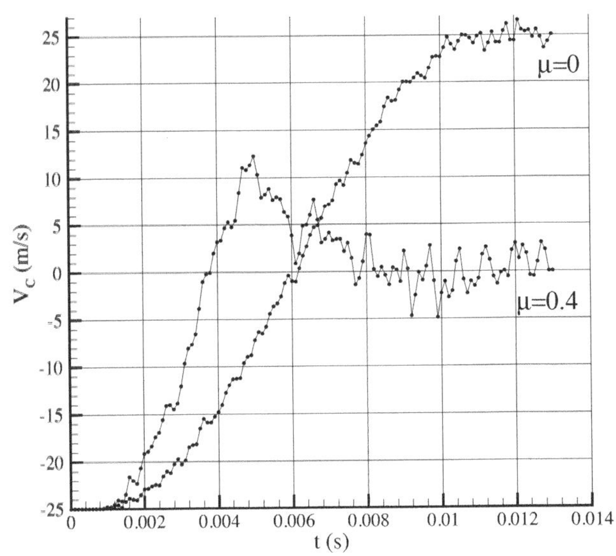

Figure 12.19. *Velocity of the disk center versus time*

• Verification of the inequality $g \leq 0$

Numerical results show that the inequality $g \leq 0$ in [5.3] is satisfied at each node of the contact surface and at each step: at the steps where contact occurs, the maximum value of the proximity g over the nodes in contact is 0.7 mm, that is, 0.7% of the disk radius.

• Contact stresses

The distribution of the contact stresses over both space and time is more or less complex. It is shown in Figure 12.20 at the instant when the disk has descended to its lowest point.

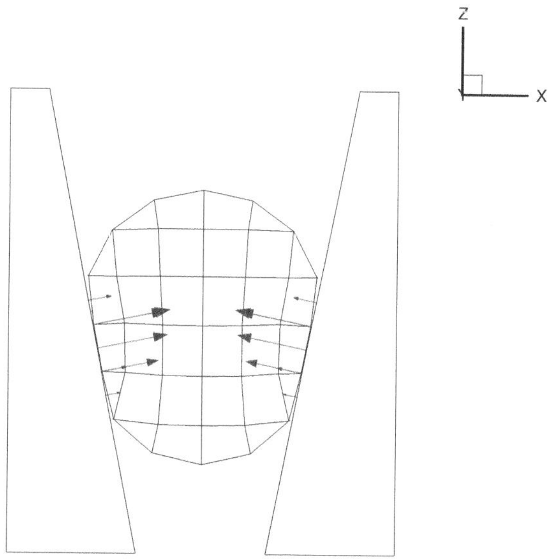

Figure 12.20. *Disk falling without friction into a funnel. Contact stresses at the instant when the disk descends to its lowest point. For a color version of this figure, see www.iste.co.uk/levan/contact.zip*

• Verification of the inequality $T_N \geq 0$

Numerical results show that the inequality $T_N \geq 0$ in [5.3] is well satisfied.

In the course of motion, inevitably there are several elements that straddle between the contact and non-contact zones. In these elements, there may be non-zero contact stresses at the nodes which are not in contact; however, on the one hand, these abnormal stresses are very small and, on the other hand, they are perfectly justifiable, as seen in section 8.9.

• Energies

The time evolution of the energies of the system (kinetic, elastic strain and total energies) is shown in Figure 12.21 in the frictionless case. Since gravity is not considered here, gravitational potential energy does not play any role. In the first phase of contact, kinetic energy drops to zero in favor of the elastic strain energy, since the disk slows down as it becomes more and more deformed. In the second phase of contact, when the disk rebounds backwards, the kinetic energy increases, whereas the strain energy decreases. Once contact has ended, the disk nearly recovers its original circular form; the strain energy is very small without being completely zero, which means that there are small oscillations of the disk after the rebound. A significant amount of the elastic strain energy that is absorbed during the shock is restored as kinetic energy. During the whole process, the total energy of the system remains constant within 0.7%.

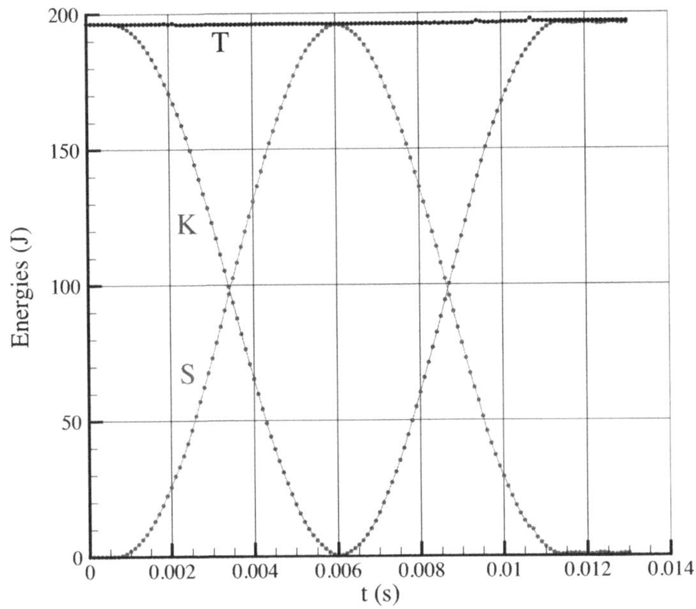

Figure 12.21. *Disk falling without friction into a funnel. Energy evolution. K = kinetic energy, S = strain energy, $T = K + S$ = total energy. For a color version of this figure, see www.iste.co.uk/levan/contact.zip*

12.3.2. *Case with friction* $\mu = 0.4$

The calculation with friction is performed with the penalty parameters $\epsilon_N = \epsilon_T = 5 \times 10^7$ N/m^3. At each step, the iterative solution using the Newton–Raphson scheme converges in seven to eight iterations on average.

• Motion

The position and velocity of the disk center **C** are represented in Figures 12.18 and 12.19 versus time. In the presence of friction $\mu = 0.4$, the disk does not sink as far into the funnel as it does when there is friction, and, at the instant $t = 3.8$ ms when the disk center has descended to its lowest height, $z_C = -68$ mm, the disk practically comes to a standstill.

The disk then slightly rebounds backwards, but fails to detach from the funnel. Instead, it continues to slightly vibrate about a blocked position, with the height of the center **C** oscillating about the value $z_C = -47$ mm. Figures 12.18 and 12.19 show that the oscillation amplitude of the center is small, while the center velocity can be quite large, about the order of $1/6$ of the initial velocity of 25 m/s. However, by anticipating Figure 12.23 later on, we know that after the instant $t = 8$ ms, the kinetic energy of the disk is of the order of $1/100$ of the initial kinetic energy. We can thus say that the disk practically does not move anymore and remains in an equilibrium between the two rough walls. All of these results are in agreement with those reported in Wriggers et al. [142] and Magnain [91].

• Verification of the inequality $g \leq 0$

The numerical results show that the inequality $g \leq 0$ in [5.3] is well satisfied at each node on the contact surface and at each step: at the steps where contact takes place, it is found that the maximum value of the proximity over the nodes in contact is 0.6 mm, that is, 0.6% of the disk radius.

• Contact stresses

With friction $\mu = 0.4$, we can roughly distinguish between three contact stress profiles:

(i) at an instant when the disk sinks into the funnel and there is slip at all nodes in contact, the contact stress vectors are directed upward in the opposite direction to the disk motion (see Figure 12.22(a));

(ii) at an instant when the disk slightly rebounds off, and when there is slip at all nodes in contact, the contract stress vectors are directed downwards (see Figure 12.22(b));

(iii) at other instants when there is stick at certain nodes in contact (adhesion), the contact stress profile is complicated but their resultant is indeed zero in both the **y** and **z** directions, guaranteeing the global equilibrium of the disk (see Figure 12.22(c)).

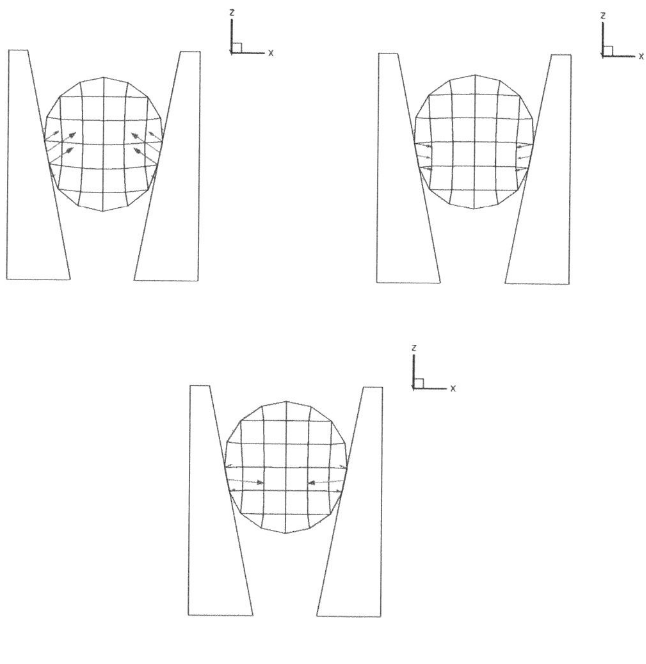

Figure 12.22. *Disk falling with friction ($\mu = 0.4$) into a funnel. Contact stresses at a few representative instants. For a color version of this figure, see www.iste.co.uk/levan/contact.zip*

• Verification of the inequality $T_N \geq 0$.

Like in the frictionless case, the inequality $T_N \geq 0$ in [5.3] is well satisfied, except for a few nodes which belong to elements that are in partial contact.

• Energies

In the frictional case, the evolution of the energy is more complicated than in the frictionless case, as seen in Figure 12.23. The curves have exactly the same shape as in Magnain [91], even if the working hypotheses there are different (this problem was solved in [91], assuming plane strain and the hyperelastic constitutive law of Blatz-Ko).

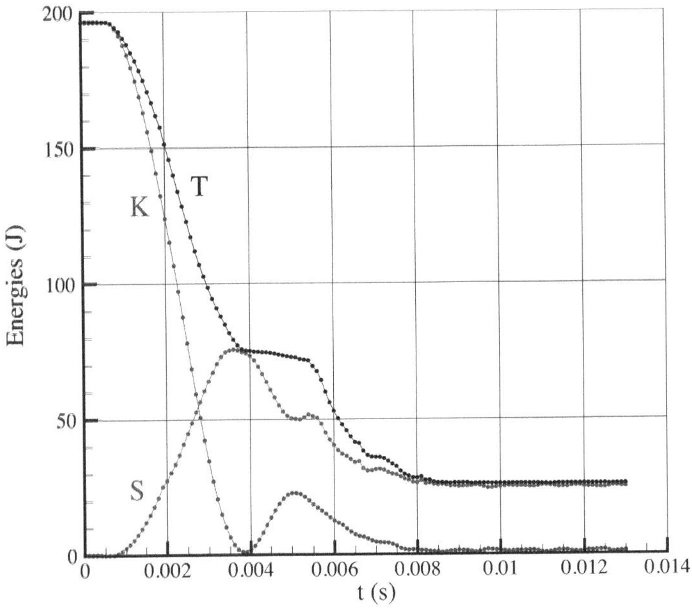

Figure 12.23. *Disk falling into a funnel with friction ($\mu = 0.4$). Energy evolution. K = kinetic energy, S = strain energy, $T = K + S$ = total energy. For a color version of this figure, see www.iste.co.uk/levan/contact.zip*

The total energy of the system drops steadily and noticeably due to friction. At the end of the computation, we find a significant dissipation of the total energy due to friction, and almost all of the remaining energy is in the form of elastic strain energy.

12.4. Final remarks

Time discretization is a difficult subject in nonlinear dynamics in general, and it is even more so when contact is involved. This subject has been the focus of numerous works in the literature, which, in recent years, led to considerable progress toward solving dynamic contact/impact problems. It will undoubtedly continue to mobilize the efforts of researchers to obtain increasingly efficient formulations. In Chapter 11, we did not seek to develop a new time discretization method capable of competing with the sophisticated pre-existing ones; instead, we used the Newmark algorithm to discretize the equation of motion $[8.47]_1$ with respect to time. This scheme is certainly not ideal, but it is simple enough to implement and also allows us to test how the contact formulation performs when contact is added. The examples discussed in this chapter show that the contact formulation, when combined with the time-stepping Newmark algorithm, produces quite reasonable results.

In these examples, we can see that the total energy of the system is fairly well conserved when there is no friction. But in fact this is fortuitous, since it is known that the Newmark algorithm with $\alpha = \beta = 1/2$ is not conservative when there are large deformations. When the Newmark algorithm does perform fairly conservatively, it tends to slightly overestimate the post-shock energy.

An interesting alternative to the Newmark algorithm is that of Moreau [97]. This first-order scheme is particularly well suited to dynamic contact problems, as it automatically incorporates the velocity discontinuities into the formulation and does not involve the acceleration variable at impact, which is a Dirac measure. From an energy point of view, Moreau's scheme performs better than Newmark's scheme, since it produces a slight decrease in total energy over time, and an energy that dissipates is physically more acceptable than one that somehow increases on its own. What is more, in some cases, Moreau's scheme allows for a larger time step than with the Newmark algorithm.

The energy-momentum impact algorithm proposed by Laursen et al. [78] and [77] is an even more sophisticated algorithm for dynamic contact. Based on a completely different approach to that proposed in this book, it forces the computational scheme to exactly conserve the total energy in the frictionless case. This algorithm was subsequently improved in [79] and [77], where the energy conservation procedure was combined with a formula that updates the velocities using a special quantity called the "discrete contact velocity".

Appendix A

Variations of Kinematic Quantities

In this appendix, variations of the different kinematic quantities defined on the contact surfaces will be calculated. The results obtained will be used to establish the expression of the contact tangent matrix in Chapter 9. We are going to study the quantities defined in Chapter 2 for the kinematics of contact.

Recall the following parameterizations [2.17] for the contact surfaces:

$$\begin{array}{llll} \forall i \in \{1,2\}, \text{ a domain of } \mathbb{R}^2 & \rightarrow S_{0c}^{(i)} & & \rightarrow S_c^{(i)} \\ \xi^{(i)} & \mapsto \mathbf{X}^{(i)} = \boldsymbol{\theta}^{(i)}(\xi^{(i)}) & \mapsto & \mathbf{x}^{(i)} = \boldsymbol{\phi}^{(i)}(\mathbf{X}^{(i)}, t) = \boldsymbol{\psi}^{(i)}(\xi^{(i)}, t) \end{array}$$

Consider a specific particle located on the contactor surface with pre-image ξ, initial position $\mathbf{X}$, current position $\mathbf{x}$ and whose projection onto the current target surface is $\mathbf{y}$. The previous parameterizations yield [2.18]–[2.19]:

$$\begin{array}{lll} \xi \mapsto \mathbf{X} & \mapsto & \mathbf{x} = \boldsymbol{\phi}^{(1)}(\mathbf{X}, t) \\ \eta \mapsto \mathbf{Y} = \boldsymbol{\theta}^{(2)}(\eta) & \mapsto & \boxed{\mathbf{y} = \boldsymbol{\phi}^{(2)}(\mathbf{Y}, t) = \boldsymbol{\phi}^{(2)}(\mathbf{Y}(\eta), t) \equiv \boldsymbol{\psi}^{(2)}(\eta, t)} \end{array}$$

Now, consider a given time instant t and the variations $\delta\mathbf{U}^{(1)}(\mathbf{X}^{(1)})$ and $\delta\mathbf{U}^{(2)}(\mathbf{X}^{(2)})$ of the displacement fields $\mathbf{U}^{(1)}$ and $\mathbf{U}^{(2)}$ at that instant. Restricted to the contact surfaces, these variations are parameterized as follows:

$$\begin{array}{llll} \forall i \in \{1,2\}, \text{ a domain of } \mathbb{R}^2 & \rightarrow S_{0c}^{(i)} & & \rightarrow \mathbb{R}^3 \\ \xi^{(i)} & \mapsto \mathbf{X}^{(i)} = \boldsymbol{\theta}^{(i)}(\xi^{(i)}) & \mapsto & \boxed{\delta\mathbf{U}^{(i)}(\mathbf{X}^{(i)}(\xi^{(i)})) \equiv \delta\mathbf{U}^{(i)}(\xi^{(i)})} \end{array}$$

where the abuse of notation $\delta\mathbf{U}^{(i)}(\xi^{(i)}) \equiv \delta\mathbf{U}^{(i)}(\mathbf{X}^{(i)}(\xi^{(i)}))$ should be no source of confusion.

These variations give rise to changes in the position of the contact surfaces $S_c^{(1)}$ and $S_c^{(2)}$. These surface positions after variations are shown by dotted lines in Figure A.1.

Let us consider a given particle situated on the contactor surface, with pre-image ξ and initial position $\mathbf{X}$. The variations $\delta\mathbf{U}^{(1)}$ and $\delta\mathbf{U}^{(2)}$ do not change ξ and $\mathbf{X}$, yet they do cause other kinematic quantities to change according to the following chain (Figure A.1):

$$\left\{ \begin{array}{l} \mathbf{U}^{(1)} \text{ varies } \rightarrow \boldsymbol{\phi}^{(1)} \text{ varies } \rightarrow \mathbf{x} = \boldsymbol{\phi}^{(1)}(\mathbf{X}, t) \text{ varies} \\ \mathbf{U}^{(2)} \text{ varies } \rightarrow \boldsymbol{\phi}^{(2)} \text{ varies} \end{array} \right\}$$

$$\rightarrow \mathbf{y} = \text{ function}(\mathbf{x}, \boldsymbol{\phi}^{(2)}(., t)) \text{ varies } \rightarrow \left\{ \begin{array}{l} \eta \text{ varies} \\ \mathbf{a}_1, \mathbf{a}_2, \boldsymbol{\nu} \text{ vary} \end{array} \right.$$

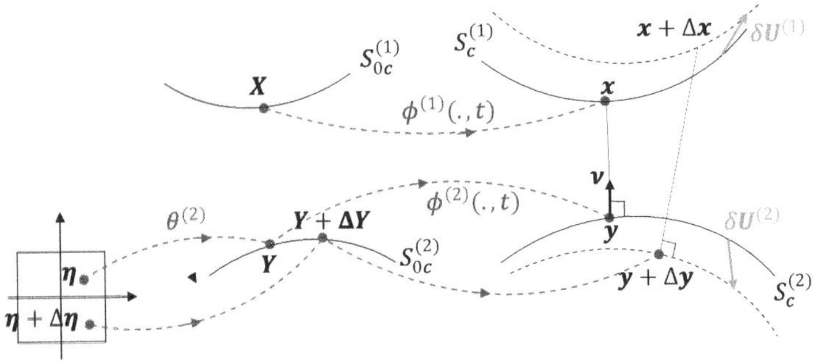

Figure A.1. *Variations $\delta U^{(1)}$ and $\delta U^{(2)}$, and how the other kinematic quantities change. For a color version of this figure, see www.iste.co.uk/levan/contact.zip*

Then, from the definition $g = -\nu.(\mathbf{x} - \mathbf{y})$, it can be deduced that g also varies.

We will calculate the variations of the quantities listed above. According to the calculus of variations, the variations $\delta\mathbf{U}^{(1)}$ and $\delta\mathbf{U}^{(2)}$ are considered as basic (i.e. first-order) infinitesimals. Let $f(\mathbf{U}^{(1)}, \mathbf{U}^{(2)})$ be a functional depending on the displacement fields $\mathbf{U}^{(1)}$ and $\mathbf{U}^{(2)}$. The variations $\delta\mathbf{U}^{(1)}$, $\delta\mathbf{U}^{(2)}$ cause a change in f, denoted by Δf, which is measured by *the difference* between the two expressions of f, one after and the other before its variation:

$$\Delta f \equiv f(\mathbf{U}^{(1)} + \delta\mathbf{U}^{(1)}, \mathbf{U}^{(2)} + \delta\mathbf{U}^{(2)}) - f(\mathbf{U}^{(1)}, \mathbf{U}^{(2)})$$

This difference is not necessarily of the same order as $\delta\mathbf{U}^{(1)}$ and $\delta\mathbf{U}^{(2)}$. By definition, the *variation* of the functional f, corresponding to the variations $\delta\mathbf{U}^{(1)}$, $\delta\mathbf{U}^{(2)}$, and denoted by δf, is the part of Δf which is first-order like $\delta\mathbf{U}^{(1)}, \delta\mathbf{U}^{(2)}$.

Note that Figure A.1 contains both the variations $\delta\mathbf{U}^{(1)}$, $\delta\mathbf{U}^{(2)}$ (written with the symbol δ) and the resulting differences (written with the symbol Δ).

The calculations developed in this appendix are similar to those in section 2.4, which was dedicated to the derivatives with respect to time of various kinematic quantities, namely, $\dot{\mathbf{x}}$, $\dot{\mathbf{y}}$, $\dot{\mathbf{a}}_\alpha$, $\dot{\nu}$, $\dot{g}$ and $\dot{\eta}^\alpha$.

Let us begin with a straightforward result.

Theorem.– $\forall i \in \{1,2\}$, the variation of the deformation at a given instant t, $\boldsymbol{\phi}^{(i)}(.,t)$, is given by

$$\forall \mathbf{X}^{(i)} \textit{ fixed}, \quad \boxed{\delta\boldsymbol{\phi}^{(i)}(\mathbf{X}^{(i)}) = \delta\mathbf{U}^{(i)}(\mathbf{X}^{(i)})} \qquad \text{[A.1]}$$

Proof. To obtain the variation $\delta\boldsymbol{\phi}^{(i)}$, let us first calculate the change $\Delta\boldsymbol{\phi}^{(i)}$ before then extracting the first-order term.

- Before the variation $\delta\mathbf{U}^{(i)}(\mathbf{X}^{(i)})$, we have $\boldsymbol{\phi}^{(i)}(\mathbf{X}^{(i)}, t) = \mathbf{X}^{(i)} + \mathbf{U}^{(i)}(\mathbf{X}^{(i)}, t)$.
- After the variation, $\boldsymbol{\phi}^{(i)}$ undergoes a change $\Delta\boldsymbol{\phi}^{(i)}$ such that

$$\left(\boldsymbol{\phi}^{(i)} + \Delta\boldsymbol{\phi}^{(i)}\right)(\mathbf{X}^{(i)}, t) = \mathbf{X}^{(i)} + \left(\mathbf{U}^{(i)} + \delta\mathbf{U}^{(i)}\right)(\mathbf{X}^{(i)}, t) = \mathbf{X}^{(i)} + \mathbf{U}^{(i)}(\mathbf{X}^{(i)}, t) + \delta\mathbf{U}^{(i)}(\mathbf{X}^{(i)})$$

By taking the difference between the two previous relations, we obtain $\Delta\boldsymbol{\phi}^{(i)} = \delta\mathbf{U}^{(i)}(\mathbf{X})$.

The variation $\delta\boldsymbol{\phi}^{(i)}$, defined to be the first-order part of $\Delta\boldsymbol{\phi}^{(i)}$, is equal to the change $\Delta\boldsymbol{\phi}^{(i)}$ itself. ■

The following result is a particular case of the previous one, when $i = 1$ and $\mathbf{X}^{(1)}$ is equal to a particular point $\mathbf{X}$ of $S_{0c}^{(1)}$:

THEOREM.– The variation of the position $\mathbf{x}$ is

$$\boxed{\delta\mathbf{x} = \delta\mathbf{U}^{(1)}(\mathbf{X})} \quad [A.2]$$

PROOF.

- Before the variation $\delta\mathbf{U}^{(1)}(\mathbf{X})$, we have $\mathbf{x} = \boldsymbol{\phi}^{(1)}(\mathbf{X}, t)$."
- After the variation, the position $\mathbf{x}$ undergoes a change $\Delta\mathbf{x}$ such that $\mathbf{x} + \Delta\mathbf{x} = \left(\boldsymbol{\phi}^{(1)} + \delta\boldsymbol{\phi}^{(1)}\right)(\mathbf{X}, t) = \boldsymbol{\phi}^{(1)}(\mathbf{X}, t) + \delta\boldsymbol{\phi}^{(1)}(\mathbf{X})$.

Then, by taking the difference and applying [A.1], we get: $\Delta\mathbf{x} = \delta\boldsymbol{\phi}^{(1)}(\mathbf{X}) = \delta\mathbf{U}^{(1)}(\mathbf{X})$.
The variation $\delta\mathbf{x}$, defined as the first-order part of $\Delta\mathbf{x}$, is equal to the change $\Delta\mathbf{x}$ itself. ■

The following result is not as straightforward as the previous ones.

THEOREM.– The variation of the projection point $\mathbf{y}$ is

$$\boxed{\delta\mathbf{y} = \delta\mathbf{U}^{(2)}(\mathbf{Y}(\mathbf{X}, t)) + \mathbf{a}_\alpha\, \delta\eta^\alpha} \quad [A.3]$$

where $\delta\eta^\alpha$ will be established in [A.5] later.

PROOF. Recall the notation from [2.19]: $\boldsymbol{\psi}^{(2)}(\,.\,, t) = \boldsymbol{\phi}^{(2)}(\,.\,, t) \circ \boldsymbol{\theta}^{(2)}(\,.\,)$, and furthermore denote:

- $\delta\boldsymbol{\psi}^{(2)}(\,.\,) \equiv \delta\boldsymbol{\phi}^{(2)}(\,.\,) \circ \boldsymbol{\theta}^{(2)}(\,.\,)$;
- $\Delta\boldsymbol{\eta}$ to be the change in $\boldsymbol{\eta}$, resulting from the variations $\delta\mathbf{U}^{(1)}$ and $\delta\mathbf{U}^{(2)}$.

We have

$$\begin{cases} \mathbf{y} &= \boldsymbol{\phi}^{(2)}(\mathbf{Y}, t) = \boldsymbol{\psi}^{(2)}(\boldsymbol{\eta}, t) \\ \mathbf{y} + \Delta\mathbf{y} &= \left(\boldsymbol{\psi}^{(2)} + \delta\boldsymbol{\psi}^{(2)}\right)\underline{(\boldsymbol{\eta} + \Delta\boldsymbol{\eta}, t)} = \boldsymbol{\psi}^{(2)}(\boldsymbol{\eta} + \Delta\boldsymbol{\eta}, t) + \delta\boldsymbol{\psi}^{(2)}(\boldsymbol{\eta} + \Delta\boldsymbol{\eta}) \end{cases}$$

Hence the change $\Delta\mathbf{y}$, by taking the difference of the two previous relations:

$$\Delta\mathbf{y} = \underbrace{\boldsymbol{\psi}^{(2)}(\boldsymbol{\eta} + \Delta\boldsymbol{\eta}, t) - \boldsymbol{\psi}^{(2)}(\boldsymbol{\eta}, t)}_{\boldsymbol{\psi}^{(2)}_{,\alpha}\Delta\eta^\alpha \,+\, \text{h.o.t.}} + \underbrace{\delta\boldsymbol{\psi}^{(2)}(\boldsymbol{\eta} + \Delta\boldsymbol{\eta})}_{\delta\boldsymbol{\psi}^{(2)}(\boldsymbol{\eta}) + \delta\boldsymbol{\psi}^{(2)}_{,\alpha}\Delta\eta^\alpha \,+\, \text{h.o.t.}}$$

where h.o.t. stands for *higher-order terms*. Thus, Q.E.D., after noting that $\boldsymbol{\psi}^{(2)}_{,\alpha} = \mathbf{a}_\alpha$, $\delta\boldsymbol{\psi}^{(2)}(\boldsymbol{\eta}) = \delta\mathbf{U}^{(2)}(\boldsymbol{\eta})$, and that the expression $\delta\boldsymbol{\psi}^{(2)}_{,\alpha}\Delta\eta^\alpha + \text{h.o.t.}$ is itself of order higher than one. ■

THEOREM.– $\forall\alpha \in \{1, 2\}$, the variation of the vector $\mathbf{a}_\alpha$ of the natural basis at $\mathbf{y}$ is given by

$$\boxed{\delta\mathbf{a}_\alpha = \delta\mathbf{U}^{(2)}_{,\alpha}(\boldsymbol{\eta}) + \boldsymbol{\psi}^{(2)}_{,\alpha\beta}\, \delta\eta^\beta} \quad [A.4]$$

where $\delta\eta^\alpha$ is given by [A.5], and $(\bullet)_{,\alpha} \equiv \dfrac{\partial(\bullet)}{\partial\xi^{(2)\alpha}}$.

PROOF. Using the notation $\Delta\eta$ introduced in the previous proof, we have

$$\begin{cases} \mathbf{a}_\alpha & = \left.\dfrac{\partial \psi^{(2)}}{\partial \xi^{(2)\alpha}}\right|_{\eta} \\ \mathbf{a}_\alpha + \Delta\mathbf{a}_\alpha & = \left.\dfrac{\partial\left(\psi^{(2)} + \delta\psi^{(2)}\right)}{\partial \xi^{(2)\alpha}}\right|_{\eta+\Delta\eta} = \left.\dfrac{\partial \psi^{(2)}}{\partial \xi^{(2)\alpha}}\right|_{\eta+\Delta\eta} + \left.\dfrac{\partial \delta\psi^{(2)}}{\partial \xi^{(2)\alpha}}\right|_{\eta+\Delta\eta} \end{cases}$$

From which, by taking their difference:

$$\Delta\mathbf{a}_\alpha = \underbrace{\left.\frac{\partial \psi^{(2)}}{\partial \xi^{(2)\alpha}}\right|_{\eta+\Delta\eta} - \left.\frac{\partial \psi^{(2)}}{\partial \xi^{(2)\alpha}}\right|_{\eta}}_{\frac{\partial^2 \psi^{(2)}}{\partial \xi^{(2)\alpha}\partial \xi^{(2)\beta}}\Delta\eta^\beta + \text{h.o.t.}} + \underbrace{\left.\frac{\partial \delta\psi^{(2)}}{\partial \xi^{(2)\alpha}}\right|_{\eta+\Delta\eta}}_{\delta\psi^{(2)}_{,\alpha}(\boldsymbol{\eta}) + \text{h.o.t.}}$$

By noticing that $\delta\psi^{(2)}_{,\alpha}(\boldsymbol{\eta}) = \delta\mathbf{U}^{(2)}_{,\alpha}(\boldsymbol{\eta})$ from [A.1], the proof is complete. ■

REMARK.– The following implication, which gives a different result to [A.4], is *false*:

$$\mathbf{a}_\alpha = \mathbf{a}_\alpha(\boldsymbol{\eta}, t) \qquad \Rightarrow \qquad \delta\mathbf{a}_\alpha = \mathbf{a}_{\alpha,\beta}\,\delta\eta^\beta = \psi^{(2)}_{,\alpha\beta}\,\delta\eta^\beta$$

Indeed, the previous statement would be true if the function $(\boldsymbol{\eta}, t) \mapsto \mathbf{a}_\alpha(\boldsymbol{\eta}, t)$ remained the same and if only $\boldsymbol{\eta}$ varied. Here, not only does $\boldsymbol{\eta}$ vary, but the function $\psi^{(2)}$ varies, becoming $\psi^{(2)} + \delta\psi^{(2)}$. It is the variation $\delta\psi^{(2)}$ which generated the term $\delta\mathbf{U}^{(2)}_{,\alpha}(\boldsymbol{\eta})$ in [A.4]. ■

THEOREM.– $\forall\alpha \in \{1, 2\}$, the variation of the coordinate η^α can be expressed in terms of the variations $\delta\mathbf{U}^{(1)}$ and $\delta\mathbf{U}^{(2)}$ by

$$\boxed{\delta\eta^\alpha = \bar{a}^{\alpha\beta}\mathbf{a}_\beta.\left[\delta\mathbf{U}^{(1)}(\mathbf{X}) - \delta\mathbf{U}^{(2)}(\mathbf{Y}(\mathbf{X}, t))\right] - g\bar{a}^{\alpha\beta}\boldsymbol{\nu}.\delta\mathbf{U}^{(2)}_{,\beta}(\boldsymbol{\eta})} \qquad \text{[A.5]}$$

where the notation $\bar{a}^{\alpha\beta}$ was introduced in [2.26].

PROOF. To determine the variations $\delta\eta^\alpha$, again let us consider the relation [2.20]:

$$\forall\alpha \in \{1, 2\}, \quad \left(\mathbf{x} - \psi^{(2)}(\boldsymbol{\eta}, t)\right).\frac{\partial \psi^{(2)}}{\partial \xi^{(2)\alpha}}(\boldsymbol{\eta}, t) = 0 \qquad \text{i.e.} \qquad (\mathbf{x} - \mathbf{y}).\mathbf{a}_\alpha = 0 \qquad \text{[A.6]}$$

Given the time instant t and point $\mathbf{x}$, this relation constitutes two scalar equations for two unknowns $\boldsymbol{\eta} = (\eta_1, \eta_2)$. Suppose that these equations have already been solved and that the point $\boldsymbol{\eta}$ is known. By taking the variation of [A.6], we get

$$(\delta\mathbf{x} - \delta\mathbf{y}).\mathbf{a}_\alpha + \underbrace{(\mathbf{x} - \mathbf{y})}_{-g\boldsymbol{\nu}}.\delta\mathbf{a}_\alpha = 0$$

By using the expressions [A.2]-[A.4] of the variations $\delta\mathbf{x}$, $\delta\mathbf{y}$ and $\delta\mathbf{a}_\alpha$ established previously, we obtain

$$\left[\delta\mathbf{U}^{(1)}(\mathbf{X}) - \delta\mathbf{U}^{(2)}(\mathbf{Y})\right].\mathbf{a}_\alpha - \underbrace{\left(a_{\alpha\beta} + g\boldsymbol{\nu}.\psi^{(2)}_{,\alpha\beta}\right)}_{\bar{a}_{\alpha\beta}}\delta\eta^\beta - g\boldsymbol{\nu}.\delta\mathbf{U}^{(2)}_{,\alpha}(\boldsymbol{\eta}) = 0$$

that is:

$$(a_{\alpha\beta} + gb_{\alpha\beta})\delta\eta^{\alpha} = \mathbf{a}_{\alpha}.\left[\delta\mathbf{U}^{(1)}(\mathbf{X}) - \delta\mathbf{U}^{(2)}(\mathbf{Y}(\mathbf{X},t))\right] - g\boldsymbol{\nu}.\delta\mathbf{U}^{(2)}_{,\alpha}(\boldsymbol{\eta})$$

This is a system involving two scalar equations for two unknowns $\delta\eta^{\alpha}$, and inverting it yields [A.5]. ■

Theorem. The variation of the proximity g is given by

$$\boxed{\delta g = -\left[\delta\mathbf{U}^{(1)}(\mathbf{X}) - \delta\mathbf{U}^{(2)}(\mathbf{Y}(\mathbf{X},t))\right].\boldsymbol{\nu}} \qquad \text{where } \boldsymbol{\nu} = \boldsymbol{\nu}(\mathbf{Y}(\mathbf{X},t)) \qquad [A.7]$$

PROOF. To calculate δg, let us take the variation of relation [2.13]:

$$\begin{aligned} -\delta g\ \boldsymbol{\nu} - g\ \delta\boldsymbol{\nu} &= \delta\mathbf{x} - \delta\mathbf{y} \\ &= \delta\mathbf{U}^{(1)}(\mathbf{X}) - \delta\mathbf{U}^{(2)}(\mathbf{Y}) - \delta\eta^{\alpha}\mathbf{a}_{\alpha} \quad \text{by using [A.2] and [A.3]} \end{aligned} \qquad [A.8]$$

Since $\|\boldsymbol{\nu}\|^2 = 1$, we have $\boldsymbol{\nu}.\delta\boldsymbol{\nu} = 0$, that is to say $\delta\boldsymbol{\nu} \perp \boldsymbol{\nu}$. By projecting relation [A.8] onto $\boldsymbol{\nu}$, we obtain [A.7]. ■

THEOREM.–

1. The variation of the normal vector $\boldsymbol{\nu}$ at $\mathbf{y}$ is given by

$$\boxed{\delta\boldsymbol{\nu} = -(\boldsymbol{\nu}.\delta\mathbf{a}_{\alpha})\,\mathbf{a}^{\alpha} = -\Big[\boldsymbol{\nu}.\delta\mathbf{U}^{(2)}_{,\alpha} + \underbrace{\boldsymbol{\nu}.\boldsymbol{\psi}^{(2)}_{,\alpha\beta}}_{b_{\alpha\beta}}\ \delta\eta^{\beta}\Big]\mathbf{a}^{\alpha}} \qquad [A.9]$$

2. A second expression for $\delta\boldsymbol{\nu}$, which is purely a function of just $\delta\mathbf{U}^{(1)}, \delta\mathbf{U}^{(2)}$ ($\delta\eta^{\alpha}$ is no longer present), is:

$$\boxed{\delta\boldsymbol{\nu} = -\left(b_{\alpha\gamma}\bar{a}^{\gamma\beta}\mathbf{a}^{\alpha}\otimes\mathbf{a}_{\beta}\right).\left(\delta\mathbf{U}^{(1)} - \delta\mathbf{U}^{(2)}\right) - \left(\bar{a}^{\alpha\beta}\mathbf{a}_{\beta}\otimes\boldsymbol{\nu}\right).\delta\mathbf{U}^{(2)}_{,\alpha}} \qquad [A.10]$$

The term $b_{\alpha\gamma}\bar{a}^{\gamma\beta}\mathbf{a}^{\alpha}\otimes\mathbf{a}_{\beta}$ can be replaced by any other equivalent expression provided from [2.31].

PROOF.

1. Since $\delta\boldsymbol{\nu} \perp \boldsymbol{\nu}$, the vector $\delta\boldsymbol{\nu}$ lies within the tangent plane spanned by $(\mathbf{a}_1, \mathbf{a}_2)$, and we can write

$$\begin{aligned} \delta\boldsymbol{\nu} &= (\delta\boldsymbol{\nu}.\mathbf{a}_{\alpha})\,\mathbf{a}^{\alpha} \\ &= -(\boldsymbol{\nu}.\delta\mathbf{a}_{\alpha})\,\mathbf{a}^{\alpha} \quad \text{since } 0 = \delta(\boldsymbol{\nu}.\mathbf{a}_{\alpha}) = \delta\boldsymbol{\nu}.\mathbf{a}_{\alpha} + \boldsymbol{\nu}.\delta\mathbf{a}_{\alpha} \end{aligned}$$

REMARK.– In an analogous manner to the remark made after expression [A.4] of $\delta\mathbf{a}_{\alpha}$, we should note that the following implication is *false*:

$$\boldsymbol{\nu} = \boldsymbol{\nu}(\boldsymbol{\eta}, t) \qquad \Rightarrow \qquad \delta\boldsymbol{\nu} = \boldsymbol{\nu}_{,\alpha}\delta\eta^{\alpha}$$

In any case, the derivative with respect to $\xi^{(2)\alpha}$ of the expression $\boldsymbol{\nu} = \dfrac{\mathbf{a}_1 \times \mathbf{a}_2}{\|\mathbf{a}_1 \times \mathbf{a}_2\|}$ is not straightforward, and it is much easier to calculate $\delta\boldsymbol{\nu}$ instead as above. ■

2. To obtain [A.10], replace $\delta\eta^\beta$ in [A.9] with its expression [A.5]:

$$\delta\boldsymbol{\nu} = -\left(\boldsymbol{\nu}.\delta\mathbf{U}^{(2)}_{,\alpha}\right)\mathbf{a}^\alpha - b_{\alpha\beta}\bar{a}^{\beta\gamma}\left(\mathbf{a}^\alpha\otimes\mathbf{a}_\gamma\right).\left(\delta\mathbf{U}^{(1)} - \delta\mathbf{U}^{(2)}\right) + \underbrace{g b_{\alpha\beta}}_{=\bar{a}_{\alpha\beta}-a_{\alpha\beta}}\bar{a}^{\beta\gamma}\left(\boldsymbol{\nu}.\delta\mathbf{U}^{(2)}_{,\gamma}\right)\mathbf{a}^\alpha$$

Knowing that $\bar{a}_{\alpha\beta}\bar{a}^{\beta\gamma} = \delta^\gamma_\alpha$ and $a_{\alpha\beta}\mathbf{a}^\alpha = \mathbf{a}_\beta$, we get

$$\begin{aligned}\delta\boldsymbol{\nu} &= -\cancel{\left(\boldsymbol{\nu}.\delta\mathbf{U}^{(2)}_{,\alpha}\right)\mathbf{a}^\alpha} - b_{\alpha\beta}\bar{a}^{\beta\gamma}\left(\mathbf{a}^\alpha\otimes\mathbf{a}_\gamma\right).\left(\delta\mathbf{U}^{(1)} - \delta\mathbf{U}^{(2)}\right) + \cancel{\left(\boldsymbol{\nu}.\delta\mathbf{U}^{(2)}_{,\alpha}\right)\mathbf{a}^\alpha} - \bar{a}^{\beta\gamma}\left(\boldsymbol{\nu}.\delta\mathbf{U}^{(2)}_{,\gamma}\right)\mathbf{a}_\beta \\ &= -\left(b_{\alpha\beta}\bar{a}^{\beta\gamma}\mathbf{a}^\alpha\otimes\mathbf{a}_\gamma\right).\left(\delta\mathbf{U}^{(1)} - \delta\mathbf{U}^{(2)}\right) - \left(\bar{a}^{\beta\gamma}\mathbf{a}_\beta\otimes\boldsymbol{\nu}\right).\delta\mathbf{U}^{(2)}_{,\gamma}\end{aligned}$$

By rewriting $b_{\alpha\beta}\bar{a}^{\beta\gamma}\mathbf{a}^\alpha\otimes\mathbf{a}_\gamma = b_{\alpha\gamma}\bar{a}^{\gamma\beta}\mathbf{a}^\alpha\otimes\mathbf{a}_\beta$, we obtain [A.10]. ■

THEOREM.– A third expression of $\delta\boldsymbol{\nu}$ is the following:

$$\boxed{\delta\boldsymbol{\nu} = -\frac{1}{g}\left[(\mathbf{I} - \boldsymbol{\nu}\otimes\boldsymbol{\nu}).\left(\delta\mathbf{U}^{(1)} - \delta\mathbf{U}^{(2)}\right) - \mathbf{a}_\alpha\delta\eta^\alpha\right]} \qquad \text{[A.11]}$$

PROOF. We just have to consider [A.8] again and, this time, project it onto the tangent plane normal to $\boldsymbol{\nu}$ by using the projection operator $\mathbf{I} - \boldsymbol{\nu}\otimes\boldsymbol{\nu}$. ■

The third expression [A.11] of $\delta\boldsymbol{\nu}$ is equivalent to the first expression [A.9], though it has the disadvantage of containing the quotient $1/g$, which is problematic when contact occurs ($g = 0$).

Recall that we have transformed [A.9] to obtain the second expression [A.10] of $\delta\boldsymbol{\nu}$. Here, we can also transform [A.11] to obtain [A.10] again, although this way is a bit longer. By substituting the expression [A.5] of $\delta\eta^\alpha$ into [A.11], we get

$$\delta\boldsymbol{\nu} = -\frac{1}{g}\left(\mathbf{I} - \boldsymbol{\nu}\otimes\boldsymbol{\nu} - \bar{a}^{\alpha\beta}\mathbf{a}_\alpha\otimes\mathbf{a}_\beta\right).\left(\delta\mathbf{U}^{(1)} - \delta\mathbf{U}^{(2)}\right) - \left(\bar{a}^{\alpha\beta}\mathbf{a}_\beta\otimes\boldsymbol{\nu}\right).\delta\mathbf{U}^{(2)}_{,\alpha}$$

To get [A.10], we just have to prove that $\frac{1}{g}\left(\mathbf{I} - \boldsymbol{\nu}\otimes\boldsymbol{\nu} - \bar{a}^{\alpha\beta}\mathbf{a}_\alpha\otimes\mathbf{a}_\beta\right)$ is equal to $b_{\alpha\gamma}\bar{a}^{\alpha\beta}\mathbf{a}_\beta\otimes\mathbf{a}^\gamma$. This is the purpose of the following theorem.

THEOREM.–

$$\boxed{\begin{aligned}\frac{1}{g}\left(\mathbf{I} - \boldsymbol{\nu}\otimes\boldsymbol{\nu} - \bar{a}^{\alpha\beta}\mathbf{a}_\alpha\otimes\mathbf{a}_\beta\right) &= \bar{a}^{\alpha\gamma}b_{\gamma\delta}a^{\delta\beta}\mathbf{a}_\alpha\otimes\mathbf{a}_\beta \\ &= a^{\alpha\gamma}b_{\gamma\delta}\bar{a}^{\delta\beta}\mathbf{a}_\alpha\otimes\mathbf{a}_\beta\end{aligned}} \qquad \text{[A.12]}$$

These expressions could be rewritten by using the contractions $a^{\delta\beta}\mathbf{a}_\beta = \mathbf{a}^\delta$, $a^{\alpha\gamma}\mathbf{a}_\alpha = \mathbf{a}^\gamma$. Unlike the left-hand side of [A.12], here the last two sides do not create any numerical issue in the case when $g \to 0$!

PROOF.

$$\mathbf{I} = a^{\alpha\beta}\mathbf{a}_\alpha\otimes\mathbf{a}_\beta + \boldsymbol{\nu}\otimes\boldsymbol{\nu} \qquad \Rightarrow \qquad \mathbf{I} - \boldsymbol{\nu}\otimes\boldsymbol{\nu} - \bar{a}^{\alpha\beta}\mathbf{a}_\alpha\otimes\mathbf{a}_\beta = \left(a^{\alpha\beta} - \bar{a}^{\alpha\beta}\right)\mathbf{a}_\alpha\otimes\mathbf{a}_\beta \qquad \text{[A.13]}$$

For brevity, let us denote the matrix $[a^{\alpha\beta}]$ (respectively, $[\bar{a}^{\alpha\beta}]$, $[b_{\alpha\beta}]$) by $a^{\cdot\cdot}$ (respectively, by $\bar{a}^{\cdot\cdot}, b_{\cdot\cdot}$).

1. We have

$$\begin{aligned} a^{\cdot\cdot} - \bar{a}^{\cdot\cdot} &= \bar{a}^{\cdot\cdot}\left[(\bar{a}^{\cdot\cdot})^{-1} a^{\cdot\cdot} - I\right] \\ &= \bar{a}^{\cdot\cdot}\left[\bar{a}_{..} a^{\cdot\cdot} - I\right] && \text{since } \bar{a}_{..} = (\bar{a}^{\cdot\cdot})^{-1} \\ &= \bar{a}^{\cdot\cdot}\left[(a_{..} + g b_{..}) a^{\cdot\cdot} - I\right] && \text{from the definition } \bar{a}_{..} = a_{..} + g b_{..} \\ &= \bar{a}^{\cdot\cdot}\left[\cancel{a_{..} a^{\cdot\cdot}} + g b_{..} a^{\cdot\cdot} - \cancel{I}\right] \\ &= g \bar{a}^{\cdot\cdot} b_{..} a^{\cdot\cdot} \end{aligned}$$

Hence $\forall \alpha, \beta, a^{\alpha\beta} - \bar{a}^{\alpha\beta} = g \bar{a}^{\alpha\gamma} b_{\gamma\delta} a^{\delta\beta}$. By substituting this result into [A.13], we obtain $[\text{A.12}]_1$.

2. This can be rewritten alternatively as

$$\begin{aligned} a^{\cdot\cdot} - \bar{a}^{\cdot\cdot} &= \left[a^{\cdot\cdot} (\bar{a}^{\cdot\cdot})^{-1} - I\right] \bar{a}^{\cdot\cdot} \\ &= \left[a^{\cdot\cdot} \bar{a}_{..} - I\right] \bar{a}^{\cdot\cdot}; \\ &= \left[a^{\cdot\cdot} (a_{..} + g b_{..}) - I\right] \bar{a}^{\cdot\cdot} \\ &= \left[\cancel{a^{\cdot\cdot} a_{..}} + g a^{\cdot\cdot} b_{..} - \cancel{I}\right] \bar{a}^{\cdot\cdot} \\ &= g a^{\cdot\cdot} b_{..} \bar{a}^{\cdot\cdot} \end{aligned}$$

From which $\forall \alpha, \beta, a^{\alpha\beta} - \bar{a}^{\alpha\beta} = g a^{\alpha\gamma} b_{\gamma\delta} \bar{a}^{\delta\beta}$. By substituting this result into [A.13], we obtain $[\text{A.12}]_2$. ■

Clearly, the left-hand side of [A.12] is symmetrical. It was seen in the proof of [2.31] that the other sides are indeed symmetric tensors as well.

• The slip $\mathbf{V}_T \Delta t$ was defined in [8.53]. Just like the other variations that we have encountered, its variation plays a role when calculating the contact tangent matrix.

THEOREM.– The variation of the slip is

$$\boxed{\delta(\mathbf{V}_T \Delta t) = \mathbf{W}^1 . \delta \mathbf{U}^{(1)}(\mathbf{X}) + \mathbf{W}^2 . \delta \mathbf{U}^{(2)}(\mathbf{Y}) - \delta \mathbf{U}^{(2)}(\mathbf{Y}_{n-1}) - \mathbf{Z}^\alpha . \delta \mathbf{U}^{(2)}_{,\alpha}(\boldsymbol{\eta})} \quad \text{[A.14]}$$

where $\mathbf{W}^\alpha$ and $\mathbf{Z}^\alpha$, $\alpha \in \{1, 2\}$, are second-order *dimensionless* tensors defined by

$$\boxed{\begin{aligned} \mathbf{W}^1 &\equiv \bar{a}^{\alpha\beta} \mathbf{a}_\alpha \otimes \mathbf{a}_\beta \\ \mathbf{W}^2 &\equiv \mathbf{I} - \mathbf{W}^1 \\ \mathbf{Z}^\alpha &\equiv g \bar{a}^{\alpha\beta} \mathbf{a}_\beta \otimes \boldsymbol{\nu} \end{aligned}} \quad \text{[A.15]}$$

Tensors $\mathbf{W}^1$ and $\mathbf{W}^2$ are *symmetric*, whereas $\mathbf{Z}^1$ and $\mathbf{Z}^2$ are not.

PROOF. Starting with the expression [8.56] $\mathbf{V}_T \Delta t \simeq \mathbf{y} - \boldsymbol{\phi}^{(2)}(\mathbf{Y}_{n-1}, t)$, and then taking its variation, we get

$$\delta(\mathbf{V}_T \Delta t) = \delta \mathbf{y} - \delta \boldsymbol{\phi}^{(2)}(\mathbf{Y}_{n-1}) \quad \text{[A.16]}$$

REMARK.– The previous relation can be recovered by reasoning geometrically. Starting again with the expression [8.56] $\mathbf{V}_T \Delta t \simeq \mathbf{y} - \boldsymbol{\phi}^{(2)}(\mathbf{Y}_{n-1}, t)$, let us define four points **A**, **B**, **C** and **D**, as in Figure A.2.

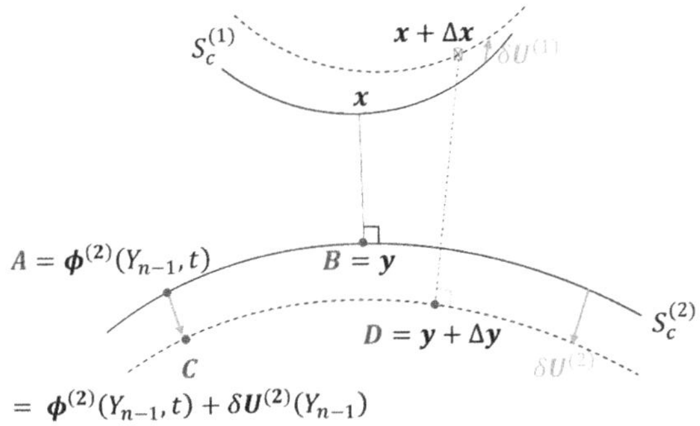

Figure A.2. *Calculation of $\delta(\mathbf{V}_T\Delta t)$. For a color version of this figure, see www.iste.co.uk/levan/contact.zip*

– Before the variation, the slip is $\mathbf{V}_T\Delta t \simeq \mathbf{y} - \boldsymbol{\phi}^{(2)}(\mathbf{Y}_{n-1}, t) = \mathbf{B} - \mathbf{A}$.

– After the variation, it becomes $\mathbf{V}_T\Delta t + \delta(\mathbf{V}_T\Delta t) = \mathbf{D} - \mathbf{C}$.

The difference of the two previous equalities yields the above-found relation [A.16]:

$$\Delta(\mathbf{V}_T\Delta t) = \underbrace{\mathbf{D} - \mathbf{B}}_{\delta\mathbf{y}} - \underbrace{(\mathbf{C} - \mathbf{A})}_{\delta\mathbf{U}^{(2)}(\mathbf{Y}_{n-1})} \quad \blacksquare$$

In the relation [A.16], we have $\delta\boldsymbol{\phi}^{(2)}(\mathbf{Y}_{n-1}) = \delta\mathbf{U}^{(2)}(\mathbf{Y}_{n-1})$ and from [A.3], $\delta\mathbf{y} = \delta\mathbf{U}^{(2)}(\mathbf{Y}(\mathbf{X}, t)) + \mathbf{a}_\alpha\delta\eta^\alpha$. Therefore:

$$\delta(\mathbf{V}_T\Delta t) = \delta\mathbf{U}^{(2)}(\mathbf{Y}(\mathbf{X}, t)) - \delta\mathbf{U}^{(2)}(\mathbf{Y}_{n-1}) + \mathbf{a}_\alpha\delta\eta^\alpha \qquad \text{[A.17]}$$

Finally, after taking into account the expression [A.5] of $\delta\eta^\alpha$, we are done, Q.E.D. ■

References

[1] P. Alart, A. Curnier. A generalized Newton method for contact problems with friction. *J. Mécanique Théorique et Appliquée*, 7, 67–82, 1988.

[2] P. Alart, A. Curnier. A mixed formulation for frictional contact problems prone to newton like solution methods. *Comput. Methods Appl. Mech. Engrg.*, 92, 353–375, 1991.

[3] F. Armero, E. Petocz. Formulation and analysis of conserving algorithms for frictionless dynamic contact/impact problems. *Comput. Methods Appl. Mech. Engrg.*, 158, 269–300, 1998.

[4] F. Armero, I. Romero. On the formulation of high-frequency dissipative time-stepping algorithms for non-linear dynamics. Part I: Low-order methods for two model problems and nonliear elastodynamics. *Comput. Methods Appl. Mech. Engrg.*, 190, 2603–2649, 2001.

[5] F. Armero, I. Romero. On the formulation of high-frequency dissipative time-stepping algorithms for non-linear dynamics. Part II: Second order methods. *Comput. Methods Appl. Mech. Engrg.*, 190, 6783–6824, 2001.

[6] K.J. Bathe. The inf-sup condition and its evaluation for mixed finite element methods. *Comput. Struct.*, 79, 243–252, 2001.

[7] K.J. Bathe. Conserving energy and momentum in nonlinear dynamics: A simple implicit time integration scheme. *Comput. Struct.*, 85, 437–445, 2007.

[8] K.J. Bathe, M.M.I. Baig. On a composite implicit time integration procedure for nonlinear dynamics. *Comput. Struct.*, 83, 2513–2524, 2005.

[9] K.J. Bathe, F. Brezzi. Stability of finite element mixed interpolations for contact problems. *Rendiconti Lincei – Matematica E Applicazioni*, 12, 167–183, 2001.

[10] F.B. Belgacem, P. Hild, P. Laborde. The mortar finite element method for contact problems. *Mathematical and Computer Modelling*, 28(4–8), 263–271, 1998.

[11] T. Belytschko, T.J.R. Hughes. Analysis of transient algorithms with particular reference to stability behavior. In *Computational Methods for Transient Analysis*. Elsevier, Amsterdam, 1983.

[12] T. Belytschko, D. Schoeberle. On the unconditional stability of an implicit algorithm for non-linear structural dynamics. *J. Appl. Mech.*, 42, 865–869, 1975.

[13] T. Belytschko, W.K. Liu, B. Moran. *Nonlinear Finite Elements for Continua and Structures*. Wiley, New York, 2000.

[14] P. Betsch. The discrete null space method for the energy consistent integration of constrained mechanical systems. Part I: Holonomic constraints. *Comput. Methods Appl. Mech. Engrg.*, 194(50–52), 5159–5190, 2005.

[15] P. Betsch, C. Hesch. Energy-momentum conserving schemes for frictionless dynamic contact problems. In *IUTAM Symposium on Computational Methods in Contact Mechanics*. Springer, Dordrecht, 2007.

[16] P. Betsch, P. Steinmann. Conservation properties of a time FE method – Part II: Time stepping schemes for non-linear elastodynamics. *Int. J. Num. Meth. Engrg.*, 50, 1931–1955, 2001.

[17] P. Betsch, S. Uhlar. Energy-momentum conserving integration of multibody dynamics. *Multibody System Dynamics*, 17(4), 243–289, 2007.

[18] E. Bittencourt, G. Creus. Finite element analysis of three-dimensional contact and impact in large deformation problems. *Comput. Struct.*, 69, 219–234, 1998.

[19] J. Casey, P.M. Naghdi. A remark on the use of the decomposition $F = F^e F^p$ in plasticity. *J. Appl. Mech.*, 47, 672–675, 1980.

[20] S. Cescotto, R. Charlier. Frictional contact finite elements based on mixed variational principles. *Int. J. Num. Meth. Engrg.*, 36(10), 1681–1701, 1993.

[21] S.K. Chan, I.S. Tuba. A finite element method for contact problems of solid bodies – Part 1. Theory and validation. *International Journal of Mechanical Sciences*, 13, 615–625, 1971.

[22] D. Chapelle, K.J. Bathe. The inf-sup test. *Comput. Struct.*, 47(415), 537–545, 1993.

[23] A.B. Chaudhary, K.J. Bathe. A solution method for static and dynamic analysis of three-dimensional contact problems with friction. *Comput. Struct.*, 24, 855–873, 1986.

[24] V. Chawla, T.A. Laursen. Energy consistent algorihtms for frictional contact problems. *Int. J. Num. Meth. Engrg.*, 42, 799–827, 1998.

[25] W.H. Chen, P. Tsai. Finite element analysis of elastodynamic sliding contact problems with friction. *Comput. Struct.*, 22, 925–938, 1986.

[26] J. Chung, G.M. Hulbert. A time integration algorithm for structural dynamics with improved numerical dissipation: The generalized-α method. *J. Appl. Mech.*, 60, 371–375, 1993.

[27] M.A. Crisfield. *Non-linear Finite Element Analysis of Solids and Structures, Volume 1. Essentials*. Wiley, New York, 1991.

[28] M.A. Crisfield. Re-visitting the contact patch test. *Int. J. Num. Meth. Engrg.*, 48, 435–449, 2000.

[29] A. Curnier. *Computational Methods in Solid Mechanics*. Springer, Dordrecht, 1994.

[30] A. Curnier, Qi.-Ch. He, A. Klarbring. Continuum mechanics modelling of large deformation contact with friction. In M. Raous, M. Jean, J.J. Moreau (eds), *Contact Mechanics*. Plenum, New York, 1995.

[31] A. Czekanski, S.A. Meguid. Analysis of dynamic frictional contact problems using variational inequalities. *Finite Elem. Anal. Des.*, 37, 861–879, 2001.

[32] G. De Saxcé, Z.-Q. Feng. The bipotential method: A constructive approach to design the complete contact law with friction and improved numerical algorithms. *Mathematical and Computer Modelling*, 28, 225–245, 1998.

[33] D. Dostal, T. Kozubek, M. Sadowska, V. Vondrak. *Scalable Algorithms for Contact Problems*. Springer, New York, 2016.

[34] N. El-Abbasi, K.J. Bathe. Stability and patch test performance of contact discretizations and a new solution algorithm. *Comput. Struct.*, 79, 1473–1486, 2001.

[35] S. Erlicher, L. Bonaventura, L. Bursi. The analysis of the α-generalized method for non-linear dynamic problems. *Computational Mechanics*, 28, 83–104, 2002.

[36] A.L. Eterovic, K.J. Bathe. On the treatment of inequality constraints arising from contact conditions in finite element analysis. *Comput. Struct.*, 40, 203–209, 1991.

[37] Z.-Q. Feng, F. Peyraut, N. Labed. Solution of large deformation contact problems with friction between Blatz-Ko hyperelastic bodies. *Int. J. Engineering Sciences*, 41, 2213–2225, 2003.

[38] Z.-Q. Feng, B. Magnain, J.M. Cros. Solution of large deformation impact problems with friction between Blatz-Ko hyperelastic bodies. *Int. J. Engineering Sciences*, 44, 113–126, 2006.

[39] K.A. Fischer, P. Wriggers. Mortar based frictional contact formulation for higher order interpolations using the moving friction cone. *Comput. Methods Appl. Mech. Engrg.*, 195(37–40), 5020–5036, 2006.

[40] A. Francavilla, O.C. Zienkiewicz. A note on numerical computation of elastic contact problems. *Int. J. Num. Meth. Engrg.*, 9(4), 913–924, 1975.

[41] M. Fremond. *Non-Smooth Thermomechanics*. Springer, Berlin, Heidelberg, 2002.

[42] P. Germain, Q.S. Nguyen, P. Suquet. Continuum thermodynamics. *J. Appl. Mech.*, 50, 1010–1020, 1983.

[43] A.E. Giannakopoulos. The return mapping method for the integration of friction constitutive relations. *Comput. Struct.*, 32, 157–167, 1989.

[44] R. Glowinski, P. Le Tallec. *Augmented Lagrangian and Operator-Splitting Methods in Nonlinear Mechanics*. SIAM Publications, Philadelphia, 1989.

[45] W. Goldsmith. *Impact – The Theory and Physical Behaviour of Colliding Solids*. Dover Publication, New York, 1960, 2001.

[46] O. Gonzalez. Exact energy and momentum conserving algorithms for general models in nonlinear elasticity. *Comput. Methods Appl. Mech. Engrg.*, 190, 1763–1783, 2000.

[47] A.E. Green, P.M. Naghdi. A general theory of an elastic-plastic continuum. *Archive for Rational Mechanics and Analysis*, 18, 251–281, 1965.

[48] A.E. Green, P.M. Naghdi. Some remarks on elastic-plastic deformation at finite strain. *Int. J. Engineering Sciences*, 9, 1219–1229, 1971.

[49] Z. Guo, L.J. Sluys. Application of a new constitutive model for the description of rubber-like materials under monotonic loading. *Int. J. Solids Struct.*, 43, 2799–2819, 2006.

[50] J.O. Hallquist, G.L. Goudreau, D.J. Benson. Sliding interfaces with contact-impact in large-scale Lagrangian computations. *Comput. Methods Appl. Mech. Engrg.*, 51(1–3), 107–137, 1985.

[51] B. Halphen, Q.S. Nguyen. Sur les matériaux standard généralisés. *Journal de Mécanique*, 14, 1, 39–63, 1975.

[52] M. Harnau, A. Konyukhov, K. Schweizerhof. Algorithmic aspects in large deformation contact analysis using 'solid-shell' elements. *Comput. Struct.*, 83(21–22), 1804–1823, 2005.

[53] S. Hartmann, J. Oliver, R. Weyler, J.C. Cante, J.A. Hernández. A contact domain method for large deformation frictional contact problems. Part 2: Numerical aspects. *Comput. Methods Appl. Mech. Engrg.*, 198(33–36), 2607–2631, 2009.

[54] J.-H. Heegaard, A. Curnier. An augmented Lagrangian method for discrete large-slip contact problems. *Int. J. Num. Meth. Engrg.*, 36, 569–593, 1993.

[55] P. Heintz, P. Hansbo. Stabilized lagrange multiplier methods for bilateral elastic contact with friction. *Comput. Methods Appl. Mech. Engrg.*, 195(33), 4323–4333, 2006.

[56] M. Hestenes. Survey paper – Multiplier and gradient methods. *J. of Optimization Theory and Applications*, 4, 303–320, 1969.

[57] H.M. Hilber, T.J.R. Hughes, R.L. Taylor. Improved numerical dissipation for time integration algorithms in structural dynamics. *Earthquake Engineering and Structural Dynamics*, 5, 283–292, 1977.

[58] G.A. Holzapfel. *Nonlinear Solid Mechanics: A Continuum Approach for Engineering*. Wiley, New York, 2000.

[59] J.C. Houbolt. A recurrence matrix solution for the dynamic response of elastic aircraft. *J. of the Aeronautical Sciences*, 17, 540–550, 1950.

[60] S. Hüeber, B.I. Wohlmuth. A primal–dual active set strategy for non-linear multibody contact problems. *Comput. Methods Appl. Mech. Engrg.*, 194(27–29), 3147–3166, 2005.

[61] S. Hüeber, B.I. Wohlmuth. Mortar methods for contact problems. In P. Wriggers, U. Nackenhorst (eds), *Analysis and Simulation of Contact Problems*. Springer, Berlin, Heidelberg, 39–47, 2006.

[62] T.J.R. Hughes. A note on the stability of Newmark's algorithm in nonlinear structural dynamics. *Int. J. Num. Meth. Engrg.*, 11, 383–386, 1977.

[63] T.J.R. Hughes. *The Finite Element Method – Linear Static and Dynamic Finite Element Analysis*. Dover Publications, New York, 2000.

[64] T.J.R. Hughes, R.L. Taylor, J.L. Sackman, A. Curnier, W. Kanoknukulchai. A finite element method for a class of contact-impact problems. *Comput. Methods Appl. Mech. Engrg.*, 8, 249–276, 1976.

[65] A. Ibrahimbegovic. *Nonlinear Solid Mechanics: Theoretical Formulations and Finite Element Solution Methods*. Springer, Dordrecht, 2009.

[66] M. Jean. The non-smooth contact dynamics methods. *Comput. Methods Appl. Mech. Engrg.*, 177, 235–257, 1999.

[67] K.L. Johnson. *Contact Mechanics*. Cambridge University Press, Cambridge, 1985.

[68] R.E. Jones, P. Papadopoulos. A yield-limited Lagrange multiplier formulation for frictional contact. *Int. J. Num. Meth. Engrg.*, 48(8), 1127–1149, 2000.

[69] W. Ju, R.L. Taylor. A perturbed Lagange formulation for the finite element solution of nonlinear frictional contact problems. *J. Theoretical and Applied Mechanics*, 7, 1–14, 1988.

[70] H.B. Khenous, P. Laborde, Y. Renard. Mass redistribution method for finite element contact problems in elastodynamics. *European Journal of Mechanics – A/Solids*, 27(5), 918–932, 2008.

[71] N. Kikuchi, J.T. Oden. *Contact Problems in Elasticity: A Study on Variational Inequalities and Finite Element Methods*. SIAM Publications, Philadelphia, 1988.

[72] A. Klarbring. The rigid punch problem in nonlinear elasticity: Formulation, variational priciples an linearization. *J. of Technical Physics*, 32, 45–60, 1991.

[73] A. Klarbring. Large displacement frictional contact: A continuum framework for finite element discretisation. *European Jounal of Mechanics – A/Solids*, 14, 237–253, 1995.

[74] A. Konyukhov, K. Schweizerhof. Contact formulation via a velocity description allowing efficiency improvements in frictionless contact analysis. *Computational Mechanics*, 33(3), 165–173, 2004.

[75] A. Konyukhov, K. Schweizerhof. *Computational Contact Mechanics: Geometrically Exact Theory for Arbitrary Shaped Bodies*. Springer, Berlin, Heidelberg, 2013.

[76] T.A. Laursen. Formulation and treatment of frictional contact problems using finite elements. PhD Thesis, Stanford University, Stanford, 1992.

[77] T.A. Laursen. *Computational Contact and Impact Mechanics – Fundamentals of Modeling Interfacial Phenomena in Nonlinear Finite Element Analysis*. Springer, Berlin, Heidelberg, 2003.

[78] T.A. Laursen, V. Chawla. Design of energy conserving algorithms for frictionless dynamic contact problems. *Int. J. Num. Meth. Engrg.*, 40, 863–886, 1997.

[79] T.A. Laursen, G.R. Love. Improved implicit integrators for transient impact problems-geometric admissibility within the conserving framework. *Int. J. Num. Meth. Engrg.*, 53, 245–274, 2002.

[80] T.A. Laursen, X.N. Meng. A new solution procedure for application of energy-conserving algorithms to general constitutive models in nonlinear elastodynamics. *Comput. Methods Appl. Mech. Engrg.*, 190, 6309–6322, 2001.

[81] T.A. Laursen, V.G. Oancea. Automation and assessment of augmented Lagrangian algorithms for frictional contact problems. *J. Appl. Mech.*, 61, 956–963, 1994.

[82] T.A. Laursen, J.C. Simo. A continuum-based finite element formulation for the implicit solution of multibody, large deformation-frictional contact problems. *International Journal for Numerical Methods in Engineering*, 36(20), 3451–3485, 1993.

[83] T.A. Laursen, J.C. Simo. Algorithmic symmetrization of coulomb frictional problems using augmented Lagrangians. *Comput. Methods Appl. Mech. Engrg.*, 108, 133–146, 1993.

[84] A. Le van, H.T.T. Nguyen. Une formulation variationnelle du problème de contact avec frottement de Coulomb. *Comptes rendus mécanique*, 336, 606–611, 2008.

[85] A. Le van, H.T.T. Nguyen. A weighted residual relationship for the contact problem with Coulomb friction. *Comput. Struct.*, 87, 1580–1601, 2009.

[86] E.H. Lee. Elastic-plastic deformations at finite strains. *J. Appl. Mech.*, 36, 1–6, 1969.

[87] E.H. Lee, D.T. Liu. Finite-strain elastic-plastic theory with application to plane-wave analysis. *J. Applied Physics*, 38, 19–27, 1967.

[88] J. Lengiewicz, J. Korelc, S. Stupkiewicz. Automation of finite element formulations for large deformation contact problems. *Int. J. Num. Meth. Engrg.*, 85, 1252–1279, 2011.

[89] W.N. Liu, G. Meschke, H.A. Mang. On the approximations of the tangential slip in frictional contact analyses. *Comput. Struct.*, 78, 53–62, 2000.

[90] G.R. Love, T.A. Laursen. Improved implicit integrators for transient impact problems-dynamic frictional dissipation within an admissible conserving framework. *Comput. Methods Appl. Mech. Engrg.*, 192, 2223–2248, 2003.

[91] B. Magnain. Développement d'algorithmes et d'un code de calcul pour l'étude des problèmes de l'impact et du choc. PhD Thesis, University Évry-Val d'Essonne, Évry-Courcouronne, 2006.

[92] R. Michlowski, Z. Mroz. Associated and non-associated sliding rules in contact friction problems. *Archives of Mechanics*, 30, 259–276, 1978.

[93] A.R. Mijar, J.S. Arora. An augmented Lagrangian optimization method for contact analysis problems, 1: Formulation and algorithm. *Structural and Multidisciplinary Optimization*, 28, 99–112, 2004.

[94] R.D. Mindlin. Compliance of elastic bodies in contact. *J. Appl. Mech.*, 16, 259–268, 1949.

[95] J.J. Moreau. Les liaisons unilatérales et le principe de Gauss. *Comptes rendus hebdomadaires des séances de l'académie des sciences*, 256, 871–874, 1963.

[96] J.J. Moreau. Sur les lois de frottement, de plasticité et de viscosité. *Comptes rendus de l'académie des sciences – série A*, 271, 608–611, 1970.

[97] J.J. Moreau. Unilateral contact and dry friction in finite freedom dynamics. In J.J. Moreau, P.D. Panagiotopoulos (eds), *Nonsmooth Mechanics and Applications*. Springer, Vienna, 1988.

[98] P.M. Naghdi. A critical review of the state of finite plasticity. *J. Appl. Math. Phys.*, 41, 315–394, 1990.

[99] P.M. Naghdi, J.A. Trapp. Restrictions on constitutive equations for finitely deformed elastic-plastic materials. *Quarterly Journal of Mechanics and Applied Mathematics*, 28, 1, 25–46, 1975.

[100] N.M. Newmark. A method of computation for structural dynamics. *Journal of the Engineering Mechanics Division, ASCE*, 85, 67–94, 1959.

[101] B. Nour-Omid, P. Wriggers. A note on the optimum choice for penalty parameters. *Communications in Applied Numerical Methods*, 3, 581–585, 1987.

[102] J.T. Oden, E.B. Pires. Algorithms and numerical results for finite element approximations of contact problems with non-classical friction laws. *Comput. Struct.*, 19, 137–147, 1984.

[103] J. Oliver, S. Hartmann, J.C. Cante, R. Weyler, J.A. Hernández. A contact domain method for large deformation frictional contact problems. Part 1: Theoretical basis. *Comput. Methods Appl. Mech. Engrg.*, 198(33–36), 2591–2606, 2009.

[104] V. Padmanabhan, T.A. Laursen. A framework for developement of structure smoothing procedures in large de formation frictional contact analysis. *Finite Elem. Anal. Des.*, 37, 173–198, 2001.

[105] D. Pantuso, K.J. Bathe, P. Bouzinov. A finite element procedure for the analysis of thermo-mechanical solids in contact. *Comput. Struct.*, 75, 551–573, 2000.

[106] P. Papadopoulos, R.L. Taylor. A mixed formulation for the finite element solution of contact problems. *Comput. Methods Appl. Mech. Engrg.*, 94, 373–389, 1992.

[107] P. Papadopoulos, J.M. Solberg. A Lagrange multiplier method for the finite element solution of frictionless contact problems. *Mathematical and Computer Modelling*, 28, 373–384, 1998.

[108] H. Parisch. A consistent tangent stiffness matrix for three-dimensional non-linear contact analysis. *Int. J. Num. Meth. Engrg.*, 28(8), 1803–1812, 1989.

[109] H. Parisch, C. Lübbing. A formulation of arbitrarily shaped surface elements for three-dimensional large deformation contact with friction. *Int. J. Num. Meth. Engrg.*, 40(18), 3359–3383, 1997.

[110] D. Perić, D.R.J. Owen. Computational model for 3-D contact problems with friction based on the penalty method. *Int. J. Num. Meth. Engrg.*, 35(6), 1289–1309, 1992.

[111] G. Pietrzak. Continuum mechanics modelling and augmented Lagrangian formulation of large deformation frictional contact problems. PhD Thesis, École Polytechnique Fédérale de Lausanne, Lausanne, 1997.

[112] G. Pietrzak, A. Curnier. Large deformation frictional contact mechanics: Continuum formulation and augmented Lagrangian treatment. *Comput. Methods Appl. Mech. Engrg.*, 177, 351–381, 1999.

[113] M.J.D. Powell. A methods for nonlinear constraints in minimization problems. In R. Flecher (ed.), *Optimization*. Academic Press, London, 1969.

[114] M.A. Puso, T.A. Laursen. A mortar segment-to-segment contact method for large deformation solid mechanics. *Comput. Methods Appl. Mech. Engrg.*, 193(6), 601–629, 2004.

[115] M.A. Puso, T.A. Laursen. A mortar segment-to-segment frictional contact method for large deformations. *Comput. Methods Appl. Mech. Engrg.*, 193(45), 4891–4913, 2004.

[116] E. Riks. An incremental approach to the solution of snapping and buckling problems. *Int. J. Solids Struct.*, 15, 524–551, 1979.

[117] W. Rust. *Non-linear Finite Element Analysis in Structural Mechanics*. Springer, Cham, 2015.

[118] J.C.J. Schellekens, R. De Borst. On the numerical integration of interface elements. *Int. J. Num. Meth. Engrg.*, 36, 43–66, 1993.

[119] W. Sextro. *Dynamical Contact Problems with Friction, Models, Methods, Experiments and Applications*. Springer, Berlin, Heidelberg, 2007.

[120] D. Sheng, K. Dieter Eigenbrod, P. Wriggers. Finite element analysis of pile installation using large-slip frictional contact. *Computers and Geotechnics*, 32, 17–26, 2005.

[121] J.C. Simo, T.A. Laursen. An augmented Lagrangian treatment of contact problems involving friction. *Comput. Struct.*, 42, 97–116, 1992.

[122] J.C. Simo, N. Tarnow. The discrete energy-momentum method. conserving algorithms for nonlinear elastodynamics. *J. Appl. Math. Phys.*, 43, 757–792, 1992.

[123] J.C. Simo, R.L. Taylor. Consistent tangent operator for rate-independent plasticity. *Comput. Methods Appl. Mech. Engrg.*, 48, 101–118, 1985.

[124] J.C. Simo, R.L. Taylor. A return mapping algorithm for plane stress elastoplasticity. *Int. J. Num. Meth. Engrg.*, 22, 649–670, 1986.

[125] J.C. Simo, P. Wriggers, R.L. Taylor. A perturbed Lagrangian formulation for the finite element solution of contact problems. *Comput. Methods Appl. Mech. Engrg.*, 50, 163–180, 1985.

[126] J.M. Solberg, P. Papadopoulos. An analysis of dual formulations for the finite element solution of two-body contact problems. *Comput. Methods Appl. Mech. Engrg.*, 194(25–26), 2734–2780, 2005.

[127] D.A. Spence. An eigenvalue problem for elastic contact with finite friction. *Proceedings of the Cambridge Philosophical Society*, 73, 249–268, 1973.

[128] D.A. Spence. The Hertz contact problem with finite friction. *J. Elasticity*, 5, 297–319, 1975.

[129] G. Stadler. Path-following and augmented Lagrangian methods for contact problems in linear elasticity. *Journal of Computational and Applied Mathematics*, 203(2), 533–547, 2007.

[130] B. Storakers. Hertz contact at finite friction and arbitrary profiles. *J. Mech. Phys.*, 53, 1422–1447, 2005.

[131] S. Stupkiewicz, J. Lengiewicz, J. Korelc. Sensitivity analysis for frictional contact problems in the augmented lagrangian formulation. *Comput. Methods Appl. Mech. Engrg.*, 199, 2165–2176, 2010.

[132] R.L. Taylor, P. Papadopoulos. On a patch test for contact problems in two dimensions. In P. Wriggers, W. Wagner (eds), *Computational Methods in Nonlinear Mechanics*, Springer-Verlag, Berlin, 1991.

[133] R.L. Taylor, P. Papadopoulos. On a finite element method for dynamic contact/impact problems. *Int. J. Num. Meth. Engrg.*, 36, 2123–2140, 1993.

[134] D. Vola, E. Pratt, M. Jean, M. Raous. Consistent time discretization for a dynamical frictional contact problem and complementarity techniques. *Revue européenne des eléments finis*, 7, 149–162, 1998.

[135] G.A. Wempner. Discrete approximations related to non-linear theories of solids. *Int. J. Solids Struct.*, 7, 1581–1599, 1971.

[136] W.L. Wood, M. Bossak, O.C. Zienkiewicz. An alpha modification of Newmark's method. *Int. J. Num. Meth. Engrg.*, 15, 1562–1566, 1981.

[137] P. Wriggers. Finite element methods for contact problems with friction. *Tribology International*, 29, 651–658, 1996.

[138] P. Wriggers. *Computational Contact Mechanics*. Wiley, New York, 2002.

[139] P. Wriggers, J. C. Simo. A note on tangent stiffness for fully nonlinear contact problems. *Communications in Applied Numerical Methods*, 1(5), 199–203, 1985.

[140] P. Wriggers, G. Zavarise. A formulation for frictionless contact problems using a weak form introduced by Nitsche. *Computational Mechanics*, 41(3), 407–420, 2008.

[141] P. Wriggers, L. Krstulovic-Opara, J. Korelc. Smooth C^1-interpolations for two-dimensional frictional contact problems. *Int. J. Num. Meth. Engrg.*, 51, 1469–1495, 2001.

[142] P. Wriggers, T. Vu Van, E. Stein. Finite element formulation of large deformation impact-contact problems with friction. *Comput. Struct.*, 37, 319–331, 1990.

[143] G. Zavarise, P. Wriggers. A segment-to-segment contact strategy. *Mathematical and Computer Modelling*, 28, 497–515, 1998.

[144] G. Zavarise, P. Wriggers. A superlinear convergent augmented Lagrangian procedure for contact problems. *Engineering Computations*, 16, 88–119, 1999.

[145] G. Zaravise, P. Wriggers, B.A. Schrefler. On augmented Lagrangian algorithms for thermomechanical contact problems with friction. *Int. J. Num. Meth. Engrg.*, 38, 2929–2949, 1995.

[146] O.C. Zienkiewicz, R.L. Taylor. *The Finite Element Method*, volume 2. Butterworth-Heinemann, Oxford, 2000.

[147] O.C. Zienkiewicz, S. Qu, R.L. Taylor, S. Nakazawa. The patch test for mixed formulations. *Int. J. Num. Meth. Engrg.*, 23, 1873–1883, 1986.

Index

A, B

abnormal contact stress, 96
acceleration, 24
algorithmic
  gap, 83
  slip, 83
  stick, 83
assembly, 91, 135
augmented Lagrangian method, 15
basis
  dual, 26
  natural, 25
boundary conditions, 52

C

Cauchy
  stress
    tensor, 37
    vector, 38
  stretch tensor, 24
coefficient of friction, 47
constitutive laws, 39, 51
contact
  force vector, 92
  law, 52
    normal, 52
    tangential, 52
  pressure, 38
  stress vector, 38
  surface, 25
    parametrization, 26
  tangent matrix(-ces), 113, 117, 121, 123, 135, 136, 204
  virtual power (VP), 62, 69, 70
contactor, 25
  surface, 25
Coulomb's
  criterion, 48
  friction
    cone, 47
    law, 47
curvilinear coordinates, 25

D, E

deformation, 24
  gradient, 24
discrete/discretized
  contact law, 94, 97, 109
  expression of the slip, 101
discretization, 78, 90, 133, 135
displacement, 24
  vector, 72
dual problem, 13
elastoplastic material, 41
element
  contact
    force vectors, 83
    tangent matrix, 134
  nodal
    coordinate vector, 74
    displacement vector, 72, 74
    multiplier vector, 74
    vector, 73
    virtual velocity vector, 74
  residual contact law vector, 83
  virtual contact power, 70
equality constraint, 8
equation(s)
  equilibrium, 109
  for the dynamic contact problem, 94
  for the static contact problem, 109
  local dynamic, 37
  of motion, 96

of motion of the bodies (semi-discrete), 94
equilibrium of the contact stresses at the contact interface, 53
external force vector
not including/excluding contact force, 90, 109

F, G, H

field equations, 51
finite element, 70
frictional stress, 38
gap, 5
global nodal vector, 88
Green strain tensor, 24
Hertz-Signorini conditions, 3, 46

I, K

inequality constraints, 10
initial conditions, 53
internal force vector, 90
inverse images, 28
Karush–Kuhn–Tucker conditions/theorem, 10, 11, 46
kinematic time, 110

L, M

Lagrangian, 9, 11
load
dead, 90
follower, 90
localization matrices, 90
master, *see also* target, 25
material
elastoplastic, 41
hyperelastic, 39
neo-Hookean, 40
Saint Venant-Kirchhoff, 40
stress tensor, 37
tangent tensor, 40, 42
min-max approach, 14
Mooney model, 43
multiplier, 60, 61, 69
method, 8
vector, 77

N

Newmark scheme/algorithm, 201
Nitsche's method, 21
nodal
displacement vector, 89
multiplier vector, 89
virtual multiplier, 89
virtual velocity vector, 89
nominal
contact pressure, 38
frictional stress, 38
stress
tensor, 37
vector, 52
non-adhesion condition, 46
non-penetrability condition, 46

P

parent element, *see also* reference element, 71
penalty
method, 3
parameter, 59
perturbed Lagrangian method, 21
Piola-Kirchhoff stress tensor, 37
position vector, 72
pre-image, *see also* inverse image , 71
primal problem, 13
principle of virtual power (PVP), 55, 59
projection
calculation, 99
of a point, 23
proximity, see also gap, 27
velocity, 36

R, S

reference element, 71
relative
contact velocity, 36
normal velocity, 36
tangential velocity, 36
residual, 110, 112, 205
contact law vector, 92
saddle point, 9, 11
slave, *see also* contactor, 25
slip, 36, 47, 48, 101
velocity, 36

solution
for the acceleration, 206
for the displacement, 206
for the velocity, 206
standard relative velocity, 35
stick, 47, 48
strain-energy density function, 39
strong formulation of the contact problem, 51
system of equations for the
dynamic contact problem, 94
static contact problem, 109

T, V

tangent matrix(-ces), 204, 209
classical (contactless), 114
element contact, 134, 138
global tangent matrix, 113, 114
target, 25
surface, 25
variation, 240
of kinematic quantities, 29, 239
of the displacement vectors, 120, 123
of the element nodal multipliers, 123
of the element node displacements, 123
of the multiplier vector, 119, 124
of the nodal multipliers, 119
of the nodal virtual multiplier, 119
velocity, 24
virtual
contact power, *see* contact virtual power multiplier, 61
power (VP), 60
velocity, 61
vector, 72

W

weak
form of the contact problem, 60
formulation of the contact problem, 55
weighted residual relationship, *see also* weak form of the contact problem, 55

Other titles from

iSTE

in

Mechanical Engineering and Solid Mechanics

2023

EL HAMI Abdelkhalak
Uncertainty and Artificial Intelligence: Additive Manufacturing, Vibratory Control, Agro-composite, Mechatronics
(Artificial Intelligence in Mechanics Set – Volume 1)

EL HAMI Abdelkhalak, EID Mohamed
Reliability-based Modeling of System Performance
(Reliability of Multiphysical Systems Set – Volume 19)

EL HAMI Abdelkhalak, DELAUX David, GRZESKOWIAK Henri
Applied Reliability for Industry 1: Predictive Reliability for the Automobile, Aeronautics, Defense, Medical, Marine and Space Industries
(Reliability of Multiphysical Systems Set – Volume 16)
Applied Reliability for Industry 2: Experimental Reliability for the Automobile, Aeronautics, Defense, Medical, Marine and Space Industries
(Reliability of Multiphysical Systems Set – Volume 17)
Applied Reliability for Industry 3: Operational Reliability for the Automobile, Aeronautics, Defense, Medical, Marine and Space Industries
(Reliability of Multiphysical Systems Set – Volume 18)

LEROY Maurice
Rheology, Physical and Mechanical Behavior of Materials 1: Physical Mechanisms of Deformation and Dynamic Behavior

2022

BAYLE Franck
Product Maturity 1: Theoretical Principles and Industrial Applications (Reliability of Multiphysical Systems Set – Volume 12)
Product Maturity 2: Principles and Illustrations (Reliability of Multiphysical Systems Set – Volume 13)

EL HAMI Abdelkhalak, DELAUX David, GRZESKOWIAK Henri
Reliability and Physics-of-Healthy in Mechatronics (Reliability of Multiphysical Systems Set – Volume 15)

LANNOY André
Reliability of Nuclear Power Plants: Methods, Data and Applications (Reliability of Multiphysical Systems Set – Volume 14)

LEDOUX Michel, EL HAMI Abdelkhalak
Heat Transfer 3: Convection, Fundamentals and Monophasic Flows (Mathematical and Mechanical Engineering Set – Volume 11)
Heat Transfer 4: Convection, Two-Phase Flows and Special Problems (Mathematical and Mechanical Engineering Set – Volume 12)

PLANCHETTE Guy
Cindynics, The Science of Danger: A Wake-up Call (Reliability of Multiphysical Systems Set – Volume 11)

2021

CHALLAMEL Noël, KAPLUNOV Julius, TAKEWAKI Izuru
Modern Trends in Structural and Solid Mechanics 1: Static and Stability
Modern Trends in Structural and Solid Mechanics 2: Vibrations
Modern Trends in Structural and Solid Mechanics 3: Non-deterministic Mechanics

DAHOO Pierre Richard, POUGNET Philippe, EL HAMI Abdelkhalak
Applications and Metrology at Nanometer Scale 1: Smart Materials, Electromagnetic Waves and Uncertainties
(Reliability of Multiphysical Systems Set – Volume 9)
Applications and Metrology at Nanometer Scale 2: Measurement Systems, Quantum Engineering and RBDO Method
(Reliability of Multiphysical Systems Set – Volume 10)

LEDOUX Michel, EL HAMI Abdelkhalak
Heat Transfer 1: Conduction
(Mathematical and Mechanical Engineering Set – Volume 9)
Heat Transfer 2: Radiative Transfer
(Mathematical and Mechanical Engineering Set – Volume 10)

2020

SALENÇON Jean
Elastoplastic Modeling

2019

BAYLE Franck
Reliability of Maintained Systems Subjected to Wear Failure Mechanisms: Theory and Applications
(Reliability of Multiphysical Systems Set – Volume 8)

BEN KAHLA Rabeb, BARKAOUI Abdelwahed, MERZOUKI Tarek
Finite Element Method and Medical Imaging Techniques in Bone Biomechanics
(Mathematical and Mechanical Engineering Set – Volume 8)

IONESCU Ioan R., QUEYREAU Sylvain, PICU Catalin R., SALMAN Oguz Umut
Mechanics and Physics of Solids at Micro- and Nano-Scales

LE VAN Anh, BOUZIDI Rabah
Lagrangian Mechanics: An Advanced Analytical Approach

MICHELITSCH Thomas, PÉREZ RIASCOS Alejandro, COLLET Bernard, NOWAKOWSKI Andrzej, NICOLLEAU Franck
Fractional Dynamics on Networks and Lattices

SALENÇON Jean
Viscoelastic Modeling for Structural Analysis

VÉNIZÉLOS Georges, EL HAMI Abdelkhalak
Movement Equations 5: Dynamics of a Set of Solids
(Non-deformable Solid Mechanics Set – Volume 5)

2018

BOREL Michel, VÉNIZÉLOS Georges
Movement Equations 4: Equilibriums and Small Movements
(Non-deformable Solid Mechanics Set – Volume 4)

FROSSARD Etienne
Granular Geomaterials Dissipative Mechanics: Theory and Applications in Civil Engineering

RADI Bouchaib, EL HAMI Abdelkhalak
Advanced Numerical Methods with Matlab® 1: Function Approximation and System Resolution
(Mathematical and Mechanical Engineering Set – Volume 6)

Advanced Numerical Methods with Matlab® 2: Resolution of Nonlinear, Differential and Partial Differential Equations
(Mathematical and Mechanical Engineering Set – Volume 7)

SALENÇON Jean
Virtual Work Approach to Mechanical Modeling

2017

BOREL Michel, VÉNIZÉLOS Georges
Movement Equations 2: Mathematical and Methodological Supplements
(Non-deformable Solid Mechanics Set – Volume 2)
Movement Equations 3: Dynamics and Fundamental Principle
(Non-deformable Solid Mechanics Set – Volume 3)

BOUVET Christophe
Mechanics of Aeronautical Solids, Materials and Structures
Mechanics of Aeronautical Composite Materials

BRANCHERIE Delphine, FEISSEL Pierre, BOUVIER Salima, IBRAHIMBEGOVIC Adnan
From Microstructure Investigations to Multiscale Modeling: Bridging the Gap

CHEBEL-MORELLO Brigitte, NICOD Jean-Marc, VARNIER Christophe
From Prognostics and Health Systems Management to Predictive Maintenance 2: Knowledge, Traceability and Decision
(Reliability of Multiphysical Systems Set – Volume 7)

EL HAMI Abdelkhalak, RADI Bouchaib
Dynamics of Large Structures and Inverse Problems
(Mathematical and Mechanical Engineering Set – Volume 5)
Fluid-Structure Interactions and Uncertainties: Ansys and Fluent Tools
(Reliability of Multiphysical Systems Set – Volume 6)

KHARMANDA Ghias, EL HAMI Abdelkhalak
Biomechanics: Optimization, Uncertainties and Reliability
(Reliability of Multiphysical Systems Set – Volume 5)

LEDOUX Michel, EL HAMI Abdelkhalak
Compressible Flow Propulsion and Digital Approaches in Fluid Mechanics
(Mathematical and Mechanical Engineering Set – Volume 4)
Fluid Mechanics: Analytical Methods
(Mathematical and Mechanical Engineering Set – Volume 3)

MORI Yvon
Mechanical Vibrations: Applications to Equipment

2016

BOREL Michel, VÉNIZÉLOS Georges
Movement Equations 1: Location, Kinematics and Kinetics
(Non-deformable Solid Mechanics Set – Volume 1)

BOYARD Nicolas
Heat Transfer in Polymer Composite Materials

CARDON Alain, ITMI Mhamed
New Autonomous Systems
(Reliability of Multiphysical Systems Set – Volume 1)

DAHOO Pierre Richard, POUGNET Philippe, EL HAMI Abdelkhalak
Nanometer-scale Defect Detection Using Polarized Light
(Reliability of Multiphysical Systems Set – Volume 2)

DE SAXCÉ Géry, VALLÉE Claude
Galilean Mechanics and Thermodynamics of Continua

DORMIEUX Luc, KONDO Djimédo
Micromechanics of Fracture and Damage
(Micromechanics Set – Volume 1)

EL HAMI Abdelkhalak, RADI Bouchaib
Stochastic Dynamics of Structures
(Mathematical and Mechanical Engineering Set – Volume 2)

GOURIVEAU Rafael, MEDJAHER Kamal, ZERHOUNI Noureddine
From Prognostics and Health Systems Management to Predictive Maintenance 1: Monitoring and Prognostics
(Reliability of Multiphysical Systems Set – Volume 4)

KHARMANDA Ghias, EL HAMI Abdelkhalak
Reliability in Biomechanics
(Reliability of Multiphysical Systems Set –Volume 3)

MOLIMARD Jérôme
Experimental Mechanics of Solids and Structures

RADI Bouchaib, EL HAMI Abdelkhalak
Material Forming Processes: Simulation, Drawing, Hydroforming and Additive Manufacturing
(Mathematical and Mechanical Engineering Set – Volume 1)

2015

KARLIČIĆ Danilo, MURMU Tony, ADHIKARI Sondipon, MCCARTHY Michael
Non-local Structural Mechanics

SAB Karam, LEBÉE Arthur
Homogenization of Heterogeneous Thin and Thick Plates

2014

ATANACKOVIC M. Teodor, PILIPOVIC Stevan, STANKOVIC Bogoljub, ZORICA Dusan
Fractional Calculus with Applications in Mechanics: Vibrations and Diffusion Processes
Fractional Calculus with Applications in Mechanics: Wave Propagation, Impact and Variational Principles

CIBLAC Thierry, MOREL Jean-Claude
Sustainable Masonry: Stability and Behavior of Structures

ILANKO Sinniah, MONTERRUBIO Luis E., MOCHIDA Yusuke
The Rayleigh–Ritz Method for Structural Analysis

LALANNE Christian
Mechanical Vibration and Shock Analysis – 5-volume series – 3rd edition
Sinusoidal Vibration – Volume 1
Mechanical Shock – Volume 2
Random Vibration – Volume 3
Fatigue Damage – Volume 4
Specification Development – Volume 5

LEMAIRE Maurice
Uncertainty and Mechanics

2013

ADHIKARI Sondipon
Structural Dynamic Analysis with Generalized Damping Models: Analysis

ADHIKARI Sondipon
Structural Dynamic Analysis with Generalized Damping Models: Identification

BAILLY Patrice
Materials and Structures under Shock and Impact

BASTIEN Jérôme, BERNARDIN Frédéric, LAMARQUE Claude-Henri
Non-smooth Deterministic or Stochastic Discrete Dynamical Systems: Applications to Models with Friction or Impact

EL HAMI Abdelkhalak, RADI Bouchaib
Uncertainty and Optimization in Structural Mechanics

KIRILLOV Oleg N., PELINOVSKY Dmitry E.
Nonlinear Physical Systems: Spectral Analysis, Stability and Bifurcations

LUONGO Angelo, ZULLI Daniele
Mathematical Models of Beams and Cables

SALENÇON Jean
Yield Design

2012

DAVIM J. Paulo
Mechanical Engineering Education

DUPEUX Michel, BRACCINI Muriel
Mechanics of Solid Interfaces

ELISHAKOFF Isaac *et al.*
Carbon Nanotubes and Nanosensors: Vibration, Buckling and Ballistic Impact

GRÉDIAC Michel, HILD François
Full-Field Measurements and Identification in Solid Mechanics

GROUS Ammar
Fracture Mechanics – 3-volume series
Analysis of Reliability and Quality Control – Volume 1
Applied Reliability – Volume 2
Applied Quality Control – Volume 3

RECHO Naman
Fracture Mechanics and Crack Growth

2011

KRYSINSKI Tomasz, MALBURET François
Mechanical Instability

SOUSTELLE Michel
An Introduction to Chemical Kinetics

2010

BREITKOPF Piotr, FILOMENO COELHO Rajan
Multidisciplinary Design Optimization in Computational Mechanics

DAVIM J. Paulo
Biotribology

PAULTRE Patrick
Dynamics of Structures

SOUSTELLE Michel
Handbook of Heterogenous Kinetics

2009

BERLIOZ Alain, TROMPETTE Philippe
Solid Mechanics using the Finite Element Method

LEMAIRE Maurice
Structural Reliability

2007

GIRARD Alain, ROY Nicolas
Structural Dynamics in Industry

GUINEBRETIÈRE René
X-ray Diffraction by Polycrystalline Materials

KRYSINSKI Tomasz, MALBURET François
Mechanical Vibrations

KUNDU Tribikram
Advanced Ultrasonic Methods for Material and Structure Inspection

SIH George C., NAÏT-ABDELAZIZ Moussa, VU-KHANH Toan
Particle and Continuum Aspects of Mesomechanics